JAK-STAT Signaling in Diseases

JAK-STAT Signaling in Diseases

Edited by
Ritobrata Goswami

CRC Press
Taylor & Francis Group
Boca Raton London New York

CRC Press is an imprint of the
Taylor & Francis Group, an **informa** business

CRC Press
Taylor & Francis Group
6000 Broken Sound Parkway, Suite 300
Boca Raton, FL 33487–2742

International Standard Book Number-13: 978-1-138-48762-8 (Hardback)

Library of Congress Control Number: 2020932394

Visit the Taylor & Francis Web site at
www.taylorandfrancis.com

and the CRC Press Web site at
www.crcpress.com

Contents

Preface

JAK-STAT (Janus kinase-Signal Transducers and Activators of Transcription) signaling is one of the few conserved signaling cascades transmitting multiple signals required for homeostasis in animals and humans. In JAK-STAT signaling, extracellular ligand binding to its cognate receptor leads to pathway activation and signal transduction. This pathway is critical for hematopoiesis, immune system development, and other events. Activation of JAK leads to induction of cell proliferation, cell migration, differentiation, and apoptosis. Mutations in JAK-STAT pathway can either lead to inflammatory diseases or may impede regular homeostasis.

JAK-STAT pathways can cross-talk with other signaling pathways. JAK-STAT pathway has been targeted to develop drugs to downregulate the immune response. Few of the JAK inhibitors are in Phase III of clinical trial and have been approved by FDA. Tofacitinib that targets JAK3 has shown its effect to treat rheumatoid arthritis, psoriasis. OPB-51602, which targets STAT3, is in phase I to treat nasopharyngeal carcinoma. JAK inhibitors have an advantage over cytokine receptor blocking drugs in that they that can be taken orally as they are small molecule drugs.

JAK-STAT pathway is a very well-studied signaling pathway. The book has consolidated information on the role of JAK-STAT signaling in autoimmune and inflammatory diseases including asthma, atopic dermatitis, hepatic fibrosis, diabetes and obesity, cancer, cardiovascular disease, immunodeficiency, renal disorders, neuro-inflammatory disorders, among others. This book is unique as it encompasses different aspects of JAK-STAT signaling and how dysregulation of this signaling pathway is associated with various disorders that are prevalent worldwide leading to morbidity and mortality associated with increased expenditure on healthcare and loss of manpower. This book has an update on the therapeutic implication of JAK-STAT inhibitors different phases of clinical trial. I sincerely believe this book will stimulate interest in graduate students, academicians and scientific community. This book will be good reference point for industry professionals who are involved in translational research leading to develop products to various diseases. Positive feedback of this book from the readers will allow us to expand the disease repertoire that is associated with dysregulated JAK-STAT signaling.

Editor

Ritobrata Goswami, PhD, is presently working as an Assistant Professor in the School of Bioscience, IIT Kharagpur since 2016. Prior to his present affiliation, Dr Goswami was associated with the Division of Biological and Life Sciences, Ahmedabad University (2014-16). He received his bachelor (2005) and masters' degree (2007) in Biotechnology from West Bengal University of Technology and Birla Institute of Technology & Science, Pilani; respectively. Dr Goswami obtained his PhD at Indiana University, Indianapolis, USA (2012) followed by post-doctoral training at Icahn School of Medicine at Mount Sinai, New York, USA (2012-14). He is a life member of the Indian Immunology Society. Dr Goswami has extramural projects funded by the Government of India. Dr Goswami's research interests include the role of nutrients and hormones in the development and function of T helper cells to regulate inflammatory disorders. He is keen on identifying a regulating network of transcription factors involved in autoimmune and inflammatory disorders so that better therapeutics can be developed.

Contributors

Marcia J. Abbott
Department of Health Sciences
Crean College of Health and Behavioral Sciences, Chapman University
Orange, California

Devendra K. Agrawal
Department of Clinical and Translational Science
Creighton University School of Medicine
Omaha, Nebraska

Raffaele Altara
Institute for Experimental Medical Research
Oslo University Hospital and University of Oslo
Oslo, Norway

Noha G. Bader El Din
Department of Microbial Biotechnology Genetic Engineering and Biotechnology Division
National Research Centre
Dokki, Egypt

Chandra S. Boosani
Department of Clinical and Translational Science
Creighton University School of Medicine
Omaha, Nebraska

George W. Booz
Department of Pharmacology and Toxicology
The University of Mississippi Medical Center
Jackson, MS

Taylor Burke
Department of Clinical and Translational Science
Creighton University School of Medicine
Omaha, Nebraska

Pike-See Cheah
Department of Human Anatomy, Faculty of Medicine and Health Sciences
Universiti Putra Malaysia
Seri Kembangan, Malaysia

Sean P. Didion
Department of Pharmacology and Toxicology
The University of Mississippi Medical Center
Jackson, MS

Christina Gavegnano
Center for AIDS Research, Laboratory of Biochemical Pharmacology, Department of Pediatrics
Emory University
Atlanta, GA

Ritobrata Goswami
School of Bioscience
Indian Institute of Technology Kharagpur
Kharagpur, India

Tristan Hayes
Department of Pediatrics and Herman B Wells Center for Pediatric Research
Indiana University School of Medicine
Indianapolis, Indiana

Nicolette Nadene Houreld
Laser Research Centre, Faculty of Health Sciences
University of Johannesburg
Johannesburg, South Africa

Marwa K. Ibrahim
Department of Microbial Biotechnology, Genetic Engineering and Biotechnology Division
National Research Centre
Dokki, Egypt

Wanlin Jiang
Department of Clinical and Translational Science
Creighton University School of Medicine
Omaha, Nebraska

Mark H. Kaplan
Department of Pediatrics and Herman B Wells Center for Pediatric Research
Indiana University School of Medicine
Indianapolis, Indiana

Han-Chung Lee
Laboratory Centre
Xiamen University Malaysia
Sepang, Malaysia

King-Hwa Ling
Department of Biomedical Sciences, Faculty of Medicine and Health Sciences
Universiti Putra Malaysia
Seri Kembangan, Malaysia

Yuji Nozaki
Department of Hematology and Rheumatology
Kindai University School of Medicine
Osaka-Sayama, Japan

Duy Pham
Department of Pathology
University of Alabama at Birmingham
Birmingham, Alabama

Thomas Pincez
Division of Pediatric Hematology-Oncology
Charles-Bruneau Cancer Center, Ste-Justine Hospital
Quebec, Canada

Amina Abdul Qayum
Department of Pediatrics and Herman B Wells Center for Pediatric Research
Indiana University School of Medicine
Indianapolis, Indiana

Parvis Sadjadian
University Clinic for Hematology, Oncology, Hemostaseology and Palliative Care, Johannes Wesling Medical Center Minden
University of Bochum
Minden, Germany

Raymond F. Schinazi
Center for AIDS Research, Laboratory of Biochemical Pharmacology
Department of Pediatrics, Emory University
Atlanta, GA

Daniel Silberger
Department of Pathology
University of Alabama at Birmingham
Birmingham, Alabama

Radomir M. Slominski
Department of Dermatology
Indiana University School of Medicine
Indianapolis, Indiana

Kai-Leng Tan
Institute of Biomedical and Pharmaceutical Sciences
Guangdong University of Technology
Guangzhou, China

Thai Hoa Tran
Division of Pediatric Hematology-Oncology
Charles-Bruneau Cancer Center, Ste-Justine Hospital
Quebec, Canada

Matthew J. Turner
Department of Dermatology
Indiana University School of Medicine
Indianapolis, Indiana

1

Regulation of Cytokine Signaling by the JAK-STAT Pathway

Nicolette Nadene Houreld
Laser Research Centre, Faculty of Health Sciences
University of Johannesburg
Johannesburg, South Africa

1.1 Introduction

The Janus kinase/signal transducers and activators of transcription (JAK-STAT) pathway is a prompt pleiotropic cytoplasmic to nuclear signaling pathway used to transduce a variety of signals, activated by cytokines, hormones, and growth factors, for development and homeostasis. The JAK-STAT pathway is responsible for controlling signals of over fifty cytokines, growth factors, and hormones (Morris, Kershaw, and Babon 2018; Hammaren et al. 2019), while negative regulation is through suppressor of cytokine signaling (SOCS) proteins (bind to and inactivate JAK3), and the protein inhibitors of activated STATs (PIAS; bind to STAT dimers thereby preventing DNA binding). Cytokines are glycoproteins (ligands) secreted by cells and operate as intercellular messengers, inducing differentiation, proliferation, growth, and apoptosis of their target cells.

Signaling via the JAK-STAT pathway is instigated by binding of a ligand to its receptor. Binding results in dimerization, oligomerization, and/or conformational changes of the receptor complex, which allow JAK proteins to bind to the receptor complex intracellular domains inducing trans-autophosphorylation of the tyrosine residues (JH1), converting the receptor into a tyrosine kinase. Phosphorylated chains serve as docking sites for SH2 domain-containing signaling molecules such as STATs. Receptor-bound STATs are phosphorylated by JAK on a specific tyrosine in the C-terminal tail, allowing them to form homo- and heterodimers, which rapidly translocate into the nucleus. In the nucleus, they associate with proteins and produce transcriptional complexes/factors with extensive effects on regulation of transcription and epigenetics (Hammaren et al. 2019).

1.2 The JAKs

Janus (kinase) comes from the Roman mythological two-faced god, who looks to the future and the past. JAK relates to the two faces due to the presence of two kinase domains, namely the pseudokinase domain (JAK homology 2, JH2) and a catalytically active kinase domain (JH1). They also contain an N-terminal band, and a four-point-one, ezrin, radixin, moesin (FERM)-domain, which mediate the interaction of JAKs with their receptors. JAKs combine with the proline-rich, membrane-proximal box1/box2 domain on cytokine receptors. They also contain an Src homology 2 (SH2)-like domain, of unknown function, which lies between the pseudokinase and FERM domains (Figure 1.1a) (Schindler, Levy, and Decker 2007; Hammaren et al. 2019). JH2 has significant regulatory functions and is a source of numerous mutations that is the cause of various diseases and disorders, including hematopoietic malignancies (JAK2 mutations),

TABLE 1.1

Type I and Type II Cytokine Receptors and their Corresponding JAKs and STATs. Associated JAK-STAT Proteins for which the Data is Weaker are Shown in Brackets (adapted from (Hammaren et al. 2019)).

	Cytokine	JAKs	STATs
Type I Cytokine Receptors	IL-6	JAK1, JAK2, TYK2	STAT3, STAT1
	IL-11	JAK1, JAK2, TYK2	STAT3, STAT1
	LIF	JAK1, JAK2, TYK2	STAT3, STAT1
	CNTF	JAK1, (JAK2, TYK2)	STAT3, (STAT1)
	CLCF1	JAK1, (JAK2)	STAT3, STAT1
	CT-1	JAK1, (JAK2, TYK2)	STAT3
	OSM	JAK1, (JAK2, TYK2)	STAT3, STAT1
	IL-31	JAK1, (JAK2)	STAT3, STAT5,STAT1
	G-CSF	JAK1, (JAK2)	STAT3
	Leptin	JAK2	STAT3
	IL-12	TYK2, JAK2	STAT4
	IL-23	TYK2, JAK2	STAT3, STAT4, STAT1
	IL-27	JAK1, JAK2	STAT1, STAT3, STAT4, (STAT5)
	IL-35	JAK1, JAK2	STAT1, STAT4
	IL-2	JAK1, JAK3, (JAK2)	STAT5, (STAT3)
	IL-4	JAK1, JAK3	STAT6
	IL-7	JAK1, JAK3	STAT5, (STAT3)
	IL-9	JAK1, JAK3	STAT5, STAT3
	IL-15	JAK1, JAK3	STAT5, (STAT3)
	IL-21	JAK1, JAK3	STAT3, STAT5, (STAT1)
	TSLP	JAK1, JAK2	STAT1, STAT3, STAT4, STAT5, STAT6
	IL-13	JAK1, JAK2, TYK2	STAT6, (STAT3)
	IL-3	JAK2, JAK1	STAT5, STAT3
	IL-5	JAK2	STAT5, STAT1, STAT3
	GM-CSF	JAK2	STAT5
	EPO	JAK2	STAT5
	GH	JAK2	STAT5, (STAT3)
	PRL	JAK2	STAT5
	TPO	JAK2	STAT5
Type II Cytokine Receptors	IFN-I (type I)	JAK1, TYK2	STAT1, STAT2, STAT3, STAT4, (STAT5, STAT6)
	IFN-γ (type II)	JAK1, JAK2	STAT1
	IL-28a, IL-28b	JAK1, TYK2	STAT1, STAT2, STAT3, STAT5
	IL-29	JAK1, TYK2	STAT1, STAT2, STAT3, STAT5
	IL-10	JAK1, TYK2	STAT3, STAT1
	IL-19	JAK1, JAK2	STAT3, STAT1
	IL-20	JAK1, JAK2	STAT3, STAT1
	IL-24	JAK1, JAK2	STAT3, STAT1
	IL-22	JAK1, TYK2	STAT3, STAT1, (STAT5)
	IL-26	JAK1, TYK2	STAT3, STAT1

Interleukin (IL); Leukemia inhibitory factor (LIF); Ciliary neurotrophic factor (CNTF); Cardiotrophin-like cytokine factor 1 (CLCF1); Cardiotrophin-1 (CT-1); Oncostatin M (OSM); Granulocyte-colony stimulating factor (G-CSF); Thymic stromal lymphopoietin (TSLP); Granulocyte-macrophage colony-stimulating factor (GM-CSF); Erythropoietin (EPO); Growth hormone (GH); Prolactin (PRL); Thrombopoietin (TPO).

1

Regulation of Cytokine Signaling by the JAK-STAT Pathway

Nicolette Nadene Houreld
Laser Research Centre, Faculty of Health Sciences
University of Johannesburg
Johannesburg, South Africa

1.1 Introduction

The Janus kinase/signal transducers and activators of transcription (JAK-STAT) pathway is a prompt pleiotropic cytoplasmic to nuclear signaling pathway used to transduce a variety of signals, activated by cytokines, hormones, and growth factors, for development and homeostasis. The JAK-STAT pathway is responsible for controlling signals of over fifty cytokines, growth factors, and hormones (Morris, Kershaw, and Babon 2018; Hammaren et al. 2019), while negative regulation is through suppressor of cytokine signaling (SOCS) proteins (bind to and inactivate JAK3), and the protein inhibitors of activated STATs (PIAS; bind to STAT dimers thereby preventing DNA binding). Cytokines are glycoproteins (ligands) secreted by cells and operate as intercellular messengers, inducing differentiation, proliferation, growth, and apoptosis of their target cells.

Signaling via the JAK-STAT pathway is instigated by binding of a ligand to its receptor. Binding results in dimerization, oligomerization, and/or conformational changes of the receptor complex, which allow JAK proteins to bind to the receptor complex intracellular domains inducing trans-autophosphorylation of the tyrosine residues (JH1), converting the receptor into a tyrosine kinase. Phosphorylated chains serve as docking sites for SH2 domain-containing signaling molecules such as STATs. Receptor-bound STATs are phosphorylated by JAK on a specific tyrosine in the C-terminal tail, allowing them to form homo- and heterodimers, which rapidly translocate into the nucleus. In the nucleus, they associate with proteins and produce transcriptional complexes/factors with extensive effects on regulation of transcription and epigenetics (Hammaren et al. 2019).

1.2 The JAKs

Janus (kinase) comes from the Roman mythological two-faced god, who looks to the future and the past. JAK relates to the two faces due to the presence of two kinase domains, namely the pseudokinase domain (JAK homology 2, JH2) and a catalytically active kinase domain (JH1). They also contain an N-terminal band, and a four-point-one, ezrin, radixin, moesin (FERM)-domain, which mediate the interaction of JAKs with their receptors. JAKs combine with the proline-rich, membrane-proximal box1/box2 domain on cytokine receptors. They also contain an Src homology 2 (SH2)-like domain, of unknown function, which lies between the pseudokinase and FERM domains (Figure 1.1a) (Schindler, Levy, and Decker 2007; Hammaren et al. 2019). JH2 has significant regulatory functions and is a source of numerous mutations that is the cause of various diseases and disorders, including hematopoietic malignancies (JAK2 mutations),

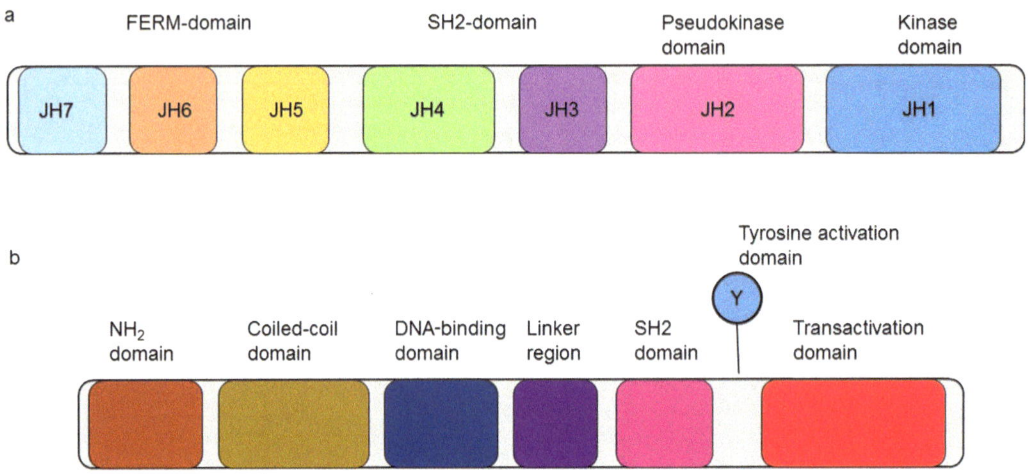

FIGURE 1.1 Structural organisation of JAK-STAT proteins. (a) JAKs share seven conserved homology domains: JH1 serves as the catalytic, kinase domain, while JH2 represents the pseudokinase domain. JH3 and half of JH4 include the nonfunctional SH2-domain, and half of JH4 to JH7 includes a FERM domain. (b) STATs share seven domains: the amino-terminal domain (NH_2), the coiled-coil domain, the DNA-binding domain, the linker domain, the SH2 domain, the tyrosine-activation domain, and the transactivation domain.

leukemia and lymphomas (all JAKs), and cancer (JAK1, JAK3) (Hammaren et al. 2019). There are four JAKs in mammals (JAK1, JAK2, JAK3, and TYK2). JAK1, JAK2, and TYK2 are ubiquitously expressed and relatively constitutive in their expression, while the expression of JAK3 is mostly confined to cells of hematopoietic origin, and its expression is more inducible.

JAK1 associates with type I (IFN-α/β), type II (IFN-γ), IL-2, and IL-6 receptors. JAK2 interacts with single-chain receptors (i.e., EPOR, GH-R, and PRL-R) and IL-3 (IL-3R, IL-5R, and GM-CSFR) cytokine families, as well as the IFN-γ receptor. Leukocyte-specific JAK3 exclusively associates with the IL-2 receptor γ-chain, and Tyk2 associates with receptors for IFN-I, IL-6, IL-10, and IL-12/23 cytokine families (Schindler, Levy, and Decker 2007).

1.3 The STATs

The STAT family includes STAT1, STAT2, STAT3, STAT5A/B, and STAT6. STAT proteins consist of seven well-defined, conserved domains: an N-terminal conserved domain (NH_2, critical for STAT function); a coiled-coil domain (involved in receptor binding, and associates with regulatory proteins); a DNA-binding domain (DBD, cooperate in binding to the promoters of target genes); a linker region (LK, spacer to maintain proper conformation between the dimerization and DNA binding domains); an SH2 domain (critical for the recruitment of STATs to activated receptor complexes and for the interaction with JAK and Src kinases); a tyrosine-activation domain (Y); and a C-terminal transactivation domain (TAD, modulates the transcriptional activation of target genes and vary considerably among STAT family members) (Jatiani et al. 2010) (Figure 1.1b).

1.4 Cytokine Receptors

Cytokines function by binding to their associated transmembrane receptor, which triggers intracellular signaling events and pathways that result in the alteration of gene expression. Most of these receptors consist of a unique ligand-binding subunit and a signal-transducing subunit. Often the signal transducing or cytoplasmic subunits are structurally similar to other cytokine receptors, particularly in regions labeled as box 1 or the proline-rich motif and the box-2 motif, and this is critical for proper receptor functioning and mediating of mitogenic signals. The ligand-binding subunit, or membrane distal region, remains uniquely different to ensure differentiation (Jatiani et al. 2010). Cytokine binding results in receptor tyrosine phosphorylation.

Cytokine receptors are divided into type I and type II receptors. Type I cytokine receptors bind to and react to cytokines with four α-helical strands and share an amino acid motif (WSXWS). Type II cytokine receptors are similar to type I, but lack the WSXWS motif. Cytokine receptors signal through the JAK-STAT pathway and other pathways that typically trigger activation of the mitogen-activated protein (MAP) kinase cascade. Different types of cells and tissues express well-defined and diverse receptor combinations that respond to cytokine combinations unique to their microenvironment. Thus, at any particular time, a single cell may respond to signals from multiple cytokine receptors (Murray 2007). Different receptor classes preferentially associate with one JAK family member, or a JAK combination.

Typically, receptors required for hematopoietic cell development and proliferation prefer JAK2; common γ-chain receptors utilize JAK1 and JAK3, while other receptors use only JAK1 (Murray 2007). All interferons (IFNs), which are essential mediators of innate immunity against bacterial and viral infection, as well as the interleukin(IL)-10 family (IL-10, IL-19, IL-20, IL-22, IL-24, IL-26), anti-inflammatory cytokines, function through type II cytokine receptors, which dimerize in multiple combinations to generate distinct downstream effects. JAK1 is imperative for signaling through these type II receptor complexes (Ferrao et al. 2016).

1.5 Activation of JAK-STAT Pathways by Cytokines

Cytokine signaling through the JAK-STAT pathway regulates numerous cellular responses, including proliferation, differentiation, motility, and cell survival. JAKs mediate signaling of around fifty to sixty different hormones, cytokines, and growth factors ranging from regulators of the immune system and hematopoiesis, such as IFNs, ILs, thrombopoietin (TPO), and erythropoietin (EPO), to regulators of development and metabolism, such as growth hormone (GH) and prolactin (PRL) (Table 1.1) (O'Shea and Plenge 2012).

Ligand binding results in receptor dimerization/oligomerization, which in turn results in the juxtapositioning of JAKs through homodimeric or heterodimeric interactions. This leads to the autophosphorylation and/or transphosphorylation of JAK, which causes the phosphorylation of receptor target tyrosine residues that serve as docking sites and allow the binding of other SH2 domain–containing signaling molecules such as STAT (Jatiani et al. 2010) (Figure 1.2). In unstimulated cells, STATs are idle, unphosphorylated cytosolic proteins. Cellular stimulation via cytokines induces phosphorylation of receptor tyrosine residues that serve as docking sites for STATs via their SH2 domains. Once phosphorylated, dimerized STATs translocate to the nucleus and drive the transcription of cytokine inducible genes. STAT activation is swift, with a maximum accumulation of phosphorylated STAT1 in the nucleus within 30 min (Lim and Cao 2006).

Recently, a number of biophysical and protein engineering studies have presented new data, which emphasizes the intricacy and complexity of signaling sparked by a cytokine–cytokine receptor complex. This permits cytokines to produce diverse biological responses in spite of using a marginal set of surface receptors and effector signaling molecules (Gorby et al. 2018). Cytokine receptors will interact with the JAK-STAT pathway though its interaction with a combination of JAKs. Cytokine-receptor interactions result in various JAK activation that exists in association with cytokine receptors, for which activation is essential for the activation of STATs. On the other hand, activated JAKs do not seem to exhibit specificity for a particular STAT, as different receptors stimulate a common STAT even though they activate distinctively

TABLE 1.1

Type I and Type II Cytokine Receptors and their Corresponding JAKs and STATs. Associated JAK-STAT Proteins for which the Data is Weaker are Shown in Brackets (adapted from (Hammaren et al. 2019)).

	Cytokine	JAKs	STATs
Type I Cytokine Receptors	IL-6	JAK1, JAK2, TYK2	STAT3, STAT1
	IL-11	JAK1, JAK2, TYK2	STAT3, STAT1
	LIF	JAK1, JAK2, TYK2	STAT3, STAT1
	CNTF	JAK1, (JAK2, TYK2)	STAT3, (STAT1)
	CLCF1	JAK1, (JAK2)	STAT3, STAT1
	CT-1	JAK1, (JAK2, TYK2)	STAT3
	OSM	JAK1, (JAK2, TYK2)	STAT3, STAT1
	IL-31	JAK1, (JAK2)	STAT3, STAT5,STAT1
	G-CSF	JAK1, (JAK2)	STAT3
	Leptin	JAK2	STAT3
	IL-12	TYK2, JAK2	STAT4
	IL-23	TYK2, JAK2	STAT3, STAT4, STAT1
	IL-27	JAK1, JAK2	STAT1, STAT3, STAT4, (STAT5)
	IL-35	JAK1, JAK2	STAT1, STAT4
	IL-2	JAK1, JAK3, (JAK2)	STAT5, (STAT3)
	IL-4	JAK1, JAK3	STAT6
	IL-7	JAK1, JAK3	STAT5, (STAT3)
	IL-9	JAK1, JAK3	STAT5, STAT3
	IL-15	JAK1, JAK3	STAT5, (STAT3)
	IL-21	JAK1, JAK3	STAT3, STAT5, (STAT1)
	TSLP	JAK1, JAK2	STAT1, STAT3, STAT4, STAT5, STAT6
	IL-13	JAK1, JAK2, TYK2	STAT6, (STAT3)
	IL-3	JAK2, JAK1	STAT5, STAT3
	IL-5	JAK2	STAT5, STAT1, STAT3
	GM-CSF	JAK2	STAT5
	EPO	JAK2	STAT5
	GH	JAK2	STAT5, (STAT3)
	PRL	JAK2	STAT5
	TPO	JAK2	STAT5
Type II Cytokine Receptors	IFN-I (type I)	JAK1, TYK2	STAT1, STAT2, STAT3, STAT4, (STAT5, STAT6)
	IFN-γ (type II)	JAK1, JAK2	STAT1
	IL-28a, IL-28b	JAK1, TYK2	STAT1, STAT2, STAT3, STAT5
	IL-29	JAK1, TYK2	STAT1, STAT2, STAT3, STAT5
	IL-10	JAK1, TYK2	STAT3, STAT1
	IL-19	JAK1, JAK2	STAT3, STAT1
	IL-20	JAK1, JAK2	STAT3, STAT1
	IL-24	JAK1, JAK2	STAT3, STAT1
	IL-22	JAK1, TYK2	STAT3, STAT1, (STAT5)
	IL-26	JAK1, TYK2	STAT3, STAT1

Interleukin (IL); Leukemia inhibitory factor (LIF); Ciliary neurotrophic factor (CNTF); Cardiotrophin-like cytokine factor 1 (CLCF1); Cardiotrophin-1 (CT-1); Oncostatin M (OSM); Granulocyte-colony stimulating factor (G-CSF); Thymic stromal lymphopoietin (TSLP); Granulocyte-macrophage colony-stimulating factor (GM-CSF); Erythropoietin (EPO); Growth hormone (GH); Prolactin (PRL); Thrombopoietin (TPO).

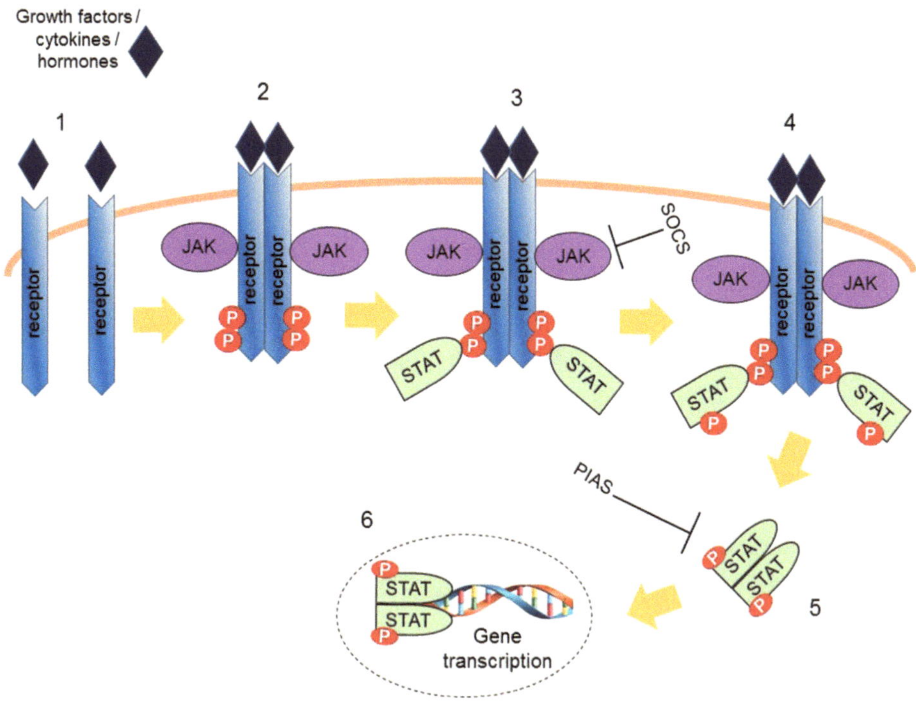

FIGURE 1.2 Overview of the JAK-STAT signaling pathway. Ligand binding (1) results in receptor dimerization, JAK activation, and receptor phosphorylation (2). STAT binds to the phosphorylated receptor (3), which becomes phosphorylated by JAK (4). STAT dimers form (5), which translocates into the nucleus and stimulates gene transcription (6). Negative regulation of JAK-STAT signaling is through suppresser of cytokine signaling (SOCS) proteins, which directly bind to and inactivate JAKs, and protein inhibitors of activated STATs (PIAS) that bind to phosphorylated STAT dimers, averting DNA binding.

different JAKs. It would appear that the specificity for the activation of STAT lies with the STAT docking sites on the receptors themselves (Jatiani et al. 2010).

Once activated STAT has translocated to the nucleus, it binds to a consensus DNA-recognition motif known as gamma-activated sites (GAS) in the promoter region of cytokine-inducible genes and activates transcription. These GAS-like elements act as ligands for a variety of STAT family members. There are three key features of STAT-binding elements (SBE), which include the core motif (with sequence TT-(X)n-AA; where X is a G/C rich palindromic sequence in length *n*), the core spacing between the palindromic A/T residues, and composition of the sequence between palindromic A/T residues. Generally, there is a preference for a specific core spacing for each class of cytokine-responsive genes, for example, GAS elements consistently occupy five-base pair core spacing, whereas genes that respond to IL-6 occupy a four-base core spacing (Wilks and Harpur 1994). STAT1 and STAT5 prefer sites with a three-base pair spacer; however, STAT5 has also been shown to bind weakly to a four base pair spacer. STAT6 prefers sites with a four-base pair spacer, and STAT4 prefers the palindromic sequence (T/A)TTCC(C/G)GGAA(T/A) where the first and last T/A sites outside of the usual motif are also necessary for binding (Morris, Kershaw, and Babon 2018).

1.6 JAK-STAT Regulation

A characteristic feature of JAK-STAT signaling is its rapid onset and subsequent decay. Activated STATs accumulate within a few hours in the nucleus, however, the signal promptly decays and the

STATs are re-exported back to the cytoplasm for the next round of signaling. This decline in activity involves down-regulation of both receptors and JAK proteins, as well as STAT-transcriptional activity (Schindler, Levy, and Decker 2007). Three well-defined mechanisms of signal decay include dephosphorylation by protein phosphotyrosine phosphatases, nuclear export, and feedback inhibition via SOCS, as well as inhibition by PIAS. Post-translational modifications (PTMs) of STAT proteins institute another important regulatory mechanism. STATS also undergo post-translational covalent modifications such as ubiquitination (post-translational modification by adding ubiquitin to the protein sequence).

Protein phosphotyrosine phosphatases (PTPs) negatively regulate the JAK-STAT pathway by dephosphorylating tyrosine residues. There are six PTPs that regulate JAK-STAT. Receptor and JAK dephosphorylation has been carried out by phosphatases, and include SHP-1, SHP-2, and CD45, while STAT dephosphorylation is carried out by SHP-2, PTP1B, TC-PTP, and PTP-BL (Schindler, Levy, and Decker 2007; Seif et al. 2017). Since PTPs are constitutively expressed, they tend to restrain the amplitude of the signaling cascade rather than controlling its duration (Morris, Kershaw, and Babon 2018). The process of nuclear translocation is complex and there is a balance between STAT nuclear import and export. This seems to be controlled by multiple nuclear export sequence (NES) and nuclear localization sequence (NLS) elements. During activation of cellular signaling, there is a shift toward STAT accumulation in the nucleus. This balance is shifted toward nuclear export when there is signal decay (Schindler, Levy, and Decker 2007). Proteolytic processing of STAT via cleavage of the C-terminal acts as a general mechanism for the negative regulation of STAT protein function (Hendry and John 2004). This may occur in the nucleus or in the cytoplasm, and this protease may cleave activated (phosphorylated) or inactivated STAT. STAT are rapidly inactivated by dephosphorylation with a half-life of phosphorylated STAT1 at less than 15 min (Lim and Cao 2006).

Post-translational modifications regulate every aspect of transcription factor function and coordinate access of RNA polymerases to promoter templates. Site-specific, DNA-binding transcription factors repress, activate, enhance, or silence complexes and associated enzymatic activities (Filtz, Vogel, and Leid 2014). Cytokine signal transduction is dependent on the ability of individual receptors to recruit signal transcription factors (STFs), which in turn activate the transcription of the genes which are related to that particular cytokine (Wilks and Harpur 1994). This process is very complex and relies on affinity differences of STFs for GAS elements. There are two main subclasses of DNA response elements in gene promoters, and include interferon-stimulated response element (ISRE) and GAS-like sequences. ISRE is located upstream of interferon alpha/beta (IFN-α/β) inducible genes, while GAS is located upstream of IFN gamma (IFN-γ)-inducible genes. GAS-like elements, which bind to STFs induced by other cytokines, all consist of the same basic DNA motif with slight changes in individual genes (Wilks and Harpur 1994). Owing to activation of other transcription factors and crosstalk between different signal transducing pathways, determination of the complete set of genes upregulated by STAT proteins is difficult. However, it is known that several hundred to thousands of genes can be upregulated or down-regulated in response to each STAT protein (Morris, Kershaw, and Babon 2018).

The SOCS family members are the primary drivers of signal attenuation and are negative-feedback inhibitors of signaling induced by cytokines and other stimuli that act via the JAK-STAT pathway. There are eight SOCS proteins: SOCS1 to 7 and CIS. These proteins all contain an SH2 domain and a short, C-terminal domain, known as the SOCS box. The SOCS box is associated with an adapter complex, elongin BC, which allows for the recruitment of an E3 ubiquitin ligase scaffold (Cullin5) to catalyse and ubiquitinate signaling intermediates recruited by their SH2 domains (Liau et al. 2018). Ubiquitination is mediated by an ubiquitin-activating enzyme (E1), ubiquitin-conjugating enzyme (E2), and ubiquitin ligase (E3), which mediates between E2 and the substrate. The resulting poly-ubiquitinated conjugates are rapidly identified and degraded by 26S proteasome. SOCS proteins form E3 ubiquitin ligase complexes with Cullin-2, Elongin B, Elongin C, and Rbx1 and ubiquitinate JAK and receptor molecules, followed by their degradation and internalization, respectively (Hatakeyama 2012). SOCS1 and SOCS3 are capable of directly binding to and inhibiting the kinase activity of JAK, and is due to a short motif, kinase inhibitory region (KIR), upstream of the SH2 domain. SOCS1 and SOCS3 also have the ability to directly inhibit JAK1, JAK2, and TYK2 (but not JAK3) by blocking the substrate-binding groove on JAK, thereby acting as a pseudosubstrate. SOCS1 is the most influential SOCS family member, and is the principal regulator of numerous cytokines involved in the immune response, in particular IFN-γ (Liau et al. 2018). Genetic deletion of SOCS1 is fatal, and its down-regulation plays a role in the progression of cancer. Each SOCS protein is also specific for only a subset of cytokines; for example, SOCS3 is prompted by IL-2, -3, -4, -6, -7, -9, -10, -11, -12, -13, -21, and -22, as well as granulocyte-colony stimulating factor (G-CSF), granulocyte-macrophage colony-stimulating factor (GM-CSF), leukemia inhibitory factor (LIF), PRL, IFN-α, IFN-γ, GH, EPO, thyroperoxidase, oncostatin M (OSM), calcitonin-1 (CT1), ciliary neurotrophic factor (CNTF), and leptin (Morris, Kershaw, and Babon 2018).

There are also members of the PIAS family, which are able to inhibit the JAK-STAT pathway by binding to and inhibiting STAT dimers, preventing DNA binding and gene transcription. However, the exact mechanism by which PIAS proteins do their regulative functions remains unrevealed. There are several PIAS family members, including PIAS1, PIAS3 (KchAP), PIASy, and PIASx (ARIP3) (O'Shea and Plenge 2012). PIAS family members possess several domains, including a serine/threonine rich domain located at the C-terminus (responsible for target binding), a Zn-binding RING–finger-like domain (RLD, responsible for SUMO transfer), and a conserved-SAP domain (scaffold attachment factor A/B, Acinus, PIAS) near the N-terminus (important part for target binding via scaffold/matrix attachment regions, S/MARs) (Seif et al. 2017).

1.7 Failure of JAK Regulation Leads to Disease

Deregulation of cytokines and/or their downstream signaling pathways are at the root of many human disorders and diseases, including asthma, severe combined immunodeficiency (SCID), and various cancers (Gorby et al. 2018; Hammaren et al. 2019). A number of these disorders are because of point mutations, as in the case of SCID (mutation and loos in function of JAK3). Oncogenic JAK2 rearrangements to multiple fusion gene partners have been identified in acute lymphoblastic leukemia (ALL), atypical chronic myeloid leukemia (CML), acute myeloid leukemia (AML), myeloproliferative neoplasms (MPN), and/or Hodgkin lymphoma (Hammaren et al. 2019).

A number of gain-of-function mutations that have been identified in JAK1, JAK2, and JAK3 are responsible for approximately 20% of T-ALL cases and to a lesser extent in B-ALL or hepatocellular adenoma (Hammaren et al. 2019). In some cases, mutations in other components and JAK-STAT regulatory molecules can be the cause of the disease, as is the case in 1%–5% of primary myelofibrosis (PMF) and essential thrombocythemia (ET). In these cases, there is a point mutation in the thrombopoietin receptor (TPOR) gene MPL, which enables ligand-independent activation of JAK2 signaling (Leroy et al. 2016). Frameshift mutations are also frequent, accounting for approximately 13% of MPN cases, causing thrombopoietin-independent TPOR–JAK2 signaling. In this instance, a frameshift mutation in calreticulin (CALR), an endoplasmic reticulum protein, results in a highly charged protein capable of binding to TPOR, thus activating the receptor (Hammaren et al. 2019). Apart from mutations resulting in JAK-STAT hyperactivity, JAK-STAT loss-of function mutations have also been described. Loss-of-function mutations in JAK3 or JAK3-associated receptors IL-7R and common gamma chain (IL-2Rγ) cause autosomal recessive SCID. TYK2 deficiency has been reported in primary immunodeficiency (PID) (Hammaren et al. 2019).

1.8 Conclusion

There is a shared interaction between external actions and internal reactions that enables a cell to remain viable and survive. Binding of extracellular cytokines and growth factors to their receptor sets off a cascade of responses and activates cellular signaling pathways, leading to gene transcription and ultimately cellular proliferation, differentiation, activation/inhibition, and survival/apoptosis. The JAK-STAT pathway plays a fundamental role in the transfer of extracellular signals from membrane receptors to the nucleus. Over fifty cytokines make use of the JAK-STAT pathway by binding to type I or type II receptors to carry out their effects. Several regulators modulate the function of the JAK-STAT pathway, such as protein tyrosine phosphatases, SOCS, and PIAS family members. There is still a lack of knowledge concerning the complete molecular understanding of JAK activation and inhibition. Any dysregulation in the JAK-STAT pathway may lead to various pathologies. Better knowledge of these mechanisms may be the answer in the development of therapeutic strategies to target the JAK-STAT pathway to treat various immunological disorders and cancers.

REFERENCES

Ferrao, R., H. J. Wallweber, H. Ho, et al. 2016. The structural basis for class II cytokine receptor recognition by JAK1. *Structure* 24 (6):897–905.

Filtz, T. M., W. K. Vogel, and M. Leid. 2014. Regulation of transcription factor activity by interconnected post-translational modifications. *Trends Pharmacol Sci* 35 (2):76–85.

Gorby, C., J. Martinez-Fabregas, S. Wilmes, and I. Moraga. 2018. Mapping determinants of cytokine signaling via protein engineering. *Front Immunol* 9:2143.

Hammaren, H. M., A. T. Virtanen, J. Raivola, and O. Silvennoinen. 2019. The regulation of JAKs in cytokine signaling and its breakdown in disease. *Cytokine* 118:48–63.

Hatakeyama, S. 2012. Ubiquitin-mediated regulation of JAK-STAT signaling in embryonic stem cells. *JAKSTAT* 1 (3):168–75.

Hendry, L., and S. John. 2004. Regulation of STAT signalling by proteolytic processing. *Eur J Biochem* 271 (23–24):4613–20.

Jatiani, S. S., S. J. Baker, L. R. Silverman, and E. P. Reddy. 2010. Jak/STAT pathways in cytokine signaling and myeloproliferative disorders: approaches for targeted therapies. *Genes Cancer* 1 (10):979–93.

Leroy, E., J. P. Defour, T. Sato, et al. 2016. His499 regulates dimerization and prevents oncogenic activation by asparagine mutations of the human thrombopoietin receptor. *J Biol Chem* 291 (6):2974–87.

Liau, N. P. D., A. Laktyushin, I. S. Lucet, et al. 2018. The molecular basis of JAK/STAT inhibition by SOCS1. *Nat Commun* 9 (1):1558.

Lim, Cheh Peng, and Xinmin Cao. 2006. Structure, function, and regulation of STAT proteins. *Mol Biosyst* 2 (11):536–50.

Morris, R., N. J. Kershaw, and J. J. Babon. 2018. The molecular details of cytokine signaling via the JAK/STAT pathway. *Protein Sci* 27 (12):1984–2009.

Murray, P. J. 2007. The JAK-STAT signaling pathway: input and output integration. *J Immunol* 178 (5):2623–9.

O'Shea, J. J., and R. Plenge. 2012. JAK and STAT signaling molecules in immunoregulation and immune-mediated disease. *Immunity* 36 (4):542–50.

Schindler, C., D. E. Levy, and T. Decker. 2007. JAK-STAT signaling: from interferons to cytokines. *J Biol Chem* 282 (28):20059–63.

Seif, F., M. Khoshmirsafa, H. Aazami, M. Mohsenzadegan, G. Sedighi, and M. Bahar. 2017. The role of JAK-STAT signaling pathway and its regulators in the fate of T helper cells. *Cell Commun Signal* 15 (1):23.

Wilks, A. F., and A. G. Harpur. 1994. Cytokine signal transduction and the JAK family of protein tyrosine kinases. *Bioessays* 16 (5):313–20.

2

The Structure-Function Bonhomie of JAK-STAT Molecules

Ritobrata Goswami
School of Bioscience
Indian Institute of Technology Kharagpur
Kharagpur, India

2.1 Introduction

The existence of life requires the ability to comprehend and respond to external signals. These signals are intercepted by numerous cell-surface receptors that utilize a range of intracellular signaling pathways to communicate with the nucleus at the cellular level. The culmination of these signaling ensue an appropriate response mediated by peripheral sensory organs via the central nervous system. The Janus kinase (JAK)-signal transducers and activators of transcription (STAT) pathway used by plants and animals, including flies, is one of the few signal transduction pathways that transduce a multitude of signals required for development and homeostasis (Rawlings, Rosler, and Harrison 2004). Activation of JAK induces various physiological events including, but not limited to, cell proliferation, migration, differentiation, and apoptosis. JAK-STAT pathway's association is evident in 12 major cancers (Vogelstein et al. 2013). JAK-STAT pathway is one of the major signaling pathways employed by multiple growth factors and cytokines (Liongue and Ward 2013). These cellular events are critical to hematopoiesis, development of immune system, and subsequent functions. JAK-STAT signaling is dependent upon tyrosine phosphorylation and harbors comparatively simple signal transduction pathway requiring few components. Activation of JAK-STAT pathway ensues when growth factors or cytokines bind to their cognate receptors leading to multimerization of receptor subunits (Rawlings, Rosler, and Harrison 2004). The multimerization of receptor subunits can form homodimers that lead to close proximity of the receptor-associated JAK molecules. Via transphosphorylation, JAKs phosphorylate each other on tyrosine residues, a cue for their activation. Activated JAKs then phosphorylate on tyrosine residues of cytokine or growth-factor receptors paving the way for binding of proteins bearing SH2 domains. With the help of SH2 domains, STAT molecules bind to the phosphorylated tyrosine residues on the cognate receptor. STAT molecules dissociate from the receptor after being tyrosine-phosphorylated by JAKs. Following activation, STAT molecules form either homo or hetero-dimers and SH2 domain of each STAT molecule binds to the phosphotyrosine residue of its corresponding STAT. Subsequently, the STAT dimer translocate to the nucleus, bind to the genes bearing consensus STAT-binding sequence, and activate gene transcription (Figure 2.1). Thus, the JAK-STAT signaling cascade is a direct mechanism to translate an extracellular signal into a transcriptional response. The canonical JAK-STAT proteins are inactivated by negative regulators, including SH2-containing protein tyrosine phosphatase (SHP) and suppressor of cytokine signaling (SOCS) proteins (Liongue et al. 2012). While phosphorylation can be denoted as an "on" switch (binary 1), dephosphorylation can be attributed as an "off" switch (binary 0). In this chapter, the structure–function relationship of JAK-STAT signaling has been discussed.

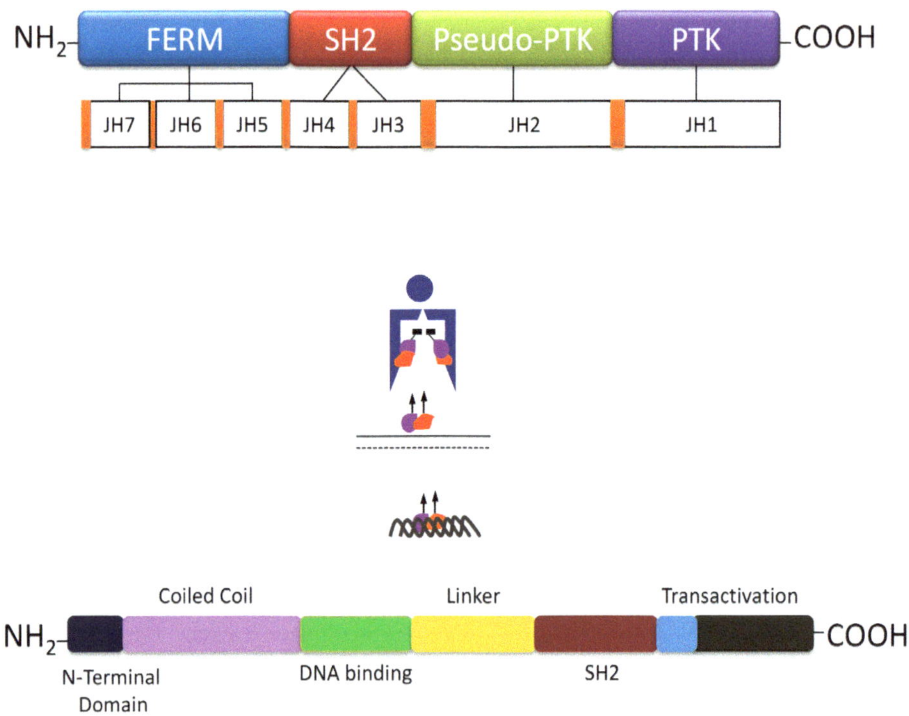

FIGURE 2.1 Brief overview of JAK-STAT domain structure. The JAK-STAT signaling pathway has been shown in between the domain structures JAK (shown on top) and STAT (shown at the bottom).

2.2 The Beginning

The journey of JAK-STAT pathway started with the discovery of "interferons" way back in 1957 by Issacs and Lindenmann (Isaacs and Lindenmann 1957). As the moniker suggests, interferons (IFNs) "interfered" with the replication process of virus. Three decades later, studies involving the induction of mRNA and translation of proteins by interferons staged the set for identifying members of JAK-STAT signaling pathway. JAK family has four members: JAK1, JAK2, JAK3, and TYK2. STAT family has seven members: STAT1, STAT2, STAT3, STAT4, STAT5A, STAT5B, and STAT6. During 1984–1988, IFN-α-induced transcriptional stimulation of specific genes was characterized and IFN-dependent promoters were identified (Stark and Darnell 2012). The first members of the JAK family were identified in 1989, while cDNA clones for the first STAT family member was discovered in 1992 (Wilks 1989, Shuai et al. 1992). The Janus of Janus kinase comes from two-faced Roman God of doorways, "Janus", since JAK possesses two near-identical phosphate-transferring domains (Wilks 1989). During 1989–1991, JAK1, JAK2, and TYK2 were identified, while JAK3 sequencing was completed in 1994 (Stark and Darnell 2012). Sequencing of STAT3, 4, 5A, 5B, and 6 were completed in the same year (Larner et al. 1993, Ruff-Jamison, Chen, and Cohen 1993, Sadowski et al. 1993, Silvennoinen et al. 1993, Akira et al. 1994, Zhong, Wen, and Darnell 1994a, 1994b, Jacobson et al. 1995). The functional and structural domains of STATs were described during 1995–1998 (Hoey and Schindler 1998). STATs were first crystalized in 1998 followed by identification of target genes (Becker, Groner, and Muller 1998). The onset of twenty-first century was when the underlying importance of the signaling pathway was linked to several disorders that were associated with JAK-STAT mutations and gained momentum.

2.3 Structure and Activation of JAK

Cytokine or growth-factor receptors lack intrinsic kinase domain, thus depending on JAK for transmitting signal to the cytoplasmic components. Structurally, JAKs have seven domains known as JAK homology regions (JH1-7) (Bousoik and Montazeri Aliabadi 2018) (Figure 2.1). Functionally, JAKs have been shown to possess four domains: catalytically active tyrosine kinase domain, the pseudotyrosine kinase domain, the Src homology 2 (SH2) domain, and the FERM domain (Bousoik and Montazeri Aliabadi 2018). JH1 encodes the kinase domain, while JH2 has a pseudo-kinase domain that controls the kinase activity of JH1 (Manning et al. 2002, LaFave and Levine 2012). The JH2 domain activity has been observed specifically for JAK2 and has been suggested to possess auto-inhibitory function (Ungureanu et al. 2011). The SH2 domain is encoded by JH3 and JH4 regions, while JH5-7 is responsible for receptor binding (Cornejo, Boggon, and Mercher 2009). The JH3-7 regions form the N-terminal of JAKs, while JH1-2 forms the C-terminus of the domain (Yamaoka et al. 2004). JAK activation occurs upon ligand-mediated receptor multimerization because two JAKs are brought into close proximity, allowing trans-phosphorylation. Binding of the ligand to cytokine receptor revamps the receptor–JAK dimers, which brings the JAKs close enough to transphosphorylate the partner JAK in the dimer at JH1. The activated JAKs subsequently phosphorylate additional targets, including both the receptors and the major substrates, STATs (Yamaoka et al. 2004). The specificity of ligand-mediated STAT activation is dependent on the interactions between SH2 domain and phosphotyrosine residues. The current understanding of JAK domains will be elaborated in the subsequent sections.

2.4 Structure of STAT and Translocation to the Nucleus

STATs function both as signal transducers and transcription factors. STATs are latent transcription factors that reside in the cytoplasm until activated. Like JAKs, STATs also possess distinct conserved domains: the N-terminal coiled-coil domain, DNA-binding domain, a linker, SH2 domain, and C-terminal transactivation domain (Figures 2.1 and 2.2). The transactivation domain interacts with multiple partners forming transcriptional complexes. Tyrosine at position 705 is important when it comes to STAT activation (Yu and Jove 2004). Other than STAT2 and STAT6, a conserved serine at position 727 is another site of phosphorylation that controls STAT transcriptional activity (Decker and Kovarik 2000). The DNA-binding domain is the central region that controls DNA-binding selectivity for each STAT molecule. Between position 600 and 700 amino acid residues, the SH2 domain is located that aids the dimer formation between two activated STAT molecules via SH2-phosphotyrosine interaction (Horvath, Wen, and Darnell 1995). The linker region is located between the amino acid residues 500 and 575 (Subramaniam et al. 2013).

After activation by tyrosine phosphorylation, STATs become dimerized and translocate into the nucleus, where they act as transcription factors. While most STATs form homodimers, evidences support heterodimer formation of STAT1/2, STAT1/3, and STAT5A/B (Delgoffe and Vignali 2013). STAT1 has been reported to exist as pre-formed homodimers, and phosphorylation induces reorientation. STAT molecules face restricted movement while translocating from cytoplasm to nucleus by the nuclear envelope owing to molecular weight >40 kDa (Mertens et al. 2006). Thus, STATs require facilitated transport mediated by specific receptor molecules that are known as importins (adaptor proteins) bearing α and β subunits. Using nuclear localization sequence (NLS), STATs bind to importin α (Gorlich and Mattaj 1996, Jans, Xiao, and Lam 2000). Following interaction of both the importin subunits, STAT–importin complex gets docked at the nuclear pore complex (NPC) (Gorlich and Mattaj 1996). The NPCs form high-order octagonal channels that are composed of proteins called nucleoporins, some of which contain hydrophobic phenylalanine/glycine (FG)-rich repeat motifs (Wolf and Mofrad 2008). The energy

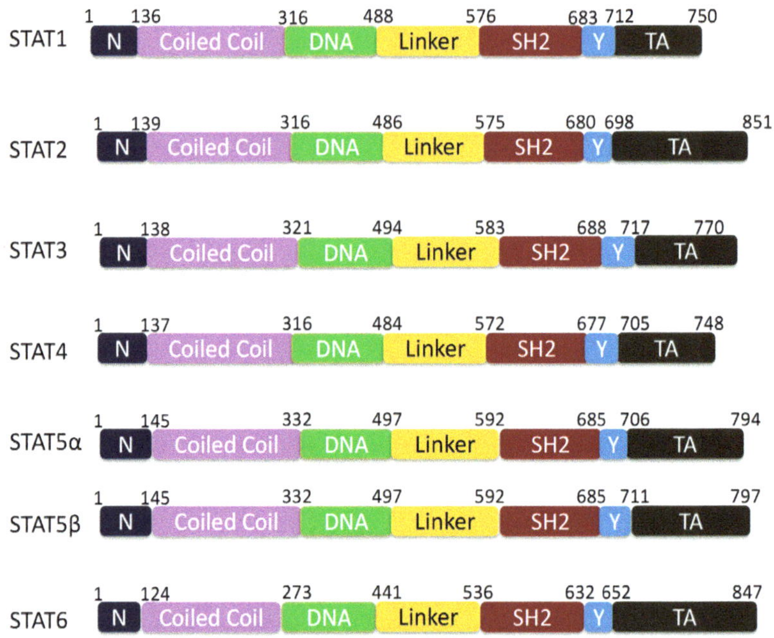

FIGURE 2.2 STAT domain structure. Minor differences in various human STAT molecules has been depicted.

for the translocation is provided by Ran, a GTPase. The intracellular RanGTP/RanGDP gradient essentially drives this active transport, which allows the accumulation of cargo against a concentration gradient. In the cytosol, the concentration of the GTP form of Ran is low due to nucleotide hydrolysis by RanGAP, which is cytoplasmically localized by RanGTPase-activating protein (Reich 2013). A differential requirement of importin α has been attributed for STAT translocation (Goldfarb et al. 2004). Out of the six human importin α molecules, STAT1 requires importin α5, while STAT5 requires importin α3 (Sekimoto et al. 1997, Liu, McBride, and Reich 2005). One study even argued that STATs may not be having functional NLS (Subramaniam, Torres, and Johnson 2001). Furthermore, it has also been suggested that NLS-binding site on STAT1 and STAT3 differs from its binding site in other proteins. STAT3, 5, and 6 could also be translocated to the nucleus in unphosphorylated form (Reich 2013). Carrier-dependent export of unphosphorylated STATs requires the specific export receptor, CRM1 (Fornerod et al. 1997). Translocation of unphosphorylated STATs requires direct interaction with nucleoporins and is carrier and energy-free, devoid of any help from importins. This process also does not require any activation by cytokines.

2.5 JAK Domains

As mentioned previously, JAKs contain specific domains. In this section, we discuss about the domains found in JAKs.

2.5.1 Protein Tyrosine Kinase Domain

The JH1 domain is the PTK domain of the JAK protein. Similar to other protein tyrosine kinases, JAKs possess an "activation loop" that regulates kinase activity, which is a major site of autophosphorylation (Toms et al. 2013). Using crystal structures of JAK2 and JAK3, it has been demonstrated that the PTK has an open conformation representing the active conformation of the protein. JAK2 and JAK3 crystal structures have been solved at 2A° and 2.5A°, respectively (Boggon et al. 2005, Lucet et al. 2006). JAK2 PTK domain has a small N-terminal lobe, comprised of one α-helix and five anti-parallel β strands, and large C-terminal lobe comprised of eight α-helices (Taylor et al. 1992). The JAK2 kinase domain is relatively an "open" conformation, but the ATP-binding site is relatively "closed" when compared to other kinases (Wilks 2008). The two lobes of the JAK2 kinase domain are slightly twisted with respect to other kinases. The ATP-binding site is relatively less accessible as a consequence. A loop structure located between amino acids 1056–1078, termed the JAK2 kinase insertion loop, is rather a unique feature not observed in any other kinases (Wilks 2008). JAK3 has ~60% sequence identity with JAK2 and most of the amino acid residues are conserved in the ATP-binding cleft (Boggon et al. 2005).

2.5.2 Pseudokinase Domain

JAKs are named after Roman God "Janus" as they have a kinase and "kinase-like" pseudokinase domain. The pseudokinase domain has been suggested to have critical positive and negative regulatory roles (Saharinen, Takaluoma, and Silvennoinen 2000, Saharinen, Vihinen, and Silvennoinen 2003). Despite lacking key residues that are shown to be important for catalytic activity, this domain has retained the capability to bind nucleotides in the presence of divalent cations and mediates essential functions in regulating catalytic activity. The pseudokinase domain shares high degree of sequence similarity with the kinase domain, but residues required for phosphotransferase activity are modified from the canonical motifs. Among the four JAK molecules, only the JAK2 pseudokinase domain possesses weak catalytic activity and autophosphorylates itself *in cis* on two auto-inhibitory phosphorylation sites, Ser523 and Tyr570 (Argetsinger et al. 2004, Ishida-Takahashi et al. 2006, Ungureanu et al. 2011). Evidence of JAK2-specific mechanism of activation comes from the fact that those residues are not conserved among the other JAKs. Two models of JAK activation have been hypothesized. The pseudokinase domain binds and inhibits the kinase domain either *in cis* or *in trans* (Babon et al. 2014). The JAK pseudokinase domains act as protein interaction modules, whose primary function is to bind and inhibit the JH1 domain activation. However, many studies have reported conflicting findings on the mechanism of inhibition. Furthermore, it is suggested that nucleotide binding to these pseudokinase domains primarily function to modulate the overall conformation of the JAK (Hammaren et al. 2015). The amino acid residues forming the "activation loop" in the JAK2, JAK1, and TYK2 pseudokinase domain form a well-defined conformation that resembles inactive protein kinase structures. These structures have been hypothesized to block the putative substrate binding site, subverting the fact that a rearrangement of JAK2 pseudokinase activation loop is required for phosphoryl-transfer activity. The importance to arrange a specific conformation to allow intermolecular interactions is evident from crucial stabilization mediated by ATP binding on purified recombinant JAK1 and JAK2 pseudokinase domains (Hammaren et al. 2015). In polycythemia vera (PV) patients, mutations in the JAK2 pseudokinase-SH2 linker region, including K539I, were recently characterized that reportedly led to constitutively activated JAK2 (Zhao et al. 2009). Two distinct mutational hotspots within the JAK2 pseudokinase domain have been identified based on functional studies of activation (Babon et al. 2014). Close to V617F, mutations point to the N-terminal lobe of the domain. The pseudokinase-SH2 linker region was found to confer factor-independent growth. However, there was no detectable increase in the catalytic activity (Babon et al. 2014). Interestingly, the crystal structure of the JAK2 pseudokinase domain possessing the activating mutant, V617F, indicates toward increased structural integrity within this domain leading to pathogenesis (Babon et al. 2014). In contrast, another structure of the JAK1 pseudokinase domain demonstrated that the αC region length can be variable regardless of the presence of the subsequent activating mutation, V658F (Toms et al. 2013). Examples exist to support JAK phosphorylation on other sites that appear to be important for regulation of catalytic activity. For example, Y785 in JAK3 and Y813 in JAK2 denote other important sites of autophosphorylation (Kurzer et al. 2004). These sites recruit SH2-domain-containing adapter molecules such as SH2-Bb. For JAK2, this could be a mechanism

2.6 STAT Domains

As described previously, STAT proteins have a modular structure with six well-defined domains (Baker, Rane, and Reddy 2007). The functions of STATs are mediated by N-terminal region since small deletions in this region do away with the STAT phosphorylation (Baker, Rane, and Reddy 2007). The N-terminal region of STAT is well conserved among family members and cooperates with the DNA-binding domain (Baker, Rane, and Reddy 2007). Additionally, this domain also has major roles in nuclear import, export, and receptor binding. Coiled-coil domain of STAT interacts with regulatory proteins, adopting α-helical conformation in doing so (Kisseleva et al. 2002). This domain functions in receptor binding. Similar to the N-terminal domain, the DNA-binding domain is also highly conserved among STAT family members (Baker, Rane, and Reddy 2007). Apart from STAT2, all STAT homodimers differentially bind more than 10 related γ-activated sequence elements characterized by the consensus sequence, TTNCNNNAA (Horvath, Wen, and Darnell 1995, Xu, Sun, and Hoey 1996). The linker domain functions as a spacer to retain true conformation between the dimerization and DNA-binding domains (Baker, Rane, and Reddy 2007). The SH2 domain, which is also found in JAKs, is highly conserved among the STAT molecules. In playing a key role in STAT signaling, the SH2 domain recruits STATs to activated receptor complexes. Moreover, the SH2 domain is also meant for interaction with JAK and Src kinases (Baker, Rane, and Reddy 2007). The same domain is also required for dimerization (both homo- and heterodimerization) of STAT molecules. This event leads to nuclear localization and DNA-binding activities (Baker, Rane, and Reddy 2007). The transactivation domain unsurprisingly modulates the transcriptional activation of target genes but varies among family members. Modulation of transcriptional activation is evident when the C-terminally truncated isoforms of STAT3, 4, and 5 minus the portions of their transactivation domains, act as dominant-negatives (Schindler and Strehlow 2000).

In an unstimulated cell, STATs are monomeric and remain in unphosphorylated state. Cytokine stimulation induces phosphorylation of tyrosine residues on the receptor that serve as docking sites for STATs via their SH2 domains. Once bound to the receptor, all members of the STAT family become tyrosine phosphorylated in response to cytokine stimulation at a conserved C-terminal tyrosine residue. For example, phosphorylation of Y694 in STAT5 can be mediated by growth-factor receptors as well as by JAK and Src kinases, depending on the nature of the cell type and the ligand/receptor interactions (Nosaka et al. 1999). This phosphorylation event induces STAT homo- and heterodimerization via the interaction of the SH2 domain of one STAT molecule with the phosphotyrosine residue of another. Once phosphorylated, the dimerized STATs are then able to translocate to the nucleus. In addition to tyrosine phosphorylation, several STATs are regulated by serine phosphorylation at a conserved PSMP motif, which is located in the transactivation domain (Jatiani et al. 2010). C-terminal serine phosphorylation is stimulated by several cytokines and is mediated by serine/threonine kinases that include extracellular responsive kinase, p38, JNK, and protein kinase C-δ (Khwaja 2006). This phosphorylation event positively regulates the transactivation potential of these proteins. One study has revealed that the specificity of signaling by STAT1 depends on SH2 and C-terminal domains that regulate Ser727 phosphorylation (Kovarik et al. 2001). This culminates in differential regulation of specific target gene expression. To acquire complete activation of STAT1, phosphorylation needs to happen at both Y701 and S727 residues (Kovarik et al. 2001). Interestingly, S727 phosphorylation of STAT1 in IFN-γ-treated mouse fibroblasts requires an intact SH2 domain and phosphorylation of Y701 without any requirement for p38 MAPK, ERK-1 and -2, and JNK (Kovarik et al. 2001). In contrast, UV-induced STAT1 phosphorylation on S727 does not need SH2 domain–phosphotyrosine interactions, but require p38 MAPK (Kovarik et al. 2001). Mutation of S727 differentially affected IFN-γ target genes, at the level of both basal and induced expression. Furthermore, S727 of STAT3 is phosphorylated in response to stimuli that differ from those for STAT1 S727, and transfer of the STAT3 C terminus to STAT1 changed the pathway specificity of STAT1 Ser727 phosphorylation to that of STAT3 (Kovarik et al. 2001). This series of experiments show that STAT C terminus contributes to the specificity of cellular responses by linking individual STATs to different kinase pathways and doing so through an intrinsically different requirement for serine phosphorylation at different target gene promoters.

2.7 Negative Regulation of JAK-STAT Pathway

It is evident that cytokines or growth factors signal through JAK-STAT pathway. Unsurprisingly, multiple studies over several years have shown that dysregulated cytokine receptor signaling is the foundation of many mammalian diseases. Normal cytokine signaling is dependent on the complex tuning of cytokine-dependent signals. There are mechanisms that work against these signals to allow a cell to be attuned to additional extracellular signals. There are multiple mechanisms by which cytokine signaling via JAK-STAT pathway is negatively regulated. In this section, we deliberate upon the negative regulation of JAKs via phosphatase, inhibition of STATs by protein inhibitors of activated STATs (PIAS) and suppressor of cytokine signaling (SOCS) proteins.

2.7.1 Effect of Protein Tyrosine Phosphatases (PTPS)

Tyrosine phosphorylation leads to the signaling that emanates ensuing cytokine–cytokine receptor interaction. The event of JAKs phosphorylating each other on tyrosine residues in activation loops is known as transphosphorylation. Due to the critical importance of JAK tyrosine phosphorylation for signal transduction, dephosphorylation of tyrosine residues is a mechanism employed by cells to attenuate kinase activity of JAKs. Several protein phosphatases act as modulators of JAK-STAT signaling. They include the SH2 domain-containing tyrosine phosphatases (SHPs), protein tyrosine phosphatase 1B (PTP1B), T-cell protein tyrosine phosphatase (TC-PTP), and protein tyrosine phosphatase-basophil like (PTP-BL). Among these, SHPs are well characterized. The classical PTPs are Cys-based class I PTPs that are true tyrosine-specific PTPs (Alonso et al. 2004). Based on their cellular localization, the classical PTPs are divided into transmembrane (TM) and non-transmembrane (non-TM) families (Tonks 2006). Few TM PTPs have been identified in T cells. Non-TM PTPs are better characterized than their TM counterparts (Lorenz 2009). Non-TM PTPs contain a single catalytic PTP domain flanked by N-/C-terminal extensions required for localization (Lorenz 2009).

SHPs were the first phosphatases to be identified. The human genome project has identified more than 100 putative PTPs (Alonso et al. 2004). The major SHPs are SHP-1 and SHP-2. SHP-1 is primarily expressed in hematopoietic cells, while SHP-2 is ubiquitously expressed (Wu et al. 2003). A spliced variant of SHP-1 has been identified as SHP-1L (Jin, Yu, and Burakoff 1999). Both SHP-1 and SHP-2 are composed of a central catalytic domain, two N-terminal SH2 domains, and a C-terminus (Poole and Jones 2005). The central catalytic domain contains the typical PTP signature motif VHCSAGIGRTG (Poole and Jones 2005). These phosphatases are named due to the two SH2 domains located N-terminal to the tyrosine phosphatase domain. The SH2 domains aid in SHP recruitment to tyrosine-phosphorylated intracellular portions of many receptors that aid in localization and activity regulation (Pao et al. 2007). SHPs inhibit JAK-STAT signaling in two ways: by impeding interactions with other effectors and by activating the elimination of the phosphotyrosine motifs acting as docking sites for effectors (Seif et al. 2017). The SHP-2 SH2 domains display overlapping binding specificities with the SOCS SH2 domain, indicating that both negative regulators compete for binding sites on phosphorylated receptors. The interaction of the SHP SH2 domains with their binding partners is a key step toward the activation of the phosphatase activity. At the basal state, this activity is repressed as N-terminal SH2-domain occupies the phosphatase active site of PTP in the intramolecular fashion (Hof et al. 1998). The release of this suppression leads to an activation of the phosphatase upon engagement of the SH2 domains. SHP-1 is a major negative regulator of receptor signaling in hematopoietic and epidermal cells (Lorenz 2009). Apart from the SH2 domains, the C-terminus could regulate the PTP-1 activity of SHP-1. Upon stimulation tyrosine phosphorylation of the C-terminal Y536 in SHP-1, PTP activity is enhanced (Zhang et al. 2003). Phosphorylation of Tyr564 leads to a modest increase of SHP-1 activity (Zhang et al. 2003). The role of SHP-2 in growth factor signaling remains less understood. Additional studies are required to completely elucidate the role of SHP-2 in hematopoiesis. The intrinsic differences in the catalytic domain allow SHP-2 to portray different substrate specificities. SHP-2 associates with multiple growth-factor receptors, but few of those interactions are true SHP-2 substrates. Additionally, SHP-2 could enact as an adapter and has several phosphotyrosine motifs that can recruit downstream effectors. Phosphorylated tyrosines have been hypothesized to generate docking sites for SH2-domain containing proteins, which could be the basis of either an adapter function for SHP-2 or a way for preventing auto-dephosphorylation. SHP-1L lacks the potential Y564 phosphorylation site, but retains the Y536 site of SHP-1 (Lorenz 2009). SHP-1L, however, has a proline-rich motif in the C-terminus, which could mediate binding to SH3-domain containing proteins (Lorenz 2009). A similar proline-rich motif is absent in SHP-1, but can be observed in the C-terminus of SHP-2 (Lorenz 2009). Critical biological functions of SHP-1 and SHP-2 have been demonstrated using mice models.

PTP1B is a negative regulator of the insulin and leptin signaling pathways. PTP1B comprises of a single catalytic domain with N- or C-terminal extensions (Fischer, Charbonneau, and Tonks 1991). Similar to SHP-1L, PTP1B has a proline-rich region adjacent to the C-terminal and a hydrophobic region (Liu, Hill, and Chernoff 1996). Substrates are recruited by the interaction of the proline-rich region with the SH3 domains. The hydrophobic sequence is inserted into the endoplasmic reticulum membrane and ties up PTP1B to the cytoplasmic side of the ER (Frangioni et al. 1992). The substrate selectivity of PTP1B is controlled partially by the modular protein interaction and the spatial restriction imposed by subcellular compartmentalization (Tonks 2003). PTP1B also acts as a negative regulator of brown fat adipogenesis (Matsuo et al. 2011).

The phosphatases, PTP1B and TC-PTP, share 74% identity between their catalytic domains (Tonks, Diltz, and Fischer 1988). TC-PTP was characterized almost three decades ago. cDNA of PTPN2 codes for TC-PTP. TC-PTP, ubiquitously expressed, shows its highest expression in hematopoietic tissues (Ibarra-Sanchez et al. 2001). TC-PTP could be induced by either a mitogen (Con A) or an anti-inflammatory cytokine (IL-10) (Rajendrakumar, Radha, and Swarup 1993, Williams et al. 2004). Expression of TC-PTP results in two isoforms due to alternative splicing. The two forms share the conservative catalytic PTP domain of 272 amino acids, but differ in the C-terminal (Kamatkar et al. 1996). The well-characterized form of TC-PTP has 387 amino acids and has MW of 45-kDa (TC45), while the less-abundant form consists of 415 amino acids with a MW of 48-kDa and expressed in the ER and the nuclear membrane (Iversen et al. 2002, Bourdeau, Dube, and Tremblay 2005). The C-terminal of TC45 aids in nuclear localization and also binds to DNA. However, in general, the C-terminal of TC-PTP negatively regulates the enzyme function. TC-PTP and PTP1B have complementary specificities for dephosphorylation of JAKs and STATs. Minor differences between JAK1/JAK3 and JAK2/TYK2 sequences determine the substrate specificity of PTP1B and TC-PTP. JAK1 and JAK3 activation loops contain either a Thr or Val residue C-terminal to the tandem phosphotyrosines. In contrast, JAK2 and TYK2 comprise of (E/D)-pY-pY-(R/K) motif in their activation loops (Myers et al. 2001). This specific motif has been revealed to be the preferred substrate of PTP1B. Both PTP1B and TC-PTP participate in STAT dephosphorylation (Bohmer and Friedrich 2014). Biological roles of PTP1B and TC-PTP have been characterized using mouse knockout studies. Mice deficient in TC-PTP gene exhibit multiple defects in the lymphoid population (You-Ten et al. 1997). TC-PTP-deficient mice die neonatally from a systemic inflammatory disease characterized by mononuclear cell infiltration. PTP1B-deficient mice demonstrate hypersensitivity to various growth factors (You-Ten et al. 1997, Hendriks et al. 2008). The other phosphatase, PTP-BL, dephosphorylates STAT1, STAT3, STAT4, STAT5, and STAT6 both *in vitro* and *in vivo.*

2.7.2 Effect of Protein Inhibitor of Activated STATs

PIAS proteins are activation-suppressing proteins for STATs incorporating transcriptional regulation of genes. The PIAS family of proteins consists of five proteins in mammals: PIAS1, PIASxα,

PIASxβ, PIAS3, and PIASy (Chung et al. 1997). PIAS1 has been named as it has been shown to inhibit STAT1, while PIAS3 interacts with STAT3 (Chung et al. 1997, Liu et al. 1998). Interestingly, PIAS proteins neither act only as inhibitors nor show any specificity for STATs. PIAS can regulate transcription either positively or negatively. The domain structure of PIAS include an N-terminal SAP domain (SAF-A/B, Acinus, and PIAS), apoptotic chromatin-condensation inducer in the nucleus, the PINIT (Pro-Ile-Asn-Ile-Thr) motif, a C3HC4-motif type RING finger-like zinc-binding domain, an acidic amino acid domain (AD), SUMO-interacting motif, and a variable C-terminal Ser/Thr-rich (S/T) domain (Aravind and Koonin 2000, Jackson 2001, Jimenez-Lara, Heine, and Gronemeyer 2002, Duval et al. 2003). The SAP domain, the PINIT motif, and the RING domain contribute to the nuclear localization of PIAS proteins (Duval et al. 2003). The SAP domain is important for sequence/structure-specific DNA binding (Aravind and Koonin 2000). PIAS proteins could modulate JAK-STAT signaling by various means including SUMOylation that attenuate transcription factor activity (Niu et al. 2018). The RING domain might interact with other proteins, which is important for SUMO-E3-ligase function (Sachdev et al. 2001). The function of AD and S/T domain remain less explored. SUMO ligase-independent modulation of transcription factor activity of PIAS proteins exists (Kotaja et al. 2002). One such mechanism is blocking the DNA-binding of a transcription factor. PIAS1 has been shown to inhibit DNA-binding activities of STAT1 (Liu et al. 1998). Similarly, PIAS3 interaction with STAT3 inhibited the DNA-binding activity of STAT3 (Chung et al. 1997). It has been hypothesized that DNA-binding activity is inhibited by PIAS, obstructing the DNA-binding site of the transcription factor within the PIAS: transcription factor complex. This could result in a conformational change or dissociation of a transcription complex due to PIAS engagement of transcription factor. Co-regulators, which can get recruited to a protein complex, are important players for PIAS-mediated regulation of transcription. PIAS proteins interact with HDAC molecules. Knockout mice studies have indicated functional redundancy between PIAS family members. It is evident in PIASy-deficient mice that reveal minor defects in Wnt and IFN signaling (Roth et al. 2004). Understanding the complex *in vivo* functions of PIAS proteins will depend on multiple experimental approaches.

2.7.3 Effect of Suppressor of Cytokine Signaling

JAK-STAT pathway is also negatively regulated by suppressor of cytokine signaling proteins or SOCS proteins. These proteins are not constitutively expressed. SOCS proteins bind to JAKs, cytokine receptors, and signaling molecules. SOCS genes encode a family of eight proteins (SOCS1-7 and cytokine-inducible SH2 protein (CIS)) (Endo et al. 1997, Naka et al. 1997, Starr et al. 1997). These proteins contain an N-terminal of variable length, a central SH2 domain that determines binding specificity, and a C-terminal termed as SOCS box (40 amino acid module) that is implicated in proteasomal degradation (Endo et al. 1997, Naka et al. 1997, Starr et al. 1997). CIS, SOCS1, 2, and 3 are well characterized, while the biological roles of SOCS4–SOCS7 remain poorly understood (Hilton et al. 1998). CIS/SOCS family proteins act as E3 ubiquitin ligase, mediating protein degradation of protein families associated with them. A 24-amino acid kinase inhibitory region (KIR) is present in the N-terminal of SOCS1 and SOCS3 (Nicholson et al. 1999). A series of studies on SOCS proteins have revealed that a wide range of cytokines that utilize JAK-STAT signaling are inhibited by SOCS. The SH2 domain of SOCS1 inhibits JAK2 by binding to Tyr1007 through insertion of its KIR into the activation loop of the kinase domain (Yasukawa et al. 1999). This prevents substrates from getting into the catalytic pocket by acting as pseudosubstrate. SOCS1 has also shown to abrogate tyrosine phosphorylation of STAT1 via direct binding to phosphorylated type I IFN receptors (Fenner et al. 2006). A model of SH2 domain: receptor interaction and KIR-mediated inhibition have been proposed. However, the SH2 domain of SOCS3 does not possess a strong affinity toward activation loops of JAKs (Yoshimura and Yasukawa 2012). The absence of SOCS1 expression due to gene methylation results in the development of a range of primary cancers (Sutherland et al. 2004). SOCS3 first binds to the receptor, followed by inhibition of JAK activity via a KIR-mediated JAK inhibition (Sasaki et al. 1999). SOCS3 also inhibits the signaling initiated by various stimuli that induce its expression. This inhibition is predominantly achieved by SOCS3 engagement of the activated receptors of these cytokines. SOCS3 has been demonstrated to specifically bind to phosphorylated tyrosines of cytokine/growth-factor receptors. A number of these receptor-docking sites also act as binding sites for another signaling regulator, SHP2. The relative contribution of SOCS and SHP2 to signal modulation and pathway crosstalk is still not completely understood. Intriguingly, SOCS3 is not phosphorylated by JAKs, but relies on other kinases, such as Src kinases and receptor-tyrosine kinases, for activation (Sommer et al. 2005). Absence of SOCS3 has been implicated to enhance susceptibility to chronic inflammatory diseases. A lack of SOCS3 correlates with the development and progression of malignancies (Riehle et al. 2008). The inhibitory mechanisms of CIS and SOCS2 differ from that of SOCS1 and SOCS3 (Krebs and Hilton 2001). Neither CIS nor SOCS2 acts on JAKs (Krebs and Hilton 2001). The biological role of CIS is not completely understood. There might be functional redundancy of CIS. Role of SOCS2 has been evaluated following the development of SOCS2-deficient mice that displays gigantism phenotype rising from dysfunction of growth hormone axis (Metcalf et al. 2000). SOCS2 partially inhibits recruitment of STAT5 by binding pTyr487 and pTyr595 on the growth hormone receptor (Greenhalgh et al. 2005). A number of studies have linked SOCS2 to cancer, including colon and prostate cancer.

SOCS are basically ubiquitin ligases that promote the degradation of their partners. The SOCS box interacts with Elongin B and Elongin C, Cullin-5, and RING-box-2 to form an E3 ligase complex termed as ECS (Elongin B/C-Cullin-SOCS box protein) (Bullock et al. 2006). The ECS binds to a ubiquitin-activating enzyme (E1) and a ubiquitin-conjugating enzyme (E2), allowing the ECS complex to recognize and bind target proteins for polyubiquitination and degradation via the 26S proteasome (Kamura et al. 2004). Proteasomal degradation has emerged as a key mechanism of signal attenuation. The nuclear protein, SLIM, has been characterized as a ubiquitin E3 ligase that targets STATs for degradation via the proteasome (Tanaka, Soriano, and Grusby 2005). The interaction of the SOCS box with the ECS complex provides a clear link between SOCS proteins and proteasomal degradation that is central to attenuating the levels of SOCS proteins within the cell (Zhang et al. 1999). It can be argued if SOCS proteins can act as adaptors to bridge their binding partners to the proteasome and induce their degradation. Various overexpression studies including SOCS proteins have supported a cross-talk between different members of the SOCS family. Physiological evidence for the importance of the SOCS box has been garnered from *in vivo* studies.

2.7.4 STAT Inhibitors as Therapeutic to Treat Inflammatory Disorders

JAK-STAT pathway is critical to the pathogenesis of inflammatory disorders mediated by the immune system. Back in the 1990s, the potential of JAK inhibitors to treat autoimmune and inflammatory disorders had been conceived (Banerjee et al. 2017). The role of JAK-STAT signaling in various inflammatory disorders has been extensively mentioned in this book. Extensive efforts have been generated to develop second-class JAK inhibitors that are more selective. Dysregulated expression of STAT molecules has been observed in various cancers. Hence, the discovery of STAT inhibitors has mainly targeted tumor cells. In the last section of the chapter, we will discuss the development of STAT inhibitors to treat inflammatory disorders. We will discuss the advent of small molecule inhibitors of STAT molecules for therapeutic purposes.

2.7.4.1 STAT1 Inhibitor

Studies have demonstrated the development of very few STAT1 inhibitors. BRD0476, a novel suppressor of pancreatic β-cell apoptosis, was developed. BRD0476 inhibited IFN-γ-induced

JAK2 and STAT1 signaling to augment β-cell survival (Chou et al. 2015). BRD0476 inhibits JAK-STAT signaling without suppressing the kinase activity of any JAK. An analog of BRD0476 was developed by adding quinolone moieties in place of naphthyl that led to more than 1000-fold solubility as well inhibited IFN-γ/STAT1 signal transduction (Scully et al. 2013). Pravastatin is another STAT1 inhibitor that attenuates the IFN-γ levels and prevents STAT1 phosphorylation by inhibiting HMG-CoA reductase (Zhou, Gao, and Sun 2009). Pravastatin treatment on Apolipoprotein E-deficient mice fed on a cholesterol-rich diet suppressed IFN-γ levels in both serum and atherosclerotic lesions, associated with reduced STAT1 activation, attenuated IRF1 expression, and induced activity of SOCS1 in aorta tissue (Zhou, Gao, and Sun 2009). ISS-840, a phosphopeptide mimetic of STAT1, disrupted STAT1 dimerization leading to moderately potent inhibition of STAT1 signaling (Gunning et al. 2007). Fludarabine, a compound used to treat hematologic malignancies, is another STAT1 activation inhibitor causing a specific depletion of STAT1 protein, but not of other STAT molecules (Feng et al. 2017). Fludarabine exerts apoptosis through increasing Bax and decreasing Bid, XIAP, and survivin expression (Meng et al. 2007).

2.7.4.2 STAT3 Inhibitor

Among the STAT-specific inhibitors, STAT3 inhibitors are most widely studied. S3I-201, a novel STAT3 inhibitor identified from NCI chemical library, displayed potent inhibition of STAT3 DNA-binding activity with IC50 of 86 μM in cell-free assays (Siddiquee et al. 2007a). S3I-201 inhibited STAT3 transcriptional activities, and induced growth inhibition and apoptosis of tumor cells expressing constitutively active STAT3 (Grandis et al. 2000). In liver fibrosis, angiogenesis and fibrogenesis were attenuated by S3I-201 (Wang et al. 2018). Derivatives of S3I-201, S3I-201.1066, S3I-1757, and BP-1-102, displayed increased potencies of STAT3 inhibition (Zhang et al. 2010, 2012, 2013). S3I-201 administration resulted in appreciable decrease in IFN-γ, T-bet, IL-17A, RORγt, Stat3, IL-21, and IL-22 levels, and increase in Foxp3 and Helios production $CD4^+$ T cells in BTBR mice, suggesting that S3I-201 could be used to treat autism (Ahmad et al. 2018). S3I-M2001 is an oxazole-based peptidomimetic of the STAT3 SH2 domain-binding phosphotyrosine peptide that selectively disrupts active STAT3:STAT3 dimers in various cancers where STAT3 expression is aberrant (Siddiquee et al. 2007b). Stattic, a nonpeptidic small molecule identified after screening of chemical libraries, has been displayed to selectively inhibit the function of the STAT3 SH2 domain at 10 μM, independent of the STAT3 activation state *in vitro* (Schust et al. 2006). STAT3 activation, dimerization, and nuclear translocation are selectively inhibited by Stattic (Schust et al. 2006). Fragments of STX-0119 (another STAT3 SH2 domain antagonist) and Stattic were chemically fused to generate HJC0123, which suppressed STAT3 phosphorylation and transcriptional activity in breast cancer cells and induced anti-tumor cell effects against breast and pancreatic cancer cells *in vitro* at IC50 values of 0.1–1.25 μM (Ashizawa et al. 2011, Chen et al. 2013b). By inhibiting STAT3, Stattic reduces the growth and increases the apoptosis of NPC and sensitizes NPC to cisplatin and IR (Pan et al. 2013). STA-21 is another small molecule inhibitor of STAT3 with a potency of 20 μM. STA-21 inhibits STAT3 dimerization and binding to DNA resulting in a transcriptional blockade of STAT3-dependent genes (Song et al. 2005). STA-21 has been shown to be useful in several inflammatory disease conditions (Park et al. 2014). STA-21 suppresses joint inflammation in a mouse CIA model through regulation of the transcription factors, Foxp3 and RORγt, in $CD8^+$ T cells and inhibits the development and progression of rheumatoid arthritis (Ahmad et al. 2017). Studies suggest that LLL12, another small-molecule inhibitor of STAT3 signaling, might suppress STAT3 activation by blocking its recruitment to the receptor, preventing phosphorylation by tyrosine kinases, and by interfering with dimerization (Bid et al. 2012). LLL12 suppressed cell viability, induced apoptosis, and repressed colony formation and migration *in vitro* in studies of glioblastoma, osteosarcoma, and breast cancer cells (Lin et al. 2010, Onimoe et al. 2012). In a drug development program involving virtual ligand screening, 2-D similarity screening, 3-D pharmacophore analysis, and SAR-based medicinal chemistry, C188-9, a potent small-molecule that targets the Src-homology

formation of phosphorylated STAT6 with its recognition DNA sequence (Sakurai et al. 2003). A novel inhibitor of STAT6, PM-242H, inhibited initiation of allergic disease induced by airway fungal challenge, reversed established allergic airway disease in mice, and blocked salmeterol-dependent enhanced allergic airway disease (Knight et al. 2015). *In vitro* structure–activity relationship studies led to the development of the lead compound, PM-43I (Knight et al. 2018). Conducting initial dose range, toxicity, and pharmacokinetic experiments with PM-43I potently inhibits both STAT5- and STAT6-dependent allergic airway diseases in mice (Knight et al. 2018). When cellular activity screen was performed for evaluating the binding capacity to SH2 domains of STAT molecules, PM-86I showed the highest specificity for STAT6 and no cross-reactivity to any of the additional targets at the highest dose tested (Knight et al. 2018). PM-86I can specifically inhibit STAT6-dependent adaptive Th2 immune responses when administered systemically (Knight et al. 2018). Other notable inhibitors of STAT6 signaling include AS1810722, a fused bicyclic pyrimidine derivative (IC_{50}value of 2.4 nM) (Nagashima et al. 2009). AS1810722 is potent, orally active STAT6 inhibitors that could be developed for the treatment of STAT6-dependent allergic diseases, such as asthma. AS1810722 showed potent STAT6 inhibition and a good CYP3A4 inhibition profile (Nagashima et al. 2009).

2.8 Concluding Remarks

JAK-STAT is one of the well-characterized signaling pathways involved in maintaining homeostasis. Aberrant regulation of this pathway affects the downstream signaling molecules leading to deleterious consequences including onset and pathophysiology of auto-immune disorders. This chapter has elaborated on the structure of JAK and STAT domains that can be targeted for therapeutic benefits. Subsequent chapters of the book have discussed the implication of JAK-STAT signaling in various inflammatory disorders.

REFERENCES

Ahmad, S. F., M. A. Ansari, A. Nadeem, S. A. Bakheet, M. A. Alshammari, M. R. Khan, A. M. S. Alsaad, and S. M. Attia. 2018. "S3I-201, a selective Stat3 inhibitor, restores neuroimmune function through upregulation of Treg signaling in autistic BTBR T(+) Itpr3(tf)/J mice." *Cell Signal* 52:127–36. doi: 10.1016/j.cellsig.2018.09.006.

Ahmad, S. F., M. A. Ansari, A. Nadeem, K. M. A. Zoheir, S. A. Bakheet, A. M. S. Alsaad, O. A. Al-Shabanah, and S. M. Attia. 2017. "STA-21, a STAT-3 inhibitor, attenuates the development and progression of inflammation in collagen antibody-induced arthritis." *Immunobiology* 222 (2):206–17. doi: 10.1016/j.imbio.2016.10.001.

Akira, S., Y. Nishio, M. Inoue, X. J. Wang, S. Wei, T. Matsusaka, K. Yoshida, T. Sudo, M. Naruto, and T. Kishimoto. 1994. "Molecular cloning of APRF, a novel IFN-stimulated gene factor 3 p91-related transcription factor involved in the gp130-mediated signaling pathway." *Cell* 77 (1):63–71.

Alonso, A., J. Sasin, N. Bottini, I. Friedberg, I. Friedberg, A. Osterman, A. Godzik, T. Hunter, J. Dixon, and T. Mustelin. 2004. "Protein tyrosine phosphatases in the human genome." *Cell* 117 (6):699–711. doi: 10.1016/j.cell.2004.05.018.

Aravind, L., and E. V. Koonin. 2000. "SAP – a putative DNA-binding motif involved in chromosomal organization." *Trends Biochem Sci* 25 (3):112–14.

Ardito, F., M. Giuliani, D. Perrone, G. Troiano, and L. Lo Muzio. 2017. "The crucial role of protein phosphorylation in cell signaling and its use as targeted therapy (review)." *Int J Mol Med* 40 (2):271–80. doi: 10.3892/ijmm.2017.3036.

Argetsinger, L. S., J. L. Kouadio, H. Steen, A. Stensballe, O. N. Jensen, and C. Carter-Su. 2004. "Autophosphorylation of JAK2 on tyrosines 221 and 570 regulates its activity." *Mol Cell Biol* 24 (11):4955–67. doi: 10.1128/MCB.24.11.4955-4967.2004.

Ashizawa, T., H. Miyata, H. Ishii, C. Oshita, K. Matsuno, Y. Masuda, T. Furuya, T. Okawara, M. Otsuka, N. Ogo, A. Asai, and Y. Akiyama. 2011. "Antitumor activity of a novel small molecule

STAT3 inhibitor against a human lymphoma cell line with high STAT3 activation." *Int J Oncol* 38 (5):1245–52. doi: 10.3892/ijo.2011.957.

Auzenne, E. J., J. Klostergaard, P. K. Mandal, W. S. Liao, Z. Lu, F. Gao, R. C. Bast, Jr., F. M. Robertson, and J. S. McMurray. 2012. "A phosphopeptide mimetic prodrug targeting the SH2 domain of Stat3 inhibits tumor growth and angiogenesis." *J Exp Ther Oncol* 10 (2):155–62.

Babon, J. J., I. S. Lucet, J. M. Murphy, N. A. Nicola, and L. N. Varghese. 2014. "The molecular regulation of Janus kinase (JAK) activation." *Biochem J* 462 (1):1–13. doi: 10.1042/BJ20140712.

Baker, S. J., S. G. Rane, and E. P. Reddy. 2007. "Hematopoietic cytokine receptor signaling." *Oncogene* 26 (47):6724–37. doi: 10.1038/sj.onc.1210757.

Banerjee, S., A. Biehl, M. Gadina, S. Hasni, and D. M. Schwartz. 2017. "JAK–STAT signaling as a target for inflammatory and autoimmune diseases: current and future prospects." *Drugs* 77 (5):521–46. doi: 10.1007/s40265-017-0701-9.

Becker, S., B. Groner, and C. W. Muller. 1998. "Three-dimensional structure of the Stat3beta homodimer bound to DNA." *Nature* 394 (6689):145–51. doi: 10.1038/28101.

Bharadwaj, U., T. K. Eckols, X. Xu, M. M. Kasembeli, Y. Chen, M. Adachi, Y. Song, Q. Mo, S. Y. Lai, and D. J. Tweardy. 2016. "Small-molecule inhibition of STAT3 in radioresistant head and neck squamous cell carcinoma." *Oncotarget* 7 (18):26307–30. doi: 10.18632/oncotarget.8368.

Bid, H. K., D. Oswald, C. Li, C. A. London, J. Lin, and P. J. Houghton. 2012. "Anti-angiogenic activity of a small molecule STAT3 inhibitor LLL12." *PLoS One* 7 (4):e35513. doi: 10.1371/journal.pone.0035513.

Boggon, T. J., Y. Li, P. W. Manley, and M. J. Eck. 2005. "Crystal structure of the Jak3 kinase domain in complex with a staurosporine analog." *Blood* 106 (3):996–1002. doi: 10.1182/blood-2005-02-0707.

Bohmer, F. D., and K. Friedrich. 2014. "Protein tyrosine phosphatases as wardens of STAT signaling." *JAKSTAT* 3 (1):e28087. doi: 10.4161/jkst.28087.

Bourdeau, A., N. Dube, and M. L. Tremblay. 2005. "Cytoplasmic protein tyrosine phosphatases, regulation and function: the roles of PTP1B and TC-PTP." *Curr Opin Cell Biol* 17 (2):203–09. doi: 10.1016/j.ceb.2005.02.001.

Bousoik, E., and H. Montazeri Aliabadi. 2018. ""Do we know Jack" about JAK? A closer look at JAK–STAT signaling pathway." *Front Oncol* 8:287. doi: 10.3389/fonc.2018.00287.

Bradshaw, J. M., V. Mitaxov, and G. Waksman. 1999. "Investigation of phosphotyrosine recognition by the SH2 domain of the Src kinase." *J Mol Biol* 293 (4):971–85. doi: 10.1006/jmbi.1999.3190.

Bradshaw, J. M., and G. Waksman. 2002. "Molecular recognition by SH2 domains." *Adv Protein Chem* 61:161–210.

Bullock, A. N., J. E. Debreczeni, A. M. Edwards, M. Sundstrom, and S. Knapp. 2006. "Crystal structure of the SOCS2-elongin C-elongin B complex defines a prototypical SOCS box ubiquitin ligase." *Proc Natl Acad Sci U S A* 103 (20):7637–42. doi: 10.1073/pnas.0601638103.

Chen, H., Z. Yang, C. Ding, L. Chu, Y. Zhang, K. Terry, H. Liu, Q. Shen, and J. Zhou. 2013a. "Discovery of O-alkylamino tethered niclosamide derivatives as potent and orally bioavailable anticancer agents." *ACS Med Chem Lett* 4 (2):180–85. doi: 10.1021/ml3003082.

Chen, H., Z. Yang, C. Ding, L. Chu, Y. Zhang, K. Terry, H. Liu, Q. Shen, and J. Zhou. 2013b. "Fragment-based drug design and identification of HJC0123, a novel orally bioavailable STAT3 inhibitor for cancer therapy." *Eur J Med Chem* 62:498–507. doi: 10.1016/j.ejmech.2013.01.023.

Chen, J., L. Bai, D. Bernard, Z. Nikolovska-Coleska, C. Gomez, J. Zhang, H. Yi, and S. Wang. 2010. "Structure-based design of conformationally constrained, cell-permeable STAT3 inhibitors." *ACS Med Chem Lett* 1 (2):85–89. doi: 10.1021/ml100010j.

Chiba, Y., M. Todoroki, Y. Nishida, M. Tanabe, and M. Misawa. 2009. "A novel STAT6 inhibitor AS1517499 ameliorates antigen-induced bronchial hypercontractility in mice." *Am J Respir Cell Mol Biol* 41 (5):516–24. doi: 10.1165/rcmb.2008-0163OC.

Chou, D. H., A. Vetere, A. Choudhary, S. S. Scully, M. Schenone, A. Tang, R. Gomez, S. M. Burns, M. Lundh, T. Vital, E. Comer, P. W. Faloon, V. Dancik, C. Ciarlo, J. Paulk, M. Dai, C. Reddy, H. Sun, M. Young, N. Donato, J. Jaffe, P. A. Clemons, M. Palmer, S. A. Carr, S. L. Schreiber, and B. K. Wagner. 2015. "Kinase-independent small-molecule inhibition of JAK–STAT signaling." *J Am Chem Soc* 137 (24):7929–34. doi: 10.1021/jacs.5b04284.

Chung, C. D., J. Liao, B. Liu, X. Rao, P. Jay, P. Berta, and K. Shuai. 1997. "Specific inhibition of Stat3 signal transduction by PIAS3." *Science* 278 (5344):1803–05. doi: 10.1126/science.278.5344.1803.

Cornejo, M. G., T. J. Boggon, and T. Mercher. 2009. "JAK3: a two-faced player in hematological disorders." *Int J Biochem Cell Biol* 41 (12):2376–79. doi: 10.1016/j.biocel.2009.09.004.
Cumaraswamy, A. A., A. M. Lewis, M. Geletu, A. Todic, D. B. Diaz, X. R. Cheng, C. E. Brown, R. C. Laister, D. Muench, K. Kerman, H. L. Grimes, M. D. Minden, and P. T. Gunning. 2014. "Nanomolar-potency small molecule inhibitor of STAT5 protein." *ACS Med Chem Lett* 5 (11):1202–06. doi: 10.1021/ml500165r.
Decker, T., and P. Kovarik. 2000. "Serine phosphorylation of STATs." *Oncogene* 19 (21):2628–37. doi: 10.1038/sj.onc.1203481.
Delgoffe, G. M., and D. A. Vignali. 2013. "STAT heterodimers in immunity: a mixed message or a unique signal?" *JAKSTAT* 2 (1):e23060. doi: 10.4161/jkst.23060.
Duval, D., G. Duval, C. Kedinger, O. Poch, and H. Boeuf. 2003. "The 'PINIT' motif, of a newly identified conserved domain of the PIAS protein family, is essential for nuclear retention of PIAS3L." *FEBS Lett* 554 (1–2):111–18. doi: 10.1016/s0014-5793(03)01116-5.
Endo, T. A., M. Masuhara, M. Yokouchi, R. Suzuki, H. Sakamoto, K. Mitsui, A. Matsumoto, S. Tanimura, M. Ohtsubo, H. Misawa, T. Miyazaki, N. Leonor, T. Taniguchi, T. Fujita, Y. Kanakura, S. Komiya, and A. Yoshimura. 1997. "A new protein containing an SH2 domain that inhibits JAK kinases." *Nature* 387 (6636):921–24. doi: 10.1038/43213.
Feng, Z., W. Zheng, Q. Tang, L. Cheng, H. Li, W. Ni, and X. Pan. 2017. "Fludarabine inhibits STAT1-mediated up-regulation of caspase-3 expression in dexamethasone-induced osteoblasts apoptosis and slows the progression of steroid-induced avascular necrosis of the femoral head in rats." *Apoptosis* 22 (8):1001–12. doi: 10.1007/s10495-017-1383-1.
Fenner, J. E., R. Starr, A. L. Cornish, J. G. Zhang, D. Metcalf, R. D. Schreiber, K. Sheehan, D. J. Hilton, W. S. Alexander, and P. J. Hertzog. 2006. "Suppressor of cytokine signaling 1 regulates the immune response to infection by a unique inhibition of type I interferon activity." *Nat Immunol* 7 (1):33–39. doi: 10.1038/ni1287.
Ferrao, R., and P. J. Lupardus. 2017. "The janus kinase (JAK) FERM and SH2 domains: bringing specificity to JAK-receptor interactions." *Front Endocrinol (Lausanne)* 8:71. doi: 10.3389/fendo.2017.00071.
Ferrao, R., H. J. Wallweber, H. Ho, et al. 2016. "The structural basis for class II cytokine receptor recognition by JAK1." *Structure* 24 (6):897–905. doi: 10.1016/j.str.2016.03.023.
Filippakopoulos, P., S. Muller, and S. Knap. 2009. "SH2 domains: modulators of nonreceptor tyrosine kinase activity." *Curr Opin Struct Biol* 19 (6):643–49. doi: 10.1016/j.sbi.2009.10.001.
Fischer, E. H., H. Charbonneau, and N. K. Tonks. 1991. "Protein tyrosine phosphatases: a diverse family of intracellular and transmembrane enzymes." *Science* 253 (5018):401–06. doi: 10.1126/science.1650499.
Fornerod, M., M. Ohno, M. Yoshida, and I. W. Mattaj. 1997. "CRM1 is an export receptor for leucine-rich nuclear export signals." *Cell* 90 (6):1051–60.
Frame, M. C., H. Patel, B. Serrels, D. Lietha, and M. J. Eck. 2010. "The FERM domain: organizing the structure and function of FAK." *Nat Rev Mol Cell Biol* 11 (11):802–14. doi: 10.1038/nrm2996.
Frangioni, J. V., P. H. Beahm, V. Shifrin, C. A. Jost, and B. G. Neel. 1992. "The nontransmembrane tyrosine phosphatase PTP-1B localizes to the endoplasmic reticulum via its 35 amino acid C-terminal sequence." *Cell* 68 (3):545–60. doi: 10.1016/0092-8674(92)90190-n.
Goldfarb, D. S., A. H. Corbett, D. A. Mason, M. T. Harreman, and S. A. Adam. 2004. "Importin alpha: a multipurpose nuclear-transport receptor." *Trends Cell Biol* 14 (9):505–14. doi: 10.1016/j.tcb.2004.07.016.
Gorlich, D., and I. W. Mattaj. 1996. "Nucleocytoplasmic transport." *Science* 271 (5255):1513–18. doi: 10.1126/science.271.5255.1513.
Grandis, J. R., S. D. Drenning, Q. Zeng, S. C. Watkins, M. F. Melhem, S. Endo, D. E. Johnson, L. Huang, Y. He, and J. D. Kim. 2000. "Constitutive activation of Stat3 signaling abrogates apoptosis in squamous cell carcinogenesis in vivo." *Proc Natl Acad Sci U S A* 97 (8):4227–32. doi: 10.1073/pnas.97.8.4227.
Greenhalgh, C. J., E. Rico-Bautista, M. Lorentzon, A. L. Thaus, P. O. Morgan, T. A. Willson, P. Zervoudakis, D. Metcalf, I. Street, N. A. Nicola, A. D. Nash, L. J. Fabri, G. Norstedt, C. Ohlsson, A. Flores-Morales, W. S. Alexander, and D. J. Hilton. 2005. "SOCS2 negatively

regulates growth hormone action in vitro and in vivo." *J Clin Invest* 115 (2):397–406. doi: 10.1172/JCI22710.

Gunning, P. T., W. P. Katt, M. Glenn, K. Siddiquee, J. S. Kim, R. Jove, S. M. Sebti, J. Turkson, and A. D. Hamilton. 2007. "Isoform selective inhibition of STAT1 or STAT3 homo-dimerization via peptidomimetic probes: structural recognition of STAT SH2 domains." *Bioorg Med Chem Lett* 17 (7):1875–78. doi: 10.1016/j.bmcl.2007.01.077.

Haftchenary, S., H. A. Luchman, A. O. Jouk, A. J. Veloso, B. D. Page, X. R. Cheng, S. S. Dawson, N. Grinshtein, V. M. Shahani, K. Kerman, D. R. Kaplan, C. Griffin, A. M. Aman, R. Al-Awar, S. Weiss, and P. T. Gunning. 2013. "Potent targeting of the STAT3 protein in brain cancer stem cells: a promising route for treating glioblastoma." *ACS Med Chem Lett* 4 (11):1102–07. doi: 10.1021/ml4003138.

Hammaren, H. M., D. Ungureanu, J. Grisouard, R. C. Skoda, S. R. Hubbard, and O. Silvennoinen. 2015. "ATP binding to the pseudokinase domain of JAK2 is critical for pathogenic activation." *Proc Natl Acad Sci U S A* 112 (15):4642–47. doi: 10.1073/pnas.1423201112.

Hendriks, W. J., A. Elson, S. Harroch, and A. W. Stoker. 2008. "Protein tyrosine phosphatases: functional inferences from mouse models and human diseases." *Febs J* 275 (5):816–30. doi: 10.1111/j.1742-4658.2008.06249.x.

Hilton, D. J., R. T. Richardson, W. S. Alexander, E. M. Viney, T. A. Willson, N. S. Sprigg, R. Starr, S. E. Nicholson, D. Metcalf, and N. A. Nicola. 1998. "Twenty proteins containing a C-terminal SOCS box form five structural classes." *Proc Natl Acad Sci U S A* 95 (1):114–19. doi: 10.1073/pnas.95.1.114.

Hoey, T., and U. Schindler. 1998. "STAT structure and function in signaling." *Curr Opin Genet Dev* 8 (5):582–87.

Hof, P., S. Pluskey, S. Dhe-Paganon, M. J. Eck, and S. E. Shoelson. 1998. "Crystal structure of the tyrosine phosphatase SHP-2." *Cell* 92 (4):441–50. doi: 10.1016/s0092-8674(00)80938-1.

Horvath, C. M., Z. Wen, and J. E. Darnell, Jr. 1995. "A STAT protein domain that determines DNA sequence recognition suggests a novel DNA-binding domain." *Genes Dev* 9 (8):984–94. doi: 10.1101/gad.9.8.984.

Hu, Y., C. Zhao, H. Zheng, K. Lu, D. Shi, Z. Liu, X. Dai, Y. Zhang, X. Zhang, W. Hu, and G. Liang. 2017. "A novel STAT3 inhibitor HO-3867 induces cell apoptosis by reactive oxygen species-dependent endoplasmic reticulum stress in human pancreatic cancer cells." *Anticancer Drugs* 28 (4):392–400. doi: 10.1097/CAD.0000000000000470.

Ibarra-Sanchez, M. J., J. Wagner, M. T. Ong, C. Lampron, and M. L. Tremblay. 2001. "Murine embryonic fibroblasts lacking TC-PTP display delayed G1 phase through defective NF-kappaB activation." *Oncogene* 20 (34):4728–39. doi: 10.1038/sj.onc.1204648.

Isaacs, A., and J. Lindenmann. 1957. "Virus interference. I. The interferon." *Proc R Soc Lond B Biol Sci* 147 (927):258–67.

Ishida-Takahashi, R., F. Rosario, Y. Gong, K. Kopp, Z. Stancheva, X. Chen, E. P. Feener, and M. G. Myers, Jr. 2006. "Phosphorylation of Jak2 on Ser(523) inhibits Jak2-dependent leptin receptor signaling." *Mol Cell Biol* 26 (11):4063–73. doi: 10.1128/MCB.01589-05.

Iversen, L. F., K. B. Moller, A. K. Pedersen, G. H. Peters, A. S. Petersen, H. S. Andersen, S. Branner, S. B. Mortensen, and N. P. Moller. 2002. "Structure determination of T cell protein-tyrosine phosphatase." *J Biol Chem* 277 (22):19982–90. doi: 10.1074/jbc.M200567200.

Jackson, P. K. 2001. "A new RING for SUMO: wrestling transcriptional responses into nuclear bodies with PIAS family E3 SUMO ligases." *Genes Dev* 15 (23):3053–58. doi: 10.1101/gad.955501.

Jacobson, N. G., S. J. Szabo, R. M. Weber-Nordt, Z. Zhong, R. D. Schreiber, J. E. Darnell, Jr., and K. M. Murphy. 1995. "Interleukin 12 signaling in T helper type 1 (Th1) cells involves tyrosine phosphorylation of signal transducer and activator of transcription (Stat)3 and Stat4." *J Exp Med* 181 (5):1755–62. doi: 10.1084/jem.181.5.1755.

Jans, D. A., C. Y. Xiao, and M. H. Lam. 2000. "Nuclear targeting signal recognition: a key control point in nuclear transport?" *Bioessays* 22 (6):532–44. doi: 10.1002/(SICI)1521-1878(200006)22:6<532::AID-BIES6>3.0.CO;2-O.

Jatiani, S. S., S. J. Baker, L. R. Silverman, and E. P. Reddy. 2010. "Jak/STAT pathways in cytokine signaling and myeloproliferative disorders: approaches for targeted therapies." *Genes Cancer* 1 (10):979–93. doi: 10.1177/1947601910397187.

Jiang, X., M. Wu, Z. Xu, H. Wang, H. Wang, X. Yu, Z. Li, and L. Teng. 2018. "HJC0152, a novel STAT3 inhibitor with promising anti-tumor effect in gastric cancer." *Cancer Manag Res* 10:6857–67. doi: 10.2147/CMAR.S188364.

Jimenez-Lara, A. M., M. J. Heine, and H. Gronemeyer. 2002. "PIAS3 (protein inhibitor of activated STAT-3) modulates the transcriptional activation mediated by the nuclear receptor coactivator TIF2." *FEBS Lett* 526 (1–3):142–46. doi: 10.1016/s0014-5793(02)03154-x.

Jin, Y. J., C. L. Yu, and S. J. Burakoff. 1999. "Human 70-kDa SHP-1L differs from 68-kDa SHP-1 in its C-terminal structure and catalytic activity." *J Biol Chem* 274 (40):28301–7. doi: 10.1074/jbc.274.40.28301.

Jung, K. H., W. Yoo, H. L. Stevenson, D. Deshpande, H. Shen, M. Gagea, S. Y. Yoo, J. Wang, T. K. Eckols, U. Bharadwaj, D. J. Tweardy, and L. Beretta. 2017. "Multifunctional effects of a small-molecule STAT3 inhibitor on NASH and hepatocellular carcinoma in mice." *Clin Cancer Res* 23 (18):5537–46. doi: 10.1158/1078-0432.CCR-16-2253.

Kamatkar, S., V. Radha, S. Nambirajan, R. S. Reddy, and G. Swarup. 1996. "Two splice variants of a tyrosine phosphatase differ in substrate specificity, DNA binding, and subcellular location." *J Biol Chem* 271 (43):26755–61. doi: 10.1074/jbc.271.43.26755.

Kamura, T., K. Maenaka, S. Kotoshiba, M. Matsumoto, D. Kohda, R. C. Conaway, J. W. Conaway, and K. I. Nakayama. 2004. "VHL-box and SOCS-box domains determine binding specificity for Cul2-Rbx1 and Cul5-Rbx2 modules of ubiquitin ligases." *Genes Dev* 18 (24):3055–65. doi: 10.1101/gad.1252404.

Kaneko, T., R. Joshi, S. M. Feller, and S. S. Li. 2012. "Phosphotyrosine recognition domains: the typical, the atypical and the versatile." *Cell Commun Signal* 10 (1):32. doi: 10.1186/1478-811X-10-32.

Khwaja, A. 2006. "The role of Janus kinases in haemopoiesis and haematological malignancy." *Br J Haematol* 134 (4):366–84. doi: 10.1111/j.1365-2141.2006.06206.x.

Kile, B. T., B. A. Schulman, W. S. Alexander, N. A. Nicola, H. M. Martin, and D. J. Hilton. 2002. "The SOCS box: a tale of destruction and degradation." *Trends Biochem Sci* 27 (5):235–41.

Kim, D., I. H. Lee, S. Kim, M. Choi, H. Kim, S. Ahn, P. E. Saw, H. Jeon, Y. Lee, and S. Jon. 2014. "A specific STAT3-binding peptide exerts antiproliferative effects and antitumor activity by inhibiting STAT3 phosphorylation and signaling." *Cancer Res* 74 (8):2144–51. doi: 10.1158/0008-5472.CAN-13-2187.

Kisseleva, T., S. Bhattacharya, J. Braunstein, and C. W. Schindler. 2002. "Signaling through the JAK–STAT pathway, recent advances and future challenges." *Gene* 285 (1–2):1–24.

Knight, J. M., G. Mak, J. Shaw, P. Porter, C. McDermott, L. Roberts, R. You, X. Yuan, V. O. Millien, Y. Qian, L. Z. Song, V. Frazier, C. Kim, J. J. Kim, R. A. Bond, J. D. Milner, Y. Zhang, P. K. Mandal, A. Luong, F. Kheradmand, J. S. McMurray, and D. B. Corry. 2015. "Long-acting beta agonists enhance allergic airway disease." *PLoS One* 10 (11):e0142212. doi: 10.1371/journal.pone.0142212.

Knight, J. M., P. Mandal, P. Morlacchi, G. Mak, E. Li, M. Madison, C. Landers, B. Saxton, E. Felix, B. Gilbert, J. Sederstrom, A. Varadhachary, M. M. Singh, D. Chatterjee, D. B. Corry, and J. S. McMurray. 2018. "Small molecule targeting of the STAT5/6 Src homology 2 (SH2) domains to inhibit allergic airway disease." *J Biol Chem* 293 (26):10026–40. doi: 10.1074/jbc.RA117.000567.

Kotaja, N., U. Karvonen, O. A. Janne, and J. J. Palvimo. 2002. "PIAS proteins modulate transcription factors by functioning as SUMO-1 ligases." *Mol Cell Biol* 22 (14):5222–34. doi: 10.1128/mcb.22.14.5222-5234.2002.

Kovarik, P., M. Mangold, K. Ramsauer, H. Heidari, R. Steinborn, A. Zotter, D. E. Levy, M. Muller, and T. Decker. 2001. "Specificity of signaling by STAT1 depends on SH2 and C-terminal domains that regulate Ser727 phosphorylation, differentially affecting specific target gene expression." *EMBO J* 20 (1–2):91–100. doi: 10.1093/emboj/20.1.91.

Krebs, D. L., and D. J. Hilton. 2001. "SOCS proteins: negative regulators of cytokine signaling." *Stem Cells* 19 (5):378–87. doi: 10.1634/stemcells.19-5-378.

Kurzer, J. H., L. S. Argetsinger, Y. J. Zhou, J. L. Kouadio, J. J. O'Shea, and C. Carter-Su. 2004. "Tyrosine 813 is a site of JAK2 autophosphorylation critical for activation of JAK2 by SH2-B beta." *Mol Cell Biol* 24 (10):4557–70. doi: 10.1128/mcb.24.10.4557-4570.2004.

LaFave, L. M., and R. L. Levine. 2012. "JAK2 the future: therapeutic strategies for JAK-dependent malignancies." *Trends Pharmacol Sci* 33 (11):574–82. doi: 10.1016/j.tips.2012.08.005.

Larner, A. C., M. David, G. M. Feldman, K. Igarashi, R. H. Hackett, D. S. Webb, S. M. Sweitzer, E. F. Petricoin, 3rd, and D. S. Finbloom. 1993. "Tyrosine phosphorylation of DNA binding proteins by multiple cytokines." *Science* 261 (5129):1730–33. doi: 10.1126/science.8378773.

Lee, B. J., Y. J. Kim, D. H. Cho, N. W. Sohn, and H. Kang. 2011. "Immunomodulatory effect of water extract of cinnamon on anti-CD3-induced cytokine responses and p38, JNK, ERK1/2, and STAT4 activation." *Immunopharmacol Immunotoxicol* 33 (4):714–22. doi: 10.3109/08923973.2011.564185.

Lin, L., B. Hutzen, P. K. Li, S. Ball, M. Zuo, S. DeAngelis, E. Foust, M. Sobo, L. Friedman, D. Bhasin, L. Cen, C. Li, and J. Lin. 2010. "A novel small molecule, LLL12, inhibits STAT3 phosphorylation and activities and exhibits potent growth-suppressive activity in human cancer cells." *Neoplasia* 12 (1):39–50. doi: 10.1593/neo.91196.

Liongue, C., L. A. O'Sullivan, M. C. Trengove, and A. C. Ward. 2012. "Evolution of JAK–STAT pathway components: mechanisms and role in immune system development." *PLoS One* 7 (3): e32777. doi: 10.1371/journal.pone.0032777.

Liongue, C., and A. C. Ward. 2013. "Evolution of the JAK–STAT pathway." *JAKSTAT* 2 (1):e22756. doi: 10.4161/jkst.22756.

Liu, B., J. Liao, X. Rao, S. A. Kushner, C. D. Chung, D. D. Chang, and K. Shuai. 1998. "Inhibition of Stat1-mediated gene activation by PIAS1." *Proc Natl Acad Sci U S A* 95 (18):10626–31. doi: 10.1073/pnas.95.18.10626.

Liu, B. A., K. Jablonowski, M. Raina, M. Arce, T. Pawson, and P. D. Nash. 2006. "The human and mouse complement of SH2 domain proteins-establishing the boundaries of phosphotyrosine signaling." *Mol Cell* 22 (6):851–68. doi: 10.1016/j.molcel.2006.06.001.

Liu, F., D. E. Hill, and J. Chernoff. 1996. "Direct binding of the proline-rich region of protein tyrosine phosphatase 1B to the Src homology 3 domain of p130(Cas)." *J Biol Chem* 271 (49):31290–95. doi: 10.1074/jbc.271.49.31290.

Liu, L., K. M. McBride, and N. C. Reich. 2005. "STAT3 nuclear import is independent of tyrosine phosphorylation and mediated by importin-alpha3." *Proc Natl Acad Sci U S A* 102 (23):8150–55. doi: 10.1073/pnas.0501643102.

Lorenz, U. 2009. "SHP-1 and SHP-2 in T cells: two phosphatases functioning at many levels." *Immunol Rev* 228 (1):342–59. doi: 10.1111/j.1600-065X.2008.00760.x.

Lucet, I. S., E. Fantino, M. Styles, R. Bamert, O. Patel, S. E. Broughton, M. Walter, C. J. Burns, H. Treutlein, A. F. Wilks, and J. Rossjohn. 2006. "The structural basis of Janus kinase 2 inhibition by a potent and specific pan-Janus kinase inhibitor." *Blood* 107 (1):176–83. doi: 10.1182/blood-2005-06-2413.

Lupardus, P. J., G. Skiniotis, A. J. Rice, C. Thomas, S. Fischer, T. Walz, and K. C. Garcia. 2011. "Structural snapshots of full-length Jak1, a transmembrane gp130/IL-6/IL-6Ralpha cytokine receptor complex, and the receptor-Jak1 holocomplex." *Structure* 19 (1):45–55. doi: 10.1016/j.str.2010.10.010.

MacDonagh, L., S. G. Gray, E. Breen, S. Cuffe, S. P. Finn, K. J. O'Byrne, and M. P. Barr. 2018. "BBI608 inhibits cancer stemness and reverses cisplatin resistance in NSCLC." *Cancer Lett* 428:117–26. doi: 10.1016/j.canlet.2018.04.008.

Machida, K., C. M. Thompson, K. Dierck, K. Jablonowski, S. Karkkainen, B. Liu, H. Zhang, P. D. Nash, D. K. Newman, P. Nollau, T. Pawson, G. H. Renkema, K. Saksela, M. R. Schiller, D. G. Shin, and B. J. Mayer. 2007. "High-throughput phosphotyrosine profiling using SH2 domains." *Mol Cell* 26 (6):899–915. doi: 10.1016/j.molcel.2007.05.031.

Manning, G., D. B. Whyte, R. Martinez, T. Hunter, and S. Sudarsanam. 2002. "The protein kinase complement of the human genome." *Science* 298 (5600):1912–34. doi: 10.1126/science.1075762.

Matsuo, K., A. Bettaieb, N. Nagata, I. Matsuo, H. Keilhack, and F. G. Haj. 2011. "Regulation of brown fat adipogenesis by protein tyrosine phosphatase 1B." *PLoS One* 6 (1):e16446. doi: 10.1371/journal.pone.0016446.

McNally, R., A. V. Toms, and M. J. Eck. 2016. "Crystal structure of the FERM-SH2 module of human Jak2." *PLoS One* 11 (5):e0156218. doi: 10.1371/journal.pone.0156218.

Meng, H., C. Yang, W. Ni, W. Ding, X. Yang, and W. Qian. 2007. "Antitumor activity of fludarabine against human multiple myeloma in vitro and in vivo." *Eur J Haematol* 79 (6):486–93. doi: 10.1111/j.1600-0609.2007.00968.x.

Seif, F., M. Khoshmirsafa, H. Aazami, M. Mohsenzadegan, G. Sedighi, and M. Bahar. 2017. "The role of JAK–STAT signaling pathway and its regulators in the fate of T helper cells." *Cell Commun Signal* 15 (1):23. doi: 10.1186/s12964-017-0177-y.

Sekimoto, T., N. Imamoto, K. Nakajima, T. Hirano, and Y. Yoneda. 1997. "Extracellular signal-dependent nuclear import of Stat1 is mediated by nuclear pore-targeting complex formation with NPI-1, but not Rch1." *EMBO J* 16 (23):7067–77. doi: 10.1093/emboj/16.23.7067.

Selvendiran, K., S. Ahmed, A. Dayton, M. L. Kuppusamy, B. K. Rivera, T. Kalai, K. Hideg, and P. Kuppusamy. 2011. "HO-3867, a curcumin analog, sensitizes cisplatin-resistant ovarian carcinoma, leading to therapeutic synergy through STAT3 inhibition." *Cancer Biol Ther* 12 (9):837–45. doi: 10.4161/cbt.12.9.17713.

Shuai, K., C. Schindler, V. R. Prezioso, and J. E. Darnell, Jr. 1992. "Activation of transcription by IFN-gamma: tyrosine phosphorylation of a 91-kD DNA binding protein." *Science* 258 (5089):1808–12. doi: 10.1126/science.1281555.

Siddiquee, K., S. Zhang, W. C. Guida, M. A. Blaskovich, B. Greedy, H. R. Lawrence, M. L. Yip, R. Jove, M. M. McLaughlin, N. J. Lawrence, S. M. Sebti, and J. Turkson. 2007a. "Selective chemical probe inhibitor of Stat3, identified through structure-based virtual screening, induces antitumor activity." *Proc Natl Acad Sci U S A* 104 (18):7391–96. doi: 10.1073/pnas.0609757104.

Siddiquee, K. A., P. T. Gunning, M. Glenn, W. P. Katt, S. Zhang, C. Schrock, S. M. Sebti, R. Jove, A. D. Hamilton, and J. Turkson. 2007b. "An oxazole-based small-molecule Stat3 inhibitor modulates Stat3 stability and processing and induces antitumor cell effects." *ACS Chem Biol* 2 (12):787–98. doi: 10.1021/cb7001973.

Silvennoinen, O., C. Schindler, J. Schlessinger, and D. E. Levy. 1993. "Ras-independent growth factor signaling by transcription factor tyrosine phosphorylation." *Science* 261 (5129):1736–39. doi: 10.1126/science.8378775.

Sommer, U., C. Schmid, R. M. Sobota, U. Lehmann, N. J. Stevenson, J. A. Johnston, F. Schaper, P. C. Heinrich, and S. Haan. 2005. "Mechanisms of SOCS3 phosphorylation upon interleukin-6 stimulation. Contributions of Src- and receptor-tyrosine kinases." *J Biol Chem* 280 (36):31478–88. doi: 10.1074/jbc.M506008200.

Song, H., R. Wang, S. Wang, and J. Lin. 2005. "A low-molecular-weight compound discovered through virtual database screening inhibits Stat3 function in breast cancer cells." *Proc Natl Acad Sci U S A* 102 (13):4700–05. doi: 10.1073/pnas.0409894102.

Stark, G. R., and J. E. Darnell, Jr. 2012. "The JAK–STAT pathway at twenty." *Immunity* 36 (4):503–14. doi: 10.1016/j.immuni.2012.03.013.

Starr, R., T. A. Willson, E. M. Viney, L. J. Murray, J. R. Rayner, B. J. Jenkins, T. J. Gonda, W. S. Alexander, D. Metcalf, N. A. Nicola, and D. J. Hilton. 1997. "A family of cytokine-inducible inhibitors of signalling." *Nature* 387 (6636):917–21. doi: 10.1038/43206.

Subramaniam, A., M. K. Shanmugam, E. Perumal, F. Li, A. Nachiyappan, X. Dai, S. N. Swamy, K. S. Ahn, A. P. Kumar, B. K. Tan, K. M. Hui, and G. Sethi. 2013. "Potential role of signal transducer and activator of transcription (STAT)3 signaling pathway in inflammation, survival, proliferation and invasion of hepatocellular carcinoma." *Biochim Biophys Acta* 1835 (1):46–60. doi: 10.1016/j.bbcan.2012.10.002.

Subramaniam, P. S., B. A. Torres, and H. M. Johnson. 2001. "So many ligands, so few transcription factors: a new paradigm for signaling through the STAT transcription factors." *Cytokine* 15 (4):175–87. doi: 10.1006/cyto.2001.0905.

Sutherland, K. D., G. J. Lindeman, D. Y. Choong, S. Wittlin, L. Brentzell, W. Phillips, I. G. Campbell, and J. E. Visvader. 2004. "Differential hypermethylation of SOCS genes in ovarian and breast carcinomas." *Oncogene* 23 (46):7726–33. doi: 10.1038/sj.onc.1207787.

Tanaka, T., M. A. Soriano, and M. J. Grusby. 2005. "SLIM is a nuclear ubiquitin E3 ligase that negatively regulates STAT signaling." *Immunity* 22 (6):729–36. doi: 10.1016/j.immuni.2005.04.008.

Taylor, S. S., D. R. Knighton, J. Zheng, L. F. Ten Eyck, and J. M. Sowadski. 1992. "Structural framework for the protein kinase family." *Annu Rev Cell Biol* 8:429–62. doi: 10.1146/annurev.cb.08.110192.002241.

Tepass, U. 2009. "FERM proteins in animal morphogenesis." *Curr Opin Genet Dev* 19 (4):357–67. doi: 10.1016/j.gde.2009.05.006.

Tierney, B. J., G. A. McCann, D. E. Cohn, E. Eisenhauer, M. Sudhakar, P. Kuppusamy, K. Hideg, and K. Selvendiran. 2012. "HO-3867, a STAT3 inhibitor induces apoptosis by inactivation of STAT3 activity in BRCA1-mutated ovarian cancer cells." *Cancer Biol Ther* 13 (9):766–75. doi: 10.4161/cbt.20559.

Toms, A. V., A. Deshpande, R. McNally, Y. Jeong, J. M. Rogers, C. U. Kim, S. M. Gruner, S. B. Ficarro, J. A. Marto, M. Sattler, J. D. Griffin, and M. J. Eck. 2013. "Structure of a pseudokinase-domain switch that controls oncogenic activation of Jak kinases." *Nat Struct Mol Biol* 20 (10):1221–23. doi: 10.1038/nsmb.2673.

Tonks, N. K. 2003. "PTP1B: from the sidelines to the front lines!" *FEBS Lett* 546 (1):140–48. doi: 10.1016/s0014-5793(03)00603-3.

Tonks, N. K. 2006. "Protein tyrosine phosphatases: from genes, to function, to disease." *Nat Rev Mol Cell Biol* 7 (11):833–46. doi: 10.1038/nrm2039.

Tonks, N. K., C. D. Diltz, and E. H. Fischer. 1988. "Purification of the major protein-tyrosine-phosphatases of human placenta." *J Biol Chem* 263 (14):6722–30.

Turkson, J., J. S. Kim, S. Zhang, J. Yuan, M. Huang, M. Glenn, E. Haura, S. Sebti, A. D. Hamilton, and R. Jove. 2004. "Novel peptidomimetic inhibitors of signal transducer and activator of transcription 3 dimerization and biological activity." *Mol Cancer Ther* 3 (3):261–69.

Ungureanu, D., J. Wu, T. Pekkala, Y. Niranjan, C. Young, O. N. Jensen, C. F. Xu, T. A. Neubert, R. C. Skoda, S. R. Hubbard, and O. Silvennoinen. 2011. "The pseudokinase domain of JAK2 is a dual-specificity protein kinase that negatively regulates cytokine signaling." *Nat Struct Mol Biol* 18 (9):971–76. doi: 10.1038/nsmb.2099.

Vogelstein, B., N. Papadopoulos, V. E. Velculescu, S. Zhou, L. A. Diaz, Jr., and K. W. Kinzler. 2013. "Cancer genome landscapes." *Science* 339 (6127):1546–58. doi: 10.1126/science.1235122.

Wagner, M. J., M. M. Stacey, B. A. Liu, and T. Pawson. 2013. "Molecular mechanisms of SH2- and PTB-domain-containing proteins in receptor tyrosine kinase signaling." *Cold Spring Harb Perspect Biol* 5 (12):a008987. doi: 10.1101/cshperspect.a008987.

Wallweber, H. J., C. Tam, Y. Franke, M. A. Starovasnik, and P. J. Lupardus. 2014. "Structural basis of recognition of interferon-alpha receptor by tyrosine kinase 2." *Nat Struct Mol Biol* 21 (5):443–48. doi: 10.1038/nsmb.2807.

Wang, Y., S. Wang, Y. Wu, Y. Ren, Z. Li, X. Yao, C. Zhang, N. Ye, C. Jing, J. Dong, K. Zhang, S. Sun, M. Zhao, W. Guo, X. Qu, Y. Qiao, H. Chen, L. Kong, R. Jin, X. Wang, L. Zhang, J. Zhou, Q. Shen, and X. Zhou. 2017. "Suppression of the growth and invasion of human head and neck squamous cell carcinomas via regulating STAT3 signaling and the miR-21/beta-catenin axis with HJC0152." *Mol Cancer Ther* 16 (4):578–90. doi: 10.1158/1535-7163.MCT-16-0606.

Wang, Z., J. Li, W. Xiao, J. Long, and H. Zhang. 2018. "The STAT3 inhibitor S3I-201 suppresses fibrogenesis and angiogenesis in liver fibrosis." *Lab Invest* 98 (12):1600–13. doi: 10.1038/s41374-018-0127-3.

Wilks, A. F. 1989. "Two putative protein-tyrosine kinases identified by application of the polymerase chain reaction." *Proc Natl Acad Sci U S A* 86 (5):1603–07. doi: 10.1073/pnas. 86.5.1603.

Wilks, A. F. 2008. "The JAK kinases: not just another kinase drug discovery target." *Semin Cell Dev Biol* 19 (4):319–28. doi: 10.1016/j.semcdb.2008.07.020.

Williams, L., L. Bradley, A. Smith, and B. Foxwell. 2004. "Signal transducer and activator of transcription 3 is the dominant mediator of the anti-inflammatory effects of IL-10 in human macrophages." *J Immunol* 172 (1):567–76. doi: 10.4049/jimmunol.172.1.567.

Wingelhofer, B., B. Maurer, E. C. Heyes, A. A. Cumaraswamy, A. Berger-Becvar, E. D. de Araujo, A. Orlova, P. Freund, F. Ruge, J. Park, G. Tin, S. Ahmar, C. H. Lardeau, I. Sadovnik, D. Bajusz, G. M. Keseru, F. Grebien, S. Kubicek, P. Valent, P. T. Gunning, and R. Moriggl. 2018. "Pharmacologic inhibition of STAT5 in acute myeloid leukemia." *Leukemia* 32 (5):1135–46. doi: 10.1038/s41375-017-0005-9.

Wolf, C., and M. R. Mofrad. 2008. "On the octagonal structure of the nuclear pore complex: insights from coarse-grained models." *Biophys J* 95 (4):2073–85. doi: 10.1529/biophysj.108.130336.

Wu, C., M. Sun, L. Liu, and G. W. Zhou. 2003. "The function of the protein tyrosine phosphatase SHP-1 in cancer." *Gene* 306:1–12.

Xu, X., Y. L. Sun, and T. Hoey. 1996. "Cooperative DNA binding and sequence-selective recognition conferred by the STAT amino-terminal domain." *Science* 273 (5276):794–97. doi: 10.1126/science.273.5276.794.

Yamaoka, K., P. Saharinen, M. Pesu, V. E. Holt, 3rd, O. Silvennoinen, and J. J. O'Shea. 2004. "The Janus kinases (Jaks)." *Genome Biol* 5 (12):253. doi: 10.1186/gb-2004-5-12-253.

Yasukawa, H., H. Misawa, H. Sakamoto, M. Masuhara, A. Sasaki, T. Wakioka, S. Ohtsuka, T. Imaizumi, T. Matsuda, J. N. Ihle, and A. Yoshimura. 1999. "The JAK-binding protein JAB inhibits Janus tyrosine kinase activity through binding in the activation loop." *EMBO J* 18 (5):1309–20. doi: 10.1093/emboj/18.5.1309.

Yoshimura, A., and H. Yasukawa. 2012. "JAK's SOCS: a mechanism of inhibition." *Immunity* 36 (2):157–59. doi: 10.1016/j.immuni.2012.01.010.

You-Ten, K. E., E. S. Muise, A. Itie, E. Michaliszyn, J. Wagner, S. Jothy, W. S. Lapp, and M. L. Tremblay. 1997. "Impaired bone marrow microenvironment and immune function in T cell protein tyrosine phosphatase-deficient mice." *J Exp Med* 186 (5):683–93. doi: 10.1084/jem.186.5.683.

Yu, H., and R. Jove. 2004. "The STATs of cancer–new molecular targets come of age." *Nat Rev Cancer* 4 (2):97–105. doi: 10.1038/nrc1275.

Zhang, D., A. Wlodawer, and J. Lubkowski. 2016. "Crystal structure of a complex of the intracellular domain of interferon lambda receptor 1 (IFNLR1) and the FERM/SH2 domains of human JAK1." *J Mol Biol* 428 (23):4651–68. doi: 10.1016/j.jmb.2016.10.005.

Zhang, J. G., A. Farley, S. E. Nicholson, T. A. Willson, L. M. Zugaro, R. J. Simpson, R. L. Moritz, D. Cary, R. Richardson, G. Hausmann, B. T. Kile, S. B. Kent, W. S. Alexander, D. Metcalf, D. J. Hilton, N. A. Nicola, and M. Baca. 1999. "The conserved SOCS box motif in suppressors of cytokine signaling binds to elongins B and C and may couple bound proteins to proteasomal degradation." *Proc Natl Acad Sci U S A* 96 (5):2071–76. doi: 10.1073/pnas.96.5.2071.

Zhang, X., Y. Sun, R. Pireddu, H. Yang, M. K. Urlam, H. R. Lawrence, W. C. Guida, N. J. Lawrence, and S. M. Sebti. 2013. "A novel inhibitor of STAT3 homodimerization selectively suppresses STAT3 activity and malignant transformation." *Cancer Res* 73 (6):1922–33. doi: 10.1158/0008-5472.CAN-12-3175.

Zhang, X., P. Yue, S. Fletcher, W. Zhao, P. T. Gunning, and J. Turkson. 2010. "A novel small-molecule disrupts Stat3 SH2 domain-phosphotyrosine interactions and Stat3-dependent tumor processes." *Biochem Pharmacol* 79 (10):1398–409. doi: 10.1016/j.bcp.2010.01.001.

Zhang, X., P. Yue, B. D. Page, T. Li, W. Zhao, A. T. Namanja, D. Paladino, J. Zhao, Y. Chen, P. T. Gunning, and J. Turkson. 2012. "Orally bioavailable small-molecule inhibitor of transcription factor Stat3 regresses human breast and lung cancer xenografts." *Proc Natl Acad Sci U S A* 109 (24):9623–28. doi: 10.1073/pnas.1121606109.

Zhang, Y., Z. Jin, H. Zhou, X. Ou, Y. Xu, H. Li, C. Liu, and B. Li. 2016. "Suppression of prostate cancer progression by cancer cell stemness inhibitor napabucasin." *Cancer Med* 5 (6):1251–58. doi: 10.1002/cam4.675.

Zhang, Z., K. Shen, W. Lu, and P. A. Cole. 2003. "The role of C-terminal tyrosine phosphorylation in the regulation of SHP-1 explored via expressed protein ligation." *J Biol Chem* 278 (7):4668–74. doi: 10.1074/jbc.M210028200.

Zhao, C., W. Wang, W. Yu, D. Jou, Y. Wang, H. Ma, H. Xiao, H. Qin, C. Zhang, J. Lu, S. Li, C. Li, J. Lin, and L. Lin. 2016. "A novel small molecule STAT3 inhibitor, LY5, inhibits cell viability, colony formation, and migration of colon and liver cancer cells." *Oncotarget* 7 (11):12917–26. doi: 10.18632/oncotarget.7338.

Zhao, L., H. Dong, C. C. Zhang, L. Kinch, M. Osawa, M. Iacovino, N. V. Grishin, M. Kyba, and L. J. Huang. 2009. "A JAK2 interdomain linker relays Epo receptor engagement signals to kinase activation." *J Biol Chem* 284 (39):26988–98. doi: 10.1074/jbc.M109.011387.

Zhong, Z., Z. Wen, and J. E. Darnell, Jr. 1994a. "Stat3 and Stat4: members of the family of signal transducers and activators of transcription." *Proc Natl Acad Sci U S A* 91 (11):4806–10. doi: 10.1073/pnas.91.11.4806.

Zhong, Z., Z. Wen, and J. E. Darnell, Jr. 1994b. "Stat3: a STAT family member activated by tyrosine phosphorylation in response to epidermal growth factor and interleukin-6." *Science* 264 (5155):95–98. doi: 10.1126/science.8140422.

Zhou, X. X., P. J. Gao, and B. G. Sun. 2009. "Pravastatin attenuates interferon-gamma action via modulation of STAT1 to prevent aortic atherosclerosis in apolipoprotein E-knockout mice." *Clin Exp Pharmacol Physiol* 36 (4):373–79. doi: 10.1111/j.1440-1681.2008.05067.x.

3

MicroRNA-Mediated Regulation of JAK-STAT Signaling in Non-Cancerous Human Diseases

Chandra S. Boosani and Devendra K. Agrawal
Department of Translational Research
College of Osteopathic Medicine of the Pacific
Western University of Health Sciences
Pomona, CA

Wanlin Jiang
Cardiovascular Research Center
Massachusetts General Hospital
Boston MA

Taylor Burke
Department of Clinical and Translational Science
Creighton University School of Medicine
Omaha, NE

3.1 Brief Background

Targeting a signaling pathway that is pathologically active and specific to a given disease could be more enticing to pursue development of drug targets for disease prevention. Even more is when the same pathway participates in multiple diseases and plays an important role in progression of the disease. Certainly, of that kind would be the Janus kinase (JAK)-signal transducers and activators of transcription (STAT) signaling pathway that has been identified as an important activator of intra-cellular signaling molecules, which are associated with many pathological conditions. Notably, this signaling mechanism promoting key cellular activities was identified to be directly associated with growth, proliferation, and migration in different cell types. Its role in proliferative disorders, cancer, and immune regulation was seen to be under tight control by STAT, which is activated by JAK. As a result of STAT-induced transcriptional activity, expression of several key inflammatory molecules was reported, which promote the pathological progression of the disease. Both direct and indirect activation of STAT by JAK has been reported, which denotes their robust involvement and regulation of multiple cross-connecting pathways, which underscores their participation in diverse cellular mechanisms.

Adding to the already-existing complexity in gene-regulatory mechanisms, the discovery of microRNAs and their proven role in controlling gene expression, layers another dimension of multiple gene regulations that are driven by a single miRNA (microRNA). At the same time, identification of miRNA targets also facilitated the predictions of possible cross talks between different pathways. The miRNAs are short oligomers with full amenability to modify and synthesize. In addition, the ease of their delivery into the cells and in live animals make them a more favorable choice to use in clinical research. Further, miRNAs as such have become a tool to drive or prevent their target pathways, which highlights the potential of miRNAs as another therapeutic avenue to treat human and animal diseases. Recent advancements using bioinformatic tools have enabled identification of specific loci for different miRNAs on the chromosomes, which furthered our understanding on the origin and putative source of miRNAs. The increasing number of miRNAs being identified as potential therapeutic candidates and their proven success in pre-clinical trials and animal trials show the promise and safety of miRNAs for disease treatment.

Ample literature evidence suggest that several miRNAs regulate JAK-STAT pathways in cancer, and many previous publications have previously discussed about this topic. In this chapter, we specifically tried to exclude discussing about the role of miRNAs that regulate JAK-STAT pathways in cancer due to the vast abundance and volume of literature available on this topic. At the same time, very little literature exists that show the role of miRNAs in regulating JAK-STAT pathway in other diseases (Table 3.1). Therefore, in this chapter, we remain focused on those miRNAs that were validated to be regulating JAK-STAT pathway in non-cancerous diseases. Although efforts were made to provide comprehensive information on this topic, the readers are encouraged to refer to the bibliography below and the current literature for more details.

TABLE 3.1

List of miRNAs and their Target Transcripts that Affect JAK-STAT Signaling in Non-Cancerous Human Diseases

miRNA	Target transcript	Anticipated effect of JAK-STAT pathway	Test model	Disease model	Reference
miR-19a	SOCS1, SOCS3, SOCS5	Inducing effect	HuH-7 immortal cells	Liver diseases	(Collins et al. 2013, e69090)
miR-155	STAT1	Inhibitory effect	Live mice	Liver injury	(Lv et al. 2015, 6013–6018)
miR-1264	DNMT1	Inhibitory effect	Smooth muscle cells	Coronary artery disease	(Boosani et al. 2015, 1365–1376)
miR-181a	To be characterized	Inducing effect	Dendritic cells	Heart disease	(Zhu et al. 2017, 2884–2895)
miR-150	To be characterized	Inhibitory effect	Dendritic cells	Heart disease	(Zhu et al. 2017, 2884–2895)
miR-146b	STAT5A	Inhibitory effect	Heart tissue	Cardiac hypertrophy	(Feng et al. 2014, 361–368)
miR-499	STAT2	Inhibitory effect	Heart tissue	Cardiac hypertrophy	(Feng et al. 2014, 361–368)
miR-19a	TNF-α, STAT3, SOCS3	Inducing effect	Human PBMCs	Arthritis	(Li et al. 2016 2531–2536)
miR-21	TNF-α, STAT3, SOCS3	Inducing effect	Human PBMCs	Arthritis	(Li et al. 2016 2531–2536)
miR-1246	HNFγ	Inducing effect	Chondrocytes	Arthritis	(Wu et al. 2017, 2010–2021)
miR-1246	PRKAR1A, PPP2CB	Inducing effect	Epithelial cells	Inflammation and cancer	(Bott et al. 2017, 43897–43914)
miR-203	MCL-1	Inhibitory effect	Human cartilage cells	Osteoarthritis	(Zhao et al. 2017 171–178)
miR-29b	To be characterized	Inducing effect	Osteoblasts	Rheumatoid arthritis	(Figueiredo Neto and Figueiredo 2017, 6365857)
miR-21	To be characterized	Inducing effect	Osteoblasts	Rheumatoid arthritis	(Figueiredo Neto and Figueiredo 2017, 6365857)
miR-155	SOCS1	Inducing effect	CD4+ T-cells	Immune disorders	(Yao et al. 2012, e46082)

(Continued)

TABLE 3.1 (Cont.)

miRNA	Target transcript	Anticipated effect of JAK-STAT pathway	Test model	Disease model	Reference
miR-1225-3p	GAB3	Inhibitory effect	Huh7 cells	Influenza A virus	(Cheng et al. 2018, 5975–5986)
miR-203	DR1	Promotes JAK-STAT signaling	A549	Influenza A virus	(Zhang et al. 2018, 6797-018-25073-9)
miR-324-5p	CUEDC2	Negative regulator	A549	H5N1	(Kumar et al. 2018, 10.1128/JVI.01057–18. Print 2018 Oct 1)
miR-130a	To be characterized	Inducing effect	J6/JFH-1 immortal cells	Hepatitis C viral infection	(Li et al. 2014, 121–128; Murayama et al. 2007, 8030–8040)
miR-122	To be characterized	To be characterized	Liver cells	Hepatitis C viral infection	(Jopling et al. 2005, 1577–1581)
miR-373	JAK1, IRF9	Inhibitory effect	Human liver biopsies	Hepatitis C viral infection	(Mukherjee et al. 2015, 3356–3365)
miR-155	SOCS1	Inducing effect	HepG2 cells, and B lymphocytes	HBV infection	(Lu et al. 2008, 10436–10443)
miR-103	CDK5R1	Inhibitory effect	Stem cells	Spinal cord injury	(Li et al. 2018, 292–300)
miR-210	Presumably JAK2	Inhibitory effect	In vivo in mice	Spinal cord injury	(Dai et al. 2018b, 6609–6615)
miR-125b	JAK1 STAT1	Inhibitory effect	In vivo in mice	Axon regeneration/ spinal cord injury	(Dai et al. 2018a, 582–589)
miR-155	To be characterized	Inducing effect	Retinal pigment epithelial cells	Eye diseases	(Kutty et al. 2010, 390–395)
miR-155	To be characterized	Inducing effect	Lymphatic endothelial cells	Angiogenesis	(Yee et al. 2017, 20,683–20,693)
miR-9	SOCS5	Inducing effect	Endothelial cells	Angiogenesis	(Zhuang et al. 2012, 3513–3523)
miR-155	SOCS1	Inducing effect	Microglial cells	Neurodegenerative diseases	(Cardoso et al. 2012, 73–88)
miR-146a	IL-6	Inducing effect	Microglial cells	Neurodegenerative diseases	(Saba et al. 2012, e30832)
miR-216a	JAK2	Inhibitory effects	Neuronal cells	Neuroprotection	(Tian et al. 2018, 977–988)

3.2 JAK-STAT Targeting miRNAs in Liver Diseases

A group of tumor-suppressor proteins, which together were classified as the suppressor of cytokine signaling proteins (SOCS), have been well characterized with defined functions toward inhibition of JAK kinase and STAT transcription factor in the canonical JAK-STAT signaling pathway. The inhibition of both JAK and STAT by SOCS proteins was found to be mediated through protein interactions and through epigenetic mechanisms, which are regulated by specific miRNAs. The miRNA miR-19a targets most of the SOCS transcripts, which include SOCS1, SOCS3, and SOCS5, in addition to Cullin (Collins et al. 2013, e69090). As JAK-STAT regulation is influenced by SOCS proteins and since miR-19a inhibits SOCS3, the functional role of miR-19a would be to curtail the inhibitory effects of SOCS3 on JAK and STAT activities. This was reported recently in immortal "HuH-7 cells" where IL-6 and IFN-α enhanced activation of pSTAT3 in presence of miR-19a. These results suggest that miR-19a indirectly promotes JAK-STAT-mediated expression of pro-inflammatory mediators (Collins et al. 2013, e69090). It would be more interesting to see if this microRNA miR-19a had any impact on activation of JAK proteins, since SOCS3 binds to the receptor of JAK in its transmembrane region and stereotypically prevents its interactions with JAK.

In a significantly large animal study using BALB/c mice with $n = 40$ in each group, an in vivo study was performed to evaluate the role of specific miRNAs associated with lipopolysaccharides-(LPS-) induced sepsis as liver injury model (Lv et al. 2015, 6013–6018). While LPS was reported to have effectively induced the anticipated liver injury in mice, there was a concomitant increase in the expression of the miRNA miR-155 in the group that was treated with LPS alone. In the mice group that were treated with LPS and with miR-155 specific inhibitor, SOCS1 protein was reported to be significantly upregulated. Further, in presence of the miR-155 specific inhibitor, the expression levels of STAT1 were shown to be greatly reduced. Also, the levels of tumor necrosis factor alpha (TNF-α) and interleukin-10 (IL-10) cytokines were reported to be significantly decreased in the mice group that was treated with LPS and miR-155 specific inhibitor. Since STAT1 drives the expression of the miRNA miR-155, and miR-155 targets SOCS1 and prevents its expression, the above study provides a clear mechanistic regulation between the microRNA miR-155 and SOCS1, where miR-155 inhibits SOCS1 to promote JAK-STAT signaling in LPS-induced liver damage (Kutty et al. 2010, 390–395).

Compared to the normal liver tissues and the para-tumor tissues in patients with hepatocellular carcinoma (HCC), the levels of microRNA miR-409 were reported to be drastically reduced in the tumor tissues. This downregulation of miR-409 was presumed as a possible cause associated with the progression of the HCC (Zhang et al. 2019, 146–154). The authors also reported the existence of an inverse correlation between the expression of miR-409 and the protein levels of JAK2 and STAT3. Further, in cell-culture system, when HCC cells were transfected with both the microRNA mimic and inhibitor separately, reduced cell viability and increased apoptosis was observed in the miR-409 transfected cells. As can be anticipated, the opposite results were found in cells that were transfected with the inhibitor, which blocks the functions of miR-409. Both the cell culture data and the histological observations indicate that miR-409 targets JAK-STAT pathway to prevent the progression of HCC. The reports, which showed E74-like factor 2 as a direct target of miR-409 and c-Met as a target for miR-409-3p, suggest that miR-409 may not be directly targeting JAK or STAT transcripts. However, all the current evidence point to the same common observation, which indicate that miR-409 regulates JAK-STAT pathway, and which is more likely to be an indirect cause. In another study, HCC development was found to be directly correlated with the expression of two miRNAs belonging to the same class—miR-196a and miR-196b—which regulate JAK-STAT pathway (Ren et al. 2019, 333-019-1530-4). These two miRNAs were reported to increase cell proliferation, migration, invasion, and apoptosis induction. However, each of these was found to target two different transcription factors. The microRNA, miR-196a, primary target was identified as FOXO1, while the target for miR-196b was FOXP2. Interestingly, both these miRNAs were also found to prevent SOCS2 expression, affecting the JAK2–STAT5 pathway in HCC. The above reports suggests that in normal hepatic cells these miRNAs establish a tight control of JAK-STAT pathway as a means to prevent oncogenic transformation.

3.3 Specific MicroRNAs that Regulate JAK-STAT Pathways in Heart Diseases

An indirect epigenetic mechanism, which we recently reported, was the direct targeting of the main DNA methylase (DNMT1, DNA methyltransferase 1) by miRNA miR-1264 (Boosani et al. 2015, 1365–1376). DNMT1 promotes methylation of CpG residues in the genome DNA, and this non-specific methylation of DNA would affect all genes, which have CpG islands. Since SOCS3 genomic sequence harbors a well-defined CpG island in its promoter region, it is highly subjected to methylation by DNMT1. This we recently reported in vascular smooth muscle cells wherein the miRNA miR-1264, which has strong affinity to bind to the DNMT1 transcript and inhibit its expression, and thus in absence of DNMT1, SOCS3 expression remain unaffected (Boosani et al. 2015, 1365–1376). Subsequently, SOCS3 was able to prevent the activation of JAK. The miRNA miR-1264 also prevented the DNMT1 activity induced by TNF-α and insulin-like growth factor 1 (IGF-1) when treated together in smooth muscle cells. These observations establish the existence of an epigenetic mechanism, which involves both miRNA and a DNA methylase. Also, inhibition of SOCS3 was identified to be an important cause for the progression of intimal hyperplasia and restenosis (Gupta et al. 2011, 346–352). Cumulatively, these studies clearly indicate that multiple mechanisms are involved in preventing SOCS3 expression during restenosis and the miRNA miR-1264 is a key mediator that can help restore SOCS3 expression to prevent neointima formation in the arteries.

The role of miRNAs specific to JAK-STAT pathway and its implications in immune regulations associated with myocardial infarction were reported recently using bone-marrow–derived dendritic cells from mice (Zhu et al. 2017, 2884–2895). In an interesting approach, the authors used the cultured supernatant from the necrotic cardiomyocytes as a means to mimic myocardial infarction environment on dendritic cells. The authors subsequently identified specific miRNAs that can potentially regulate inflammatory responses mediated by the dendritic cells. The supernatant from the necrotic cells were reported to upregulate maturation markers in the dendritic cells, such as CD40, CD83, and CD86, along with increased levels of proinflammatory cytokines. Concurrent increase in the expression of the miRNA, miR-181a, and simultaneous downregulation of miR-150 was also reported. Treatment with the necrotic supernatant was shown to activate the JAK-STAT pathway and also facilitated nuclear translocation of nuclear factor kappa B (NF-κB) and c-Fos. These reports highlight the significance of the miRNAs—miR-181a and miR-150—in regulating JAK-STAT signalling, which controls the hypoxic cell death in cardiomyocytes.

In a rat model to study the role of different miRNAs in cardiac hypertrophy and heart failure, transverse aortic constriction (TAC) was performed in 20 rats (Feng et al. 2014, 361–368). RNA isolated from the hearts of these rats showed differential expression of multiple miRNAs that regulated different cellular pathways. The miRNAs, which directly or indirectly targeted JAK-STAT pathway in the above study, were discussed here. Expression of the microRNAs let-7c, let-7f, miR-1, miR-300-5p, miR-347, and miR-874 were found downregulated. These miRNAs indirectly act in regulating JAK-STAT pathway by targeting different interleukins. At the same time, the expression of miRNAs let-7i, miR-22, miR-29, miR-30a, miR-101, miR-125b-5p, miR-146b, miR-214, and miR-499 were found to be increased. Notably, only miR-146b and miR-499 were found to act directly on STAT5 and STAT2, respectively. Cumulatively, this comprehensive miRNA profiling in the cardiac hypertrophy model resulting from pressure overload due to TAC, identified signature miRNA species. Although a significant role of JAK-STAT signaling was evident in this study, several other effective target transcripts and other potent miRNAs regulating different cellular pathways exist which were well characterized.

3.4 MicroRNAs Targeting JAK-STAT Pathways in Arthritis

In efforts to identify specific miRNAs that regulate JAK-STAT signaling in patients with arthritis, a recent human clinical study was conducted using peripheral blood monocytes (PBMCs) collected from 20 normal individuals and 20 patients who were suffering from symptomatic juvenile idiopathic arthritis (Li et al. 2016, 2531–2536). Although the study showed minimal analysis to identify specific miRNAs, it was interesting to note that two selected miRNAs, miR-19a and miR-21, were reported to be expressing at a very low level in the patients. Importantly, the authors conclude that TNF-α, STAT3, and SOCS3 transcripts are the targets for these two miRNAs, miR-19a and miR-21. The increased expression of TNF-α and STAT3 in the arthritic patients agrees with the other reports in the literature. However, the expression of SOCS3 can be debatable but it should be noted that SOCS3 has also been shown to encourage inflammatory responses in a few other diseases, such as in asthma (Zafra et al. 2014, e91996). Also, the study includes juvenile patients, and presumably is focused on the local ethnic group, which greatly limits extension of the findings to other population groups unless evaluated. Additional analysis in this study could have been more beneficial, such as conducting a microarray-based analysis of all the samples and evaluating the samples to identify what specific

JAK or STAT molecules are activated through phosphorylation. Another notable limitation in this study was using real-time PCR as means for quantifying the gene expression. The procedure as such depends on the primer choice, and thus the estimated gene expression based on the available RNA needs to support through additional means of evaluation, such as using a protein array or ELISA or western blotting.

LPS is widely known to initiate many proinflammatory mechanisms in different cell types. Several volumes of information in the literature shows that inflammatory mediators, which include interleukins, cytokines, and growth factors, can be upregulated by LPS, both in vitro and in animal models. TNF-α is certainly one of them, which has the potential to mediate multiple signaling mechanisms and the evidence is accumulating, which supports the role of TNF-α in inducing epigenetic mechanisms that promote disease progression. We have earlier identified a clear role of TNF-α in exacerbating neointimal hyperplasia in the coronary arteries (Gupta et al. 2011, 346–352). Recently, we have also shown that TNF-α regulates different intracellular pathways including the JAK-STAT signaling (Dhar et al. 2013, H776-85; Boosani et al. 2015, 1365–1376). LPS-induced expression of IL-1b, IL-6, IL-8, and TNF-α was identified to promote inflammation in the chondrogenic cells "ATDC5." LPS-induced expression of these inflammatory mediators were found to regulate cell viability and apoptotic pathways that are mediated by JAK1, STAT1, STAT3, Akt, and PI3K (Wu et al. 2017, 2010–2021). Notably, the regulation of JAK-STAT pathway and its cross-talk with PI3K-Akt pathway by LPS was found to be mediated through the miRNA, miR-1246. The hepatic nuclear factor gamma "HNFγ" was reported as a specific target of miR-1246 in the putative chondrocytes. The study highlights multiple functions of HNFγ, and its inhibition by the miRNA miR-1246 in presence of LPS as a potential mechanism by which JAK-STAT promotes intracellular inflammation.

In mesenchymal stem cells, the miRNA miR-1246 was reported to be expressed in higher levels and interestingly, this miRNA expression was independent of the inflammatory signal induced by TNF-α. However, the main functions of miR-1246 reported was to target and prevent the expression of two tumor-suppressor genes—cAMP-dependent protein kinase type I-alpha "PRKAR1A" and the protein phosphates 2C-beta "PPP2CB" (Bott et al. 2017, 43897–43914). An interesting aspect was that the conditioned medium from the miR-1246 transfected cells was found to induce JAK-STAT signaling in the normal and tumor epithelial cells.

To understand the role of JAK-STAT pathway and miRNAs that affect JAK-STAT signaling in osteoarthritis, LPS was used as means to induce inflammation and cellular injury in cultured human cartilage cells C28/I2 (Zhao et al. 2017, 171–178). The authors report that when chondrocytes were treated with LPS, expression of a specific miRNA miR-203 was significantly upregulated with concomitant decrease in cell viability and increase in cellular apoptosis. Further, the authors also report that overexpression of miR-203 in the chondrocytes exaggerates LPS-induced inflammatory responses in the transfected cells. Since the direct target of the miRNA, miR-203, is the myeloid cell leukemia-1 transcript (MCL-1), it was speculated that MCL-1 would play a critical role in LPS-mediated inflammation during osteoarthritis. Further, as the effector pathways for MCL-1 are Wnt/β-catenin and JAK-STAT, it was speculated that the inflammatory responses initiated during the progression of osteoarthritis are mediated through upregulation of the miRNA miR-203, which indirectly affects JAK-STAT pathway through the expression of MCL-1.

In treating rheumatoid arthritis patients, it is not just sufficient to address inflammation since bone erosion occurs which is also a complication seen in these patients. In this direction, IL-27 has been shown to promote mineralization in the osteoblasts, which can prevent bone erosion. In attempts to identify the factors that enhance the effects of IL-27, a microarray-based study revealed that two specific miRNAs—miR-29b and miR-21—augment the expression of IL-27 in osteoblasts (Figueiredo Neto and Figueiredo 2017, 6365857). The study showed that treatment with IL-27 not only promoted osteoblast differentiation, but it also inhibited the proteins that prevent osteoblast formation. Two main signaling pathways were identified to be associated with osteoblast differentiation, which include JAK-STAT and TGF-β/BMP/SMAD signaling. In another study using human bone marrow stromal cells which were induced for osteogenic differentiation, specific miRNAs were identified that were characterized as osteogenic specific miRNAs, and these were reported to be targeting JAK-STAT pathway (Vimalraj and Selvamurugan 2014, 194–202).

3.5 JAK-STAT Specific miRNAs Regulating Immune Responses

CD4+ T-cells are an important class of immune cells with multiple functions and are primarily involved in suppressing immune reaction. IL-6 and STAT3 were shown to be essentially required for the differentiation of immune cells, such as T-regulatory and Th17, cells from uncommitted CD4+ T-cells. In a recent report, inhibition of SOCS1 by the miRNAs miR-155 was shown to induce differentiation of T-regulatory cells and Th17 cells (Yao et al. 2012, e46082). This cell differentiation was further supported by the observations that when CD4+ T-cells were transfected with the miRNAs miR-155, increased expression of activated STAT3 and STAT5 were observed, which correlated with decrease in the expression of SOCS1 in the same group. These reports clarify a clear role of JAK-STAT signaling in immune cell differentiation, and SOCS1 prevents activation of JAK kinases, which is required to induce STAT3 and STAT5 activation. In another study to identify the significance and role of miRNAs in regulating proliferation and death of T-cells, a comprehensive microarray analysis was carried out which was based on the observations that a drastic decrease in JAK1 expression was noted in IL-2 depleted cells (Ranji et al. 2015, 169–183). The study identified several miRNAs that can specifically target important pathway mediators of JAK-STAT signaling to regulate immune responses.

3.6 miRNAs Regulating Viral Diseases Induced by JAK-STAT Pathways

Chronic hepatitis resulting from hepatitis C viral infections can be a major cause that can lead to hepatocellular carcinoma and other complications in humans. A recent report showed a strong correlation between the expression of the miRNA miR-130a and the replication of hepatitis C virus (Li et al. 2014, 121–128). Specifically, the expression of miRNA miR-130a was reported to significantly inhibit the virus replication in chimeric J6/JFH-1 cells. It is to be noted that the J6/JFH-1 cell line, which has the chimeric genomic regions that encode J6 structural proteins and JFH-1 nonstructural proteins, enables the J6/JFH-1 cells to replicate autonomously and produce infectious HCV particles (Murayama et al. 2007, 8030–8040). The finding that overexpression of miR-130a in these cells upregulates the expression of type I interferons and other essential proteins which initiate innate responses clearly indicates the role of interferon in stimulating genes that have been earlier reported to activate JAK-STAT pathway (Martensen and Justesen 2004, 1–19). Together, the above reports would provide strong inference, which indicates the potential role of miR-130a in inhibiting replication of hepatitis C virus in humans. Another important observation reported in the above studies is the indirect effects of the miRNA miR-130a that inhibits the expression of another miRNA miR-122. The miRNA miR-122 has been previously reported to promote proliferation of hepatitis C virus (Jopling et al. 2005, 1577–1581).

Since SOCS3 is a competitive inhibitor of JAK kinases, miRNAs that inhibit SOCS3 would have JAK–promoting effects. Conversely, those miRNAs that promote SOCS3 expression would confer indirect inhibitory effects on JAK activation. Such an indirect role of miRNA miR-122 was recently reported to affect interferon treatment, which is the current standard of care to treat hepatitis C viral infections. While overexpression of the miRNA miR-122 suppressed the interferon-stimulated response element-mediated gene expression, silencing of miR-122 was reported to enhanced interferon-induced functions, and as a consequence, the expression of SOCS3 was reportedly affected (Yoshikawa et al. 2012, 637).

In another study, elevated levels of the miRNA miR-373 was reported in hepatitis C virus infected human primary hepatocytes, and also in the liver biopsies obtained from human patients (Mukherjee et al. 2015, 3356–3365). A direct target of miR-373 indicated in the above article is the JAK1 transcript and the interferon regulating factor-9 and therefore, activation of STAT1 was found to be decreased. Subsequently, knockdown of miR-373 in the hepatocytes was found to prevent the expression of JAK1 and IRF9, and, therefore, the authors speculate that the hepatitis C virus induces the expression of miR-373 in hepatocytes and this is required for its replication inside the cell.

The Epstein–Barr virus infection of B lymphocytes has been recently shown to elevate the expression levels of the miRNA miR-155, which contributes to immortalization (Lu et al. 2008, 10436–10443). Subsequently, in HepG2 human hepatocellular carcinoma cell lines, it was reported that expression of miR-155 upregulates the interferon inducible anti-viral response. However, overexpression of miR-155 was also found to inhibit SOCS1 expression and, as a consequence, STAT1 and STAT3 activation was induced (Su et al. 2011, 354-422X-8-354). SOCS1 being a negative regulator of JAK-STAT pathway, the above reports provide a clear understanding that the miRNA miR-155 enhances antiviral immune response by activating JAK-STAT pathway, and this is required to prevent HBV infection in humans.

Diagnostic biomarkers, if detected early in the disease, can be very helpful in strategizing an effective treatment plan to cure the disease. In this line, using the blood samples from 50 patients diagnosed with HBV infections and hepatic fibrosis, circulating miRNAs were screened using miRNA microarray (Zhang et al. 2015, 5647–5654). The authors reported that 12 miRNAs were differentially expressed, of which 10 were overexpressed and 2 were downregulated. Among the target genes identified, 31 of them were specific for JAK-STAT pathway. Although the study identified putative biomarkers for early diagnosis of HBV-associated hepatic fibrosis, further validation studies are required.

As identified earlier in this manuscript, the role of miR-203 is evident in regulating different cellular pathways in many disease conditions. Apart from the above-discussed studies, the role of miR-203 in regulating viral infections was also reported. Within the 2500bp promoter region of miR-203, the binding sites for two transcription factors, NFκB and ISGF3, were identified which are known to regulate the JAK-STAT signaling pathway. ISGF3 is primarily involved in IFN-mediated regulation of JAK-STAT pathway with STAT1, STAT2, and IRF9 as key mediators. Interestingly, in the IFN-deficient cells that were infected with influenza A virus, increased expression of miR-203 was observed. These reports suggest a direct correlation between the increased expression of miR-203 with influenza A virus infection; however, the same study also identifies indirect role of JAK-STAT signaling during influenza A viral infection (Zhang et al. 2018, 6797-018-25073-9). IFNs are commonly administered to treat viral infections. In both cancerous and non-cancerous liver cells, treatment with IFN was reported to downregulate the levels of miR-1225-3p. Since viral infection induces the production of endogenous IFN levels, Huh7 and 293T cells when infected with RNA viruses, such as Hepatitis C Virus, Sendai Virus, and New castle Disease Virus, was also reported to lower the levels of miR-1225-3p (Cheng et al. 2018, 5975–5986). Since IFN directly regulates JAK-STAT pathway through its cell surface receptors, the above reports indicate IFN-mediated regulation of JAK-STAT signaling by miR-1225-3p during viral infections.

The highly pathogenic influenza A virus, H5N1, showed more than 50% mortality rate in infected people. In mice infected with H5N1, the expression of miR-324-5p was reported to be down-regulated by Poly(IC), which is a viral pathogen-associated molecular pattern protein. By targeting the RNA polymerase subunits PB1 and PB2, the miR-324-5p prevents H5N1 viral replication in the host (Kumar et al. 2018). Further transcriptome analysis of the miR-324-5p identified one of its direct target transcripts CUEDC2, which is a negative regulator of JAK-STAT signaling. It was also shown that ectopic expression of miR-324-5p induces the expression of antiviral gene expression, notably the IFNs. Taken together these reports show an essential role of different miRNAs in regulating viral infections, primarily through IFN-mediated JAK-STAT signaling pathway comprising different isoforms of JAK and STAT proteins.

3.7 Involvement of miRNAs in Other Diseases Regulated by JAK-STAT Pathways

3.7.1 Spinal Cord Injury

The transcription factor SOX2 is critically required for tissue development and homeostasis in the cells. In addition, for stem cells, SOX2 expression is required to maintain their self-renewability and pluripotency. Just recently, the miRNA miR-103 was identified to prevent two important cellular functions, autophagy and apoptosis, by regulating SOX2 expression. The miRNA miR-103 was shown to indirectly inhibit both MAPK-ERK and JAK-STAT pathways by upregulating SOX2 in PC12 cancer cells (Li et al. 2018, 292–300). Further in the same article, the authors showed that in rats with spinal cord injury, delivery of the miRNA miR-103 agomir, increased the expression of SOX2. The treatment with miR-103 agomir was also found to prevent the expression of Beclin-1, Bax, activated caspase-3 and caspase-8, and in contrast, LC3-1, p62, and BcL-2 were reported to be upregulated. Although whether miR-103 can directly regulate SOX2 expression remained unclear, it was evident that SOX2 plays a critical role in tissue repair. Further, since CDK5R1 is a known direct target of miR-103, it remains to be elucidated whether SOX2 is at a pivotal point between CDK5R1 and JAK-STAT pathways, and their possible cross talk.

A mouse model of spinal cord injury due to contusion was reported as a feasible study to evaluate axon growth and regeneration (Dai et al. 2018a, 582–589). In this mouse model, significantly reduced expression of the miRNA miR-125b was noted due to spinal cord contusion. Both JAK1 and STAT1 transcripts were identified as the direct targets of this miRNA miR-125b, suggesting the direct role of JAK-STAT signaling in regulating spinal cord injury and axon regeneration. The authors further extended the mice results to a rat model and reported that miR-125b reduces protein levels of caspase3, which is a known activator of cellular apoptosis. The results presented in the above study clearly indicate that miR-125b promotes axon growth and regeneration by inhibiting the JAK-STAT signaling pathway. The same group also evaluated the role of another miRNA miR-210 in the mice with spinal cord injury (Dai et al. 2018b, 6609–6615). A key observation made in this study was the late recovery to grasping strength in mice with spinal cord injury, compared to sham control group. The late recovery was found to correlate with the reduced levels of the miRNA miR-210. Correlating with this observation, the mice group that was administered with miR-210 mimic was found to show higher grip strength. In assessing the target gene transcripts of miR-210, increased expression of JAK2 and STAT3 proteins was observed in the control group. Their observations were further supported from the results seen in the mice with miR-210 mimic group, which showed reduced JAK2 and STAT3 levels. The above two studies point to the possibility of targeting JAK-STAT proteins to promote spinal cord injury. In these lines, JAK or STAT inhibitors can be tested in animal models to show their effects in promoting axon regeneration and recovery from spinal cord injury.

3.7.2 Eye Diseases

Eye transplant is vital to restoring vision in suitable patients. In efforts to address the inflammatory responses in the retinal pigment epithelial cells (RPE) from donor eyes, the role of several cytokines was recently evaluated (Kutty et al. 2010, 390–395). Three important cytokines, TNF-α, IL-1β, and IFN-γ, were tested in RPE cells. The authors reported that all three were effective in inducing the inflammatory responses in the RPE cells. While the effects of each cytokine were noticed in RPE cells when treated individually, the cellular inflammation was reported to be exaggerated when RPE cells were treated with all three cytokines at the same time. Importantly, the combined treatment was shown to enhance the expression of miRNA miR-155, and the RPE cells when treated with the three cytokines along with JAK1 inhibitor, the expression of miR-155 was drastically reduced. In addition, the authors also showed that STAT1 activation was elevated in presence of the three cytokines and the same was inhibited when the cells were treated with JAK1 inhibitor. These reports suggest that miR-155 promotes JAK-STAT signaling to induce inflammation in the cells. However, additional studies are required to precisely elucidate specific transcripts that are targeted by miR-155 and the molecular intermediates that promote proinflammatory signaling in these cells.

3.7.3 Endothelial Cell Migration and Angiogenesis

Within this last decade, several studies have identified that targeting the programmed death ligand 1 "PD-L1" could be an efficient strategy to prevent progression of certain cancer types, especially with its initial success in treating melanoma patients (Wolchok et al. 2013a, 1–13, 2013b, 122–133). However, in many non-cancerous diseases, upregulation of cytokines is also a characteristic feature such as in atherosclerosis where there is an increased expression of TNF-α. Whether the increased expression of cytokines and growth factors in non-cancerous diseases would have any impact on normal cells and tissues would enable strategizing a better treatment plan. In these directions, using normal human lymphatic endothelial cells, the effects of the cytokines TNF-α and IFN-γ were recently reported where the combined presence of TNF-α and IFN-γ significantly elevated the expression of PD-L1 expression (Yee et al. 2017, 20683–20693). A key finding in the above study was that TNF-α induced the expression of miRNA miR-155 and this effect was enhanced by treatment with IFN-γ. Further, the same effects on PD-L1 and miR-155 expression were also reported in normal dermal fibroblasts. Importantly, it was reported that the cytokines TNF-α and mostly the IFN-γ, initiates JAK-STAT pathways and drives the activation of STAT1 and STAT3 contributing to the induced expression of miR-155 and PD-L1.

SOCS5 is another member of SOCS family, which is a negative regulator of JAK-STAT pathway as it prevents activation of JAK. The miRNA miR-9 was reported to be increased in different tumor types (Ma et al. 2010, 247–256; Shigehara et al. 2011, e23584). However, this miRNA was found to target and inhibit SOCS5 expression by binding to its 3′UTR sequence and relieves the inhibition on JAK activation (Zhuang et al. 2012, 3513–3523). Transfection of the miRNA miR-9 into human and mouse endothelial cells was shown to induce phosphorylation of JAK1, JAK2, STAT1, and STAT3, and simultaneously inhibit SOCS5 expression. Further, miR-9 was also shown to promote endothelial cell migration and angiogenic sprouting in HUVECs.

3.7.4 Neuroprotective Functions

Besides the above-discussed biological effects of the miRNA, miR-155, its proinflammatory functions through inhibition of SOCS1 were also reported in immortal mouse microglial cells (Cardoso et al. 2012, 73–88). Upon transfection of the microglial cells with miRNA miR-155, the expression of SOCS1 was significantly downregulated. On the contrary, knockdown of the miRNA miR-155 not only upregulated the expression of SOCS1, but it also prevented induction of nitric oxide synthase and inflammatory cytokines. It is to be noted that SOCS1 prevents JAK and STAT activation. The authors report that when the microRNA miR-155 was inhibited in microglial cells and treated with LPS, the LPS induced effects on the expression of TNF-α and IL-6 were significantly reduced. These findings indicate that under chronic inflammatory conditions, inhibition of miR-155 could confer neuroprotective effects and may prevent neuronal cell death, which can be an effective treatment strategy for neurodegenerative diseases.

In the prion-infected mice brain, the levels of miRNA miR-146a was reported to be elevated. Further, TLR2 and TLR4 activation also was found to hold strong correlation with the increased expression of miR-146a (Saba et al. 2012, e30832). Interestingly, in the murine microglial cells, the authors report that upon transfection with the miRNA miR-146a, the expression of the proinflammatory cytokine IL-6 was drastically reduced. However, the expression of the master proinflammatory transcription factor, NFκB was affected in absence of miR-146a, in addition to the JAK-STAT pathway mediators. Although these studies demonstrate the effect of miR-146a in modulating the cellular inflammation, the cross talk between NFκB and JAK-STAT pathway is still warranted.

Cerebral ischemia is a major cause of permanent damage to health that can lead to disability and even death of the individual. Inflammation initiates several cellular mechanisms, which eventually cause neuronal damage. In the cultured primary neuronal cells from mouse brain and also in the mouse model for ischemic stroke induced due to middle cerebral artery occlusion, targeting JAK2 was reported to confer neuroprotection against ischemic injury (Tian et al. 2018, 977–988). The cultured neuronal cells, when subjected to oxygen-glucose deprivation and

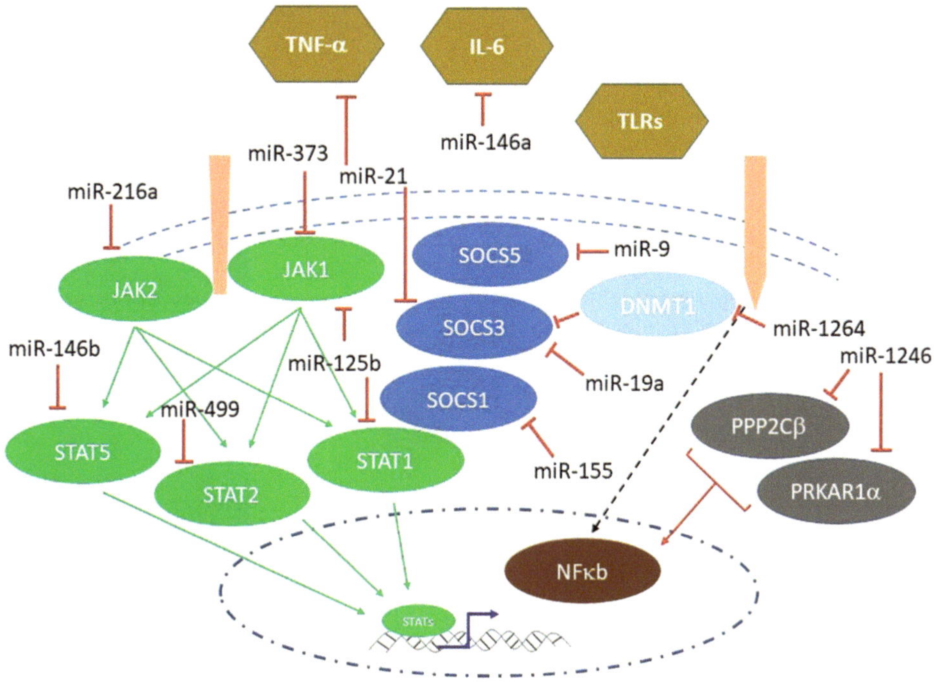

FIGURE 3.1 Regulation of JAK-STAT signaling pathway by microRNAs.

reoxygenation, appear to mimic the ischemic conditions. Expression of JAK2 and the downstream proinflammatory molecules such as iNOS, MMP9, TNF-α, and IL-1β were reported in this in vitro cell culture model. In this study, the miRNA miR-216a was identified as a direct target for JAK2 and inhibition of JAK2 in the neuronal cells transfected with the miR-216a prevented the expression of the above inflammatory mediators. Based on the above published reports, targeting JAK2-mediated signaling appears to be a beneficial approach to protect neuronal damage resulting from the ischemic injury.

Figure 3.1 shows known miRNAs that can directly target JAK and STAT transcripts and prevent their translation. Also, specific miRNAs that can potentially promote JAK-STAT signaling pathway by indirectly targeting the expression of proteins, such as the suppressors of cytokine signaling proteins, which inhibit JAK1 and JAK2, were highlighted. In addition, the miRNAs, which intercept the crosstalk between the JAK-STAT and PI3K–Akt pathways were shown in the above figure.

3.8 Summary and Future Directions

The information presented here clearly shows that JAK-STAT pathway mediators promote pathological progression of many human diseases. Several important miRNAs have been identified, which have been proven to be key regulators of JAK-STAT signaling. In Figure 3.1, we have summarized few miRNAs, which affect JAK-STAT pathway. Also, specific miRNAs were found to activate JAK-STAT pathway in more than one disease conditions. Since inflammation and cell proliferation are key cellular mechanisms that are associated with many diseases, and JAK-STAT pathway promotes both these important cellular mechanisms, it appears very promising to identify and test those miRNAs that can modulate JAK-STAT pathway. As current bioinformatic tools were proven to be effective in target identification and characterization of different miRNAs, high throughput screening and cell-based assays would facilitate validation of the putative miRNAs to identify their therapeutic benefits. Since miRNAs have gained entry into human clinical trials, their initial evaluation through pre-clinical trials in animal models is already underway in many research laboratories. It is anticipated that there would be a significant change in the therapeutic strategies to treat different human diseases using miRNA-based medicine.

ACKNOWLEDGEMENTS

This work was supported by the State of Nebraska LB692 grant to CSB by Creighton University and research grants R01 HL112597, R01 HL116042, and R01 HL120659 to DK Agrawal from the National Institutes of Health, USA. The content of this book chapter is solely the responsibility of the authors and does not necessarily represent the official views of the National Institutes of Health or the State of Nebraska.

COMPETING INTERESTS

The authors declare no other relevant affiliations or financial involvement with any organization or entity with financial interest or financial conflict with the subject matter or materials discussed in this chapter. No writing assistance was utilized in the production of this manuscript.

REFERENCES

Boosani, C. S., K. Dhar, and D. K. Agrawal. 2015. "Down-Regulation of Hsa-miR-1264 Contributes to DNMT1-Mediated Silencing of SOCS3." *Molecular Biology Reports* 42 (9): 1365–1376.

Bott, A., N. Erdem, S. Lerrer, A. Hotz-Wagenblatt, C. Breunig, K. Abnaof, A. Worner et al. 2017. "miRNA-1246 Induces Pro-Inflammatory Responses in Mesenchymal Stem/Stromal Cells by Regulating PKA and PP2A." *Oncotarget* 8 (27): 43897–43914.

Cardoso, A. L., J. R. Guedes, L. Pereira de Almeida, and M. C. Pedroso de Lima. 2012. "miR-155 Modulates Microglia-Mediated Immune Response by Down-Regulating SOCS-1 and Promoting Cytokine and Nitric Oxide Production." *Immunology* 135 (1): 73–88.

Cheng, M., Y. Niu, J. Fan, X. Chi, X. Liu, and W. Yang. 2018. "Interferon Down-Regulation of miR-1225-3p as an Antiviral Mechanism through Modulating Grb2-Associated Binding Protein 3 Expression." *The Journal of Biological Chemistry* 293 (16): 5975–5986.

Collins, A. S., C. E. McCoy, A. T. Lloyd, C. O'Farrelly, and N. J. Stevenson. 2013. "miR-19a: An Effective Regulator of SOCS3 and Enhancer of JAK–STAT Signalling." *PLoS One* 8 (7): e69090.

Dai, J., L. J. Xu, G. D. Han, H. L. Sun, G. T. Zhu, H. T. Jiang, G. Y. Yu, and X. M. Tang. 2018a. "MicroRNA-125b Promotes the Regeneration and Repair of Spinal Cord Injury through Regulation of JAK/STAT Pathway." *European Review for Medical and Pharmacological Sciences* 22 (3): 582–589.

Dai, J., G. Y. Yu, H. L. Sun, G. T. Zhu, G. D. Han, H. T. Jiang, and X. M. Tang. 2018b. "MicroRNA-210 Promotes Spinal Cord Injury Recovery by Inhibiting Inflammation Via the JAK–STAT Pathway." *European Review for Medical and Pharmacological Sciences* 22 (20): 6609–6615.

Dhar, K., K. Rakesh, D. Pankajakshan, and D. K. Agrawal. 2013. "SOCS3 Promotor Hypermethylation and STAT3-NF-kappaB Interaction Downregulate SOCS3 Expression in Human Coronary Artery Smooth Muscle Cells." *American Journal of Physiology.Heart and Circulatory Physiology* 304 (6): H776–H785.

Feng, H. J., W. Ouyang, J. H. Liu, Y. G. Sun, R. Hu, L. H. Huang, J. L. Xian, C. F. Jing, and M. J. Zhou. 2014. "Global microRNA Profiles and Signaling Pathways in the Development of Cardiac Hypertrophy." *Brazilian Journal of Medical and Biological Research = Revista Brasileira De Pesquisas Medicas E Biologicas* 47 (5): 361–368.

Figueiredo Neto, M. and M. L. Figueiredo. 2017. "Combination of Interleukin-27 and MicroRNA for Enhancing Expression of Anti-Inflammatory and Proosteogenic Genes." *Arthritis* 2017: 6365857.

Gupta, G. K., K. Dhar, M. G. Del Core, W. J. Hunter 3rd, G. I. Hatzoudis, and D. K. Agrawal. 2011. "Suppressor of Cytokine Signaling-3 and Intimal Hyperplasia in Porcine Coronary Arteries Following Coronary Intervention." *Experimental and Molecular Pathology* 91 (1): 346–352.

Jopling, C. L., M. Yi, A. M. Lancaster, S. M. Lemon, and P. Sarnow. 2005. "Modulation of Hepatitis C Virus RNA Abundance by a Liver-Specific MicroRNA." *Science (New York, N.Y.)* 309 (5740): 1577–1581.

Kumar, A., A. Kumar, H. Ingle, S. Kumar, R. Mishra, M. K. Verma, D. Biswas et al. 2018. "MicroRNA Hsa-miR-324-5p Suppresses H5N1 Virus Replication by Targeting the Viral PB1 and Host CUEDC2." *Journal of Virology* 92 (19): doi: 10.1128/JVI.01057-18. Print 2018 Oct 1.

Kutty, R. K., C. N. Nagineni, W. Samuel, C. Vijayasarathy, J. J. Hooks, and T. M. Redmond. 2010. "Inflammatory Cytokines Regulate microRNA-155 Expression in Human Retinal Pigment Epithelial Cells by Activating JAK/STAT Pathway." *Biochemical and Biophysical Research Communications* 402 (2): 390–395.

Li, G., T. Chen, Y. Zhu, X. Xiao, J. Bu, and Z. Huang. 2018. "MiR-103 Alleviates Autophagy and Apoptosis by Regulating SOX2 in LPS-Injured PC12 Cells and SCI Rats." *Iranian Journal of Basic Medical Sciences* 21 (3): 292–300.

Li, H. W., Y. Xie, F. Li, G. C. Sun, Z. Chen, and H. S. Zeng. 2016. "Effect of miR-19a and miR-21 on the JAK/STAT Signaling Pathway in the Peripheral Blood Mononuclear Cells of Patients with Systemic Juvenile Idiopathic Arthritis." *Experimental and Therapeutic Medicine* 11 (6): 2531–2536.

Li, S., X. Duan, Y. Li, B. Liu, I. McGilvray, and L. Chen. 2014. "MicroRNA-130a Inhibits HCV Replication by Restoring the Innate Immune Response." *Journal of Viral Hepatitis* 21 (2): 121–128.

Lu, F., A. Weidmer, C. G. Liu, S. Volinia, C. M. Croce, and P. M. Lieberman. 2008. "Epstein-Barr Virus-Induced miR-155 Attenuates NF-kappaB Signaling and Stabilizes Latent Virus Persistence." *Journal of Virology* 82 (21): 10436–10443.

Lv, X., Y. Zhang, Y. Cui, Y. Ren, R. Li, and Q. Rong. 2015. "Inhibition of microRNA155 Relieves Sepsisinduced Liver Injury through Inactivating the JAK/STAT Pathway." *Molecular Medicine Reports* 12 (4): 6013–6018.

Ma, L., J. Young, H. Prabhala, E. Pan, P. Mestdagh, D. Muth, J. Teruya-Feldstein et al. 2010. "miR-9, a MYC/MYCN-Activated microRNA, Regulates E-Cadherin and Cancer Metastasis." *Nature Cell Biology* 12 (3): 247–256.

Martensen, P. M. and J. Justesen. 2004. "Small ISGs Coming Forward." *Journal of Interferon & Cytokine Research: The Official Journal of the International Society for Interferon and Cytokine Research* 24 (1): 1–19.

Mukherjee, A., A. M. Di Bisceglie, and R. B. Ray. 2015. "Hepatitis C Virus-Mediated Enhancement of microRNA miR-373 Impairs the JAK/STAT Signaling Pathway." *Journal of Virology* 89 (6): 3356–3365.

Murayama, A., T. Date, K. Morikawa, D. Akazawa, M. Miyamoto, M. Kaga, K. Ishii et al. 2007. "The NS3 Helicase and NS5B-to-3′X Regions are Important for Efficient Hepatitis C Virus Strain JFH-1 Replication in Huh7 Cells." *Journal of Virology* 81 (15): 8030–8040.

Ranji, N., M. Sadeghizadeh, M. Karimipoor, M. A. Shokrgozar, R. Nakhaei Sistani, and S. H. Paylakhi. 2015. "MicroRNAs Signature in IL-2-Induced CD4+ T Cells and their Potential Targets." *Biochemical Genetics* 53 (7–8): 169–183.

Ren, W., S. Wu, Y. Wu, T. Liu, X. Zhao, and Y. Li. 2019. "MicroRNA-196a/-196b Regulate the Progression of Hepatocellular Carcinoma through Modulating the JAK/STAT Pathway Via Targeting SOCS2." *Cell Death & Disease* 10 (5): 333.

Saba, R., S. Gushue, R. L. Huzarewich, K. Manguiat, S. Medina, C. Robertson, and S. A. Booth. 2012. "MicroRNA 146a (miR-146a) is Over-Expressed during Prion Disease and Modulates the Innate Immune Response and the Microglial Activation State." *PLoS One* 7 (2): e30832.

Shigehara, K., S. Yokomuro, O. Ishibashi, Y. Mizuguchi, Y. Arima, Y. Kawahigashi, T. Kanda et al. 2011. "Real-Time PCR-Based Analysis of the Human Bile microRNAome Identifies miR-9 as a Potential Diagnostic Biomarker for Biliary Tract Cancer." *PLoS One* 6 (8): e23584.

Su, C., Z. Hou, C. Zhang, Z. Tian, and J. Zhang. 2011. "Ectopic Expression of microRNA-155 Enhances Innate Antiviral Immunity against HBV Infection in Human Hepatoma Cells." *Virology Journal* 8: 354.

Tian, Y. S., D. Zhong, Q. Q. Liu, X. L. Zhao, H. X. Sun, J. Jin, H. N. Wang, and G. Z. Li. 2018. "Upregulation of miR-216a Exerts Neuroprotective Effects against Ischemic Injury through Negatively Regulating JAK2/STAT3-Involved Apoptosis and Inflammatory Pathways." *Journal of Neurosurgery* 130 (3): 977–988.

Vimalraj, S. and N. Selvamurugan. 2014. "MicroRNAs Expression and their Regulatory Networks during Mesenchymal Stem Cells Differentiation toward Osteoblasts." *International Journal of Biological Macromolecules* 66: 194–202.

Wolchok, J. D., F. S. Hodi, J. S. Weber, J. P. Allison, W. J. Urba, C. Robert, S. J. O'Day et al. 2013a. "Development of Ipilimumab: A Novel Immunotherapeutic Approach for the Treatment of Advanced Melanoma." *Annals of the New York Academy of Sciences* 1291: 1–13.

Wolchok, J. D., H. Kluger, M. K. Callahan, M. A. Postow, N. A. Rizvi, A. M. Lesokhin, N. H. Segal et al. 2013b. "Nivolumab Plus Ipilimumab in Advanced Melanoma." *The New England Journal of Medicine* 369 (2): 122–133.

Wu, D. P., J. L. Zhang, J. Y. Wang, M. X. Cui, J. L. Jia, X. H. Liu, and Q. D. Liang. 2017. "MiR-1246 Promotes LPS-Induced Inflammatory Injury in Chondrogenic Cells ATDC5 by Targeting HNF4gamma." *Cellular Physiology and Biochemistry: International Journal of Experimental Cellular Physiology, Biochemistry, and Pharmacology* 43 (5): 2010–2021.

Yao, R., Y. L. Ma, W. Liang, H. H. Li, Z. J. Ma, X. Yu, and Y. H. Liao. 2012. "MicroRNA-155 Modulates Treg and Th17 Cells Differentiation and Th17 Cell Function by Targeting SOCS1." *PLoS One* 7 (10): e46082.

Yee, D., K. M. Shah, M. C. Coles, T. V. Sharp, and D. Lagos. 2017. "MicroRNA-155 Induction Via TNF-Alpha and IFN-Gamma Suppresses Expression of Programmed Death Ligand-1 (PD-L1) in Human Primary Cells." *The Journal of Biological Chemistry* 292 (50): 20683–20693.

Yoshikawa, T., A. Takata, M. Otsuka, T. Kishikawa, K. Kojima, H. Yoshida, and K. Koike. 2012. "Silencing of microRNA-122 Enhances Interferon-Alpha Signaling in the Liver through Regulating SOCS3 Promoter Methylation." *Scientific Reports* 2: 637.

Zafra, M. P., C. Mazzeo, C. Gamez, A. Rodriguez Marco, A. de Zulueta, V. Sanz, I. Bilbao, J. Ruiz-Cabello, J. M. Zubeldia, and V. Del Pozo. 2014. "Gene Silencing of SOCS3 by siRNA Intranasal Delivery Inhibits Asthma Phenotype in Mice." *PLoS One* 9 (3): e91996.

Zhang, C. S., Y. Lin, F. B. Sun, J. Gao, B. Han, and S. J. Li. 2019. "miR-409 Down-Regulates JAK–Stat Pathway to Inhibit Progression of Liver Cancer." *European Review for Medical and Pharmacological Sciences* 23 (1): 146–154.

Zhang, Q., M. Xu, Y. Qu, Z. Li, Q. Zhang, X. Cai, and L. Lu. 2015. "Analysis of the Differential Expression of Circulating microRNAs during the Progression of Hepatic Fibrosis in Patients with Chronic Hepatitis B Virus Infection." *Molecular Medicine Reports* 12 (4): 5647–5654.

Zhang, S., J. Li, J. Li, Y. Yang, X. Kang, Y. Li, X. Wu, Q. Zhu, Y. Zhou, and Y. Hu. 2018. "Up-Regulation of MicroRNA-203 in Influenza A Virus Infection Inhibits Viral Replication by Targeting DR1." *Scientific Reports* 8 (1): 6797.

Zhao, C., Y. Wang, H. Jin, and T. Yu. 2017. "Knockdown of microRNA-203 Alleviates LPS-Induced Injury by Targeting MCL-1 in C28/I2 Chondrocytes." *Experimental Cell Research* 359 (1): 171–178.

Zhu, J., K. Yao, J. Guo, H. Shi, L. Ma, Q. Wang, H. Liu et al. 2017. "miR-181a and miR-150 Regulate Dendritic Cell Immune Inflammatory Responses and Cardiomyocyte Apoptosis Via Targeting JAK1-STAT1/C-Fos Pathway." *Journal of Cellular and Molecular Medicine* 21 (11): 2884–2895.

Zhuang, G., X. Wu, Z. Jiang, I. Kasman, J. Yao, Y. Guan, J. Oeh et al. 2012. "Tumour-Secreted miR-9 Promotes Endothelial Cell Migration and Angiogenesis by Activating the JAK–STAT Pathway." *The EMBO Journal* 31 (17): 3513–3523.

4

JAK-STAT Signaling in Asthma and Allergic Airway Inflammation

Amina Abdul Qayum, Tristan Hayes, and Mark H. Kaplan
Department of Pediatrics and Herman B Wells Center for Pediatric Research
Indiana University School of Medicine
Indianapolis, Indiana

4.1 Introduction

Classical asthma is an inflammatory disease mediated by T_H2 and T_H9 cells characterized by increased serum IgE, eosinophilia, and responsiveness to steroids. Non-canonical asthma, which affects less than 10% of all asthmatics, is thought to be promoted by T_H17 cells and is non-atopic, characterized by steroid-resistant neutrophilia.[1] Cytokines that are well known to implement the asthmatic phenotype include IL-4, IL-5, IL-9, IL-13, IL-17, TSLP, and IL.-31[2,3] These cytokines signal through the JAK-STAT (see Table 4.1) pathway to mediate communication between cells and their environment. In this chapter, we will discuss the current understanding of JAK-STAT signaling in multiple cell types as it relates to asthma and allergic lung inflammation in animal models.

4.2 Janus Kinase

Janus kinase (JAK) proteins are a family of intracellular non-receptor tyrosine protein kinases (PTKs) that aid in signal transduction of JAK-STAT pathways. There are four known JAK kinases: JAK1, JAK2, JAK3, and TYK2. The structures of these kinases contain seven homology domains (JH1-JH7). The kinase domain (JH1) has catalytic activity while the pseudo-kinase domain (JH2) located at the carboxy terminus end, does not have catalytic activity. JH2 domains interact with and regulate the JH1 domain.[4,5] The kinase and pseudo-kinase domains are the defining feature of the JAK family of proteins. This name is derived from the Roman God of duality, Janus, who is depicted as having two heads; one looking to the past and one looking to the future. The JH3–JH4 domains contain homology to SH2 domains and putatively play a structural role and stabilize JAK protein conformation. The JH5–JH7 domains are located at the amino-terminus and serve as the FERM proteins that are critical for receptor binding of JAKs.[6,7]

When a ligand binds to its receptor, JAKs associated with that receptor phosphorylate each other as well as the cytokine receptor. The phosphorylated receptor serves as a docking site for STAT proteins. When the STAT proteins bind to the receptor, they become phosphorylated by JAKS. Activated STATs can then translocate to the nucleus where they directly bind DNA and promote RNA transcriptoin. There is also evidence for the role of nuclear JAK proteins in cell signaling. Some studies have shown that JAKs can modify other non-STAT transcription factors and modify histone proteins.[8,9] However, the main function defined for JAKs is in the activation of STAT proteins.

TABLE 4.1
Th2 Cytokine Usage of JAK-STAT Proteins

JAK/STAT	Role in Signaling	Mutation Phenotype	References
JAK1	IL-4, IL-13, IL-5, IL-9, TSLP, IL-31	KO prenatally lethal	6
JAK2	IL-5, TSLP, IL-31	KO embryonically lethal	6,7
JAK3	IL-4, IL-13, IL-9	KO causes SCID (similar in humans) Defective B and T cell maturation	12,8,11
Tyk2	(IFNs, IL-12)	KO susceptible to infections Defective STAT3 activation	10
STAT1	(IFNs), IL-5, IL-31, IL-9	Defective in Type I/II IFN signaling	18,21,19
STAT2	IL-5	KO phenotype not yet reported	18,21,19
STAT3	IL-5, IL-31, IL-9	KO embryonically lethal and mutation causes HIES	27
STAT4	(IL-12)	Defective Th1 Development Enhanced Th2 development	33,34,35
STAT5a/b	IL-5, TSLP, IL-31, IL-9	STAT5a/b: defects in hematopoietic cell development and immune cell signaling	33,34,35,38,40
STAT6	IL-4, IL-13	Defective Th2 and Th9 development Protected from AHR	46 132, 178

KO, knock out; AHR, airway hyperresponsiveness; TSLP, thymic stromal lymphopoietin; SCID, severe combined immunodeficiency; IFN, interferon. Brackets indicate canonical activators of the pathway that are not Th2 cytokines.

4.2.1 JAKs and Asthma

The importance of JAK proteins in human diseases is evident by the presentation of autoimmune and inflammatory disorders in humans with JAK mutations. Somatic missense mutations in JAKs are linked to the development of malignancies. Genome-wide association studies (GWAS) have not yet identified small nucleotide polymorphisms (SNPs) in JAK genes linked to asthma.[8]

In mice, the loss of JAK1 or JAK2 is lethal, respectively, at perinatal or embryonic stages of development. Based on this evidence, loss of function mutations in JAK1 and JAK2 are predicted to have high level of intolerance in humans.

Tyk2−/− animals show increased susceptibility to viral and intracellular infections.[10] The deficiency in Tyk2 affects Type II interferon activity and IL-12 signaling.[11] Only a few patients have been identified with Tyk2 mutations and all of them show high susceptibility to infections.[11] One patient with a homozygous recessive Tyk2 mutation was diagnosed with hyper-IgE syndrome (HIES), which is an immune deficiency characterized by high levels of IgE, eczema, and recurrent lung infections.[11] HIES patient cells are also defective in several cytokine signaling pathways (IL-6 and IL-12).[10,11] Of note is that *Tyk2−/−* mice do not develop HIES despite defects in cytokine signaling (IL-12).[10] This case serves as a good example of differences in JAK signaling between human and animal models, though it could also indicate differences between loss of the entire gene and hypomorphic mutations in patients, highlighting the complexity behind developing and testing treatments.

JAK3 is unique in that it is not expressed in all cell types like the other three JAKs. JAK3 expression is predominantly present in hematopoietic cell lineages and it only associates with IL-2 common gamma chain receptor. *JAK3−/−* animals have defective T and B cell maturation. Mutations in JAK3 cause autosomal recessive severe combined immunodeficiency syndrome (SCID). Human mutations of JAK3 are limited to defects in the immune system, which has made targeted inhibitors of JAK3 promising immunosuppressants.[8,9,12]

4.2.2 JAKinibs in Asthma

The criticality of JAKs in cytokine signaling has led to the development of numerous JAK inhibitors; both specific to certain JAKs and broad tyrosine kinase inhibitors. Some of these

TABLE 4.2
Use of JAK-STAT Protein Inhibitors in Mouse Models of Asthma

Inhibitor	Targeted Protein	Mouse Asthma Model	References
TyrA1	All JAKs	↓STAT3 activation ↓Eosinophilia	15
WHI-P97	JAK3	↓Eosinophilia ↓AHR	14
VR588	All JAKs	↓STAT3 activation ↓AHR ↓Immune cell infiltrate into lung	16
Decoy ODN	STAT1	↓BAL infiltrate, AHR, IL-5	56
C1889	STAT3	↓Th2, Th17 cell development	57
R256	JAK1, JAK3	↓Eosinophilia ↓AHR ↓Mucus production	62
AS1517499	STAT6	↓RhoA Expression	196
Leflunomide	STAT6	↓RhoA Expression	197

AHR, airway hyperresponsiveness; BAL, bronchoalveolar lavage.

inhibitors are FDA approved and are being prescribed for conditions like rheumatoid arthritis (Ruxolitinib, Tofacitinib, Oclacitinib), where they show considerable efficacy.[13] These inhibitors are in clinical trials for their potential in the treatment of other immune-related diseases as well.[13]

While JAK inhibitors have not yet been tested in patients with asthma, mouse model studies of asthma have employed JAK inhibitors (see Table 4.2). One study shows that inhibiting JAK3 specifically (WHI-P97) reduces leukotriene synthesis in mast cells and in vivo administration can prevent eosinophilia and airway hyperresponsiveness (AHR).[14]

A pan-JAK inhibitor (TyrA1) has been used which resulted in reduced STAT3 activation and eosinophilia during a mouse model of allergic asthma.[15] Another pan JAK inhibitor (VR588) also resulted in reduced STAT3 activation, AHR, and immune cell infiltration in mice during allergic asthma.[16] These studies provide clear evidence that JAKs are important in orchestrating asthma and can be targeted for inhibition. However, it is still not clear how they will affect asthmatic patients.

4.3 STATs and Asthma

STATs are latent cytoplasmic proteins classified as transcription factors. There are seven known mammalian STAT proteins (STAT1, STAT2, STAT3, STAT4, STAT5a, STAT5b, STAT6) (see Table 4.1). The structure of STAT proteins allows them to serve a dual role of both signal transduction and DNA binding. Sequentially from the amino terminus end to the carboxy terminus, these proteins have a STAT dimerization domain, coiled-coil domain, DNA binding domain, SH2 domain, and transactivation domain. STAT proteins use their SH2 domains to bind phosphorylated cytokine receptors. STATs are then phosphorylated by JAK kinases and released from the receptor. This allows STATs to form homo or hetero dimers, which are poised for translocation into the nucleus and transcriptional activation.[17]

4.3.1 STAT1 and STAT2

The overall importance of STAT1 and STAT2 for asthma is unclear. Type I, Type II, and Type III interferons signal through STAT1 and STAT2. While some evidence exists that interferons may perpetuate asthma in general, Interferons are thought to be protective in asthma development because they skew the immune system to Th1-like responses.[18–20]

Through the use of knockout animals and administration of recombinant IFNs during allergic asthma, studies have shown that IFN-λ can reduce IL-5-, IL-13-, and IL-17-positive T cells and dampen allergic asthma in mice.[19–21] One study found that indirect activation of STAT1 (through interferons or IL-12) can be protective in allergic asthma.[22] There is also evidence that STAT1 induces pulmonary chemokines indicating that under some conditions it may not play a protective role in mouse models of asthma.[23] Evidence for the upregulation of STAT1 activity in epithelial cells of asthmatic patients has also been documented.[24]

4.3.2 STAT3

Loss of function mutations in STAT3 in humans is the underlying cause of dominant negative hyper-IgE syndrome (HIES), while gain of function mutations in STAT3 is associated with the development of autoimmunity and immunodeficiencies. No association has been found between SNPs or mutations in STAT3 and occurrences of asthma.[25] It has been demonstrated that STAT3 is a positive regulator of migration and function of eosinophils from asthmatic patients.[26]

Germ-line *Stat3−/−* animals are difficult to assess because they die early in embryonic development. However, conditional mutant mice that lack STAT3 in specific tissues have been used to address the role of STAT3 in allergic inflammation. The importance of STAT3 in house dust mite (HDM) models of asthma has been demonstrated using STAT3 airway epithelial cell conditional mutant animals. These animals have reduced airway eosinophilia, lung T_H2, and T_H2 cytokine levels in a chronic asthma model.[15]

STAT3 is known to play an important role in the development of T_H2, T_H9, and T_H17 cells. Mice with *Stat3−/−* T cells fail to develop allergic asthma. STAT3 can bind to T_H2 cytokine loci and this process is critical for STAT6 induced T_H2 cell development.[27] STAT3 is also critical for inhibiting stability of T_H9 cell development.[28] In T_H17 cells, STAT3 can activate several cytokines and this process is well documented. T cells that are *Stat3−/−* have reduced RORγt, a transcription factor that is critical for T_H17 cell development. Similarly, T_H17 cells fail to develop in HIES patients.[27,29] However, in contrast to the mouse system, STAT3 is required for human T_H9 development.[29]

4.3.3 STAT4

STAT4 is expressed in lymphoid and myeloid cells.[30] It is implicated in the development and regulation of T_H1, T_H2, and follicular helper T cells.[31–35] STAT4 is capable of exerting cytokine regulatory effects in other cell types such as regulatory T cells, T_H17 cells, mast cells, and natural killer cells.[36–39] Beyond its role in cell development, STAT4 is associated with increasing the severity of airway diseases including chronic obstructive pulmonary disorder (COPD), allergic rhinitis, and asthma.[40–43] At present, however, the direct role that STAT4 plays on the development of these diseases remains under-investigated. The role of STAT4 on the regulation of several cytokines important for the development of airway diseases such as asthma, has been explored using mouse models of asthma and airway hypersensitivity.

STAT4 is a transcription factor that was initially described as being a master controller of the development of T_H1 cells. STAT4, which can be activated by IL-12, is capable of inducing effects ranging from increased IFN-γ production, lymphocyte proliferation, and enhanced natural killer cell cytotoxicity.[33,35] During viral infections, STAT4 binds to the IFN-γ gene suggesting that STAT4 alone is capable of regulating the IFN-γ pathway.[44]

As asthma is an airway disease with a complex interplay between T_H1, T_H2, and T_H17 cells, the cytokine milieu can have conflicting effects on T_H cell development. IL-12 promotes the differentiation of naïve T cells into T_H1 cells while suppressing the generation of T_H2 cytokines such as IL-4 and IL-5. However, T_H2 cytokines such as IL-4 inhibit the development of T_H1 cells. Much of this interplay can be attributed to transcriptionally activated T-bet which induces IL-12 receptor beta 2 subunit (IL-12Rβ2) expression and subsequently IL-12-dependent STAT4 activation. T-bet conversely represses the expression of several transcription factors including GATA-3. In an environment with IL-4, IL-5, and IL-13, IFN-γ production and IL-12Rβ2 gene expression are suppressed.[45–48] T_H17 cells are additional players in some asthma models and IL-12-induced STAT4 activation and the induction of T-bet can drive them to an IFN-γ producing phenotype further suppressing T_H17 cell development.[49–54]

4.3.3.1 The IL-12/STAT4/IFN-γ Pathway in Mouse Models of Asthma

In mouse models of allergen-induced asthma, direct treatment with IL-12 can inhibit the development of asthma-like symptoms, including airway hyperresponsiveness and eosinophilic infiltration of the airways.[55–57] Several studies elucidate the mechanism that drives the control of STAT4 and IL-12 on airway hyperresponsiveness in allergen-induced asthma.[58] Mice deficient in STAT4, on both the C57BL/6 and BALB/C background, develop airway hyperresponsiveness and airway eosinophilia.[59,60] When these mice were given an IL-12 blocking antibody during allergen re-challenge, airway hyperesponsiveness was reduced.[61] In this model, *Stat4−/−* mice showed decreased numbers of peribronchial eosinophils as compared to controls. As expected, when profiling the airways for cytokines, IL-12 was lower as compared to WT controls, perhaps suggesting that the lack of IL-12-induced STAT4 directly affects eosinophils.[61] IFN-γ was not altered in this model, suggesting an IFN-γ-STAT4-independent process in some models of allergen-induced asthma. One such mechanism has been described.[61–63] Interestingly, in other studies using WT mice, when IL-12 is applied directly to the airways of OVA-sensitized mice, IFN-γ levels are increased after antigen challenge. In this model, eosinophil recruitment is still inhibited.[58] Additionally, when mice are treated with T_H2 cytokine reducing compounds, such as CpG Oligodeoxynucleotide, the ratio of IFN-γ and IL-12 increases as compared to IL-5. This, in turn, delays the development of asthma-like symptoms.[64] Thus in these studies, the role that IL-12/STAT4 plays in asthma could be beneficial.

In other mouse models of allergen-induced asthma, STAT4 contributes to airway hyperresponsiveness. However, its contribution compared to IFN-γ appears dependent upon concentration of allergen used. For example, in one study, low dosage of LPS (0.1 μg) induced type 2 asthma (airway hyperresponsiveness and eosinophilic infiltration) while high dosages (10 μg) induced type 1 asthma (airway hyperresponsiveness and non-eosinophilic infiltration). In the high dose model, the asthma was not able to develop when using IFN-γ-deficient mice. Levels of lung inflammation, IgG2a production, and airway infiltration by IFN-producing cells ($CD8^+$ helper T cells) were decreased. Interestingly, when using *Stat4−/−* mice, high dosages of LPS did not induce pulmonary inflammation.[65] A separate study utilizing cockroach antigen-treated control and *Stat4−/−* mice showed that pulmonary levels of chemokines CCL5, CCL6, CCL11, and CCL17 were reduced when compared to WT mice. Additionally, and unlike the previous model, T_H2 cytokines including IL-4 and IL-13 were not reduced.[61] Allergen-induced airway hyperresponsiveness was found to be impaired by STAT4-deficiency in a methacholine-challenge airway hyperresponsiveness model. Though IL-12 levels remained the same between WT and *Stat4−/−* groups, lung infiltration by IFN-γ producing cells was only enhanced in WT mice.[66] Thus, most data suggest that IL-12-signaling and STAT4-dependent IFN-γ production are critical to controlling airway hyperresponsiveness, but in some models they can be dispensable or even required for development of disease.

4.3.3.2 STAT4 in Patient Asthma

Several population-based studies have found that IL-12 levels are reduced in asthmatics and that polymorphisms in the IL-12 gene can serve as a risk factor for the development of asthma.[67,68]

Following inhaled corticosteroid treatments, IL-12 levels are increased in these patients.[69] However, when using high dose-inhaled corticosteroids on populations with defects in the STAT4 gene, asthma symptoms do not improve.[43]

Additionally, human airway disease studies have mixed conclusions in correlating IFN-γ and disease severity.[70–76] For example, sputum IFN-γ expression is higher in severe asthmatics as compared to mild to moderate asthmatics.[65] In a study examining discordant twins, hypermethlyation of IFN-γ was associated with asthma severity.[77] Another study found hypermethlyation of the IFN-γ promoter to be associated with the increased risk of the development of occupational-related asthma.[78] These suggestions indicate that the role of IFN-γ may be more subtle and context-dependent.

In the human population, polymorphisms of STATs, including STAT4 and STAT6, are associated with airway diseases including asthma and COPD.[79,80] In studies using a Japanese population of asthmatics, patients homozygous for the STAT4 SNP (rs925847) display reduced lung function and increased bronchial airway remodeling.[43] In a study examining an asthmatic Korean population, that same SNP was associated with an increased risk of the development of allergic asthma.[81] In studies of children, STAT4 SNPs (rs16833215 and rs4853546) were associated with childhood wheezing and increased asthmatic exacerbations.[82] Finally, in studies incorporating IFN-γ, STAT4, and STAT6 polymorphisms, SNPs in any of these genes were correlated with the development of asthma.[83] Taken together, these studies suggest that defects in the STAT4-IFN-γ pathway could lead to the development of asthma and increased severity of symptoms.

4.3.4 STAT5

There are two genes encoding STAT5 proteins: *Stat5a* and *Stat5b* which are 96% similar in sequence but have some distinct biological functions. *Stat5a/b−/−* animals have prenatal lethality and severe immunodeficiency similar to *JAK3−/−*.[84] In humans, STAT5b deficiency causes autoimmune disease in part due to dysfunctional T regulatory cells.[85] Patients with severe refractory asthma (SRA) were found to have lower levels of STAT5a in their peripheral blood mononuclear cells.[86]

STAT5 proteins can be activated via JAK and Src kinases. It has been well documented that STAT5 plays an important role in the development of cells that drive airway disease such as T_H2, T_H9, ILC2s, mast cells, and eosinophils.[87–89] STAT5 is critical in mast cell development and IgE-mediated mast cell function.[87] In T_H9 cells, IL-9 gene transcription is in part dependent on STAT5[88,90,91] STAT5 activation, through IL-2 signaling, has been shown to induce T cell proliferation in asthma models.[92]

4.3.4.1 STAT5, IL-5, and IL-2

Eosinophilic lung inflammation and high peripheral blood eosinophil numbers have been associated with severity of asthma in humans. Eosinophils are also implicated in airway remodeling because they are the source of remodeling molecules such as tumor growth factors (TGF), VEGF, and metalloproteinases.[89] IL-5 is undoubtedly an important mediator of asthma because it is important for the maturation and migration of eosinophils. In the asthmatic lung milieu, IL-5 is primarily produced by T cells, eosinophils, mast cells, and innate lymphoid cells. In humans, high levels of IL-5 protein can be found in the mucosa of asthmatic airways.[93] Studies of anti-IL-5 treatment (Mepolizumab) have shown reduced disease exacerbation in severe asthmatics with eosinophilia. Anti-IL-5 treatment in these patients reduced peripheral blood and sputum eosinophil counts.[94–96]

Ovalbumin-challenged *Il5−/−* mice show decreased numbers of circulating eosinophils and airway eosinophilia.[97,98] Conversely, inhalation of IL-5 has been shown to increase eosinophils and airway hyperresponsiveness. IL-5 primarily signals through the JAK2/STAT5 pathway. It has been demonstrated that *Stat5a−/−* and *Stat5b−/−* animals have diminished antigen-induced eosinophil recruitment in the airways. Interestingly, this effect is due to both diminished IL-5 responsiveness of eosinophils and defects in antigen-specific T_H2 cell proliferation.[99,100] T_H2 cell differentiation is thought to be orchestrated by IL-2 signaling, which primarily signals through STAT5. It has been shown that constitutively active STAT5a, independent of STAT6, can increase accessibility of the IL-4 gene leading to T_H2 cell differentiation.[101] Consistent with this concept, mice that are *Stat5a−/− Stat6−/−* have severely decreased allergic airway inflammation with low eosinophilia and T_H2 cell differentiation compared to *Stat6−/−* alone.[102]

4.3.4.2 STAT5 and TSLP

TSLP is a type I cytokine that signals through JAK2/JAK1 to activate STAT5 (and other STATs to a lesser degree).[103] In the lung, TSLP is primarily expressed in epithelia and stromal cells and some nonstructural cells such as DCs.[104] TSLP supports maturation of B and T cells in addition to modulating many immune cells that play a role in development of lung inflammation like DCs, mast cells, and eosinophils. TSLPR-deficient animals have been used to demonstrate that without TSLP signaling, mice do not develop allergic asthma. Inflammation can be restored in these animals when wild type T cells are reconstituted, emphasizing the importance of TSLP signaling for T cells.[105,106] It has also been demonstrated that TSLP inhibits the development of tolerance to an allergen in the lung. TSLP can suppress development of antigen-specific human and mouse T regulatory cells.[107,108]

4.3.5 STAT6

In mice and humans, IL-4 induces the tyrosine phosphorylation and DNA binding activity of STAT6[109,110] IL-13 activates STAT6 in cells that express the receptor including macrophages.[46,111,112] Some evidence suggests that IL-15 is additionally capable of activating STAT6; however, more investigations are warranted.[113–115] Activated STAT6 has roles in the regulation of several genes including the T_H2 transcription factor GATA3, IL-4Rα receptor, eotaxin-3, CCL17, and FIZZ.1[46,115–124] STAT6 also inhibits the expression of CD40 and E-selectins.[125,126]

IL-4 and IL-13 are cytokines that induce the activation of STAT6. In asthma, these cytokines play numerous roles. For example, activated T_H2, NK, and mast cells secrete IL-4 and IL-13 to drive the activation of B-cells. Activated B-cells initiate an IgE-producing phenotype, which allows mast cell priming.[127–130] In asthmatics, both cytokines are increased in airway smooth muscle cells.[131] These cytokines promote goblet cell metaplasia which enhances mucus secretion and airway hyperresponsiveness.[127,130,132,133] Therefore, it is of interest to identify the link between IL-4 and IL-13 with STAT6 in asthma. Indeed, asthma patients treated with the IL-4/IL-13 signaling inhibitor, dupilimab, experience decreased asthma exacerbations and improved lung functions.[134]

STAT6 plays several roles in helper T-cell development. While it mainly skews the differentiation of naïve T-cells towards a T_H2 producing phenotype, it can regulate both T_H9 and T_H17 cell development.[35,46,135–137] In airway diseases, STAT6 affects numerous cell types. For example, in an Alternaria-allergen challenge model of asthma, STAT6 controlled eosinophilia.[138] Another study linked the control of eosinophilia by STAT6 to PARP1, a poly (ADP-ribose) polymerase.[139]

In asthmatics, STAT6 positively correlates with increased serum IgE levels.[140] Furthermore, several studies indicate that increased STAT6 levels also correlate with more severe forms of allergic rhinitis and asthma.[141–145] In patients who undergo glucocorticosteroid treatments, STAT6 controls cytokine secretion in cell types such as innate lymphoid cells (ILC2s).[146]

4.3.5.1 STAT6 and IL-4 in Patient Asthma

Observations of induced sputum in asthmatics reveal that STAT6 and the IL-4 receptor are expressed more highly in these patients than in control subjects.[147] Mutations in the IL-4 receptor are wide ranging and can result in increased serum IgE. Several mutations are linked with the increased incidence of atopic asthma and defective STAT6 DNA binding (Q576R and S503P).[148–153] For example, a mutation in an extracellular region of IL-4R (V50) leads to prolonged STAT6 phosphorylation.[153] Another IL-4R mutation (Ala57Thr) is associated with decreased atopy among patients in Greenland.[150]

Several STAT6 variants are associated with asthma risk.[154] The variant rs324015 has a protective effect on atopic asthma. Variant rs7180246 is associated with an increased risk of asthma in all but an Indian population.[143,154] In Caucasians, the variant rs324011 is associated with asthma risk.[154] Variant rs324011 though not associated with asthma risk, is associated with recurrent wheezing in early childhood.[155] A recently characterized STAT6 variant, rs167769, is associated with asthma risk in a multi-ancestry study.[156]

4.3.5.2 Pharmaceutical and Biological Inhibition of STAT6

Researchers have used various inhibitors to block the IL-4/STAT6 pathway. Fasudil, an approved clinical drug that can block genes such as RhoA/Rho kinase from enhancing hypermucus secretion in asthmatics, was used in mouse studies to further clarify the role of STAT6 in asthma. In an OVA-challenge mouse model, secreted IL-4 was decreased in Fasudil-treated animals compared to untreated wild types. Importantly, this correlated with a decrease in both phosphorylated STAT6 and STAT6 protein levels as compared to OVA-challenged untreated controls.[157]

Another drug, heparin, limits the amount of secreted cytokines from asthmatic cells and prevents the development of asthma in mice, sheep, and pigs.[158–161] In humans, it limits bronchoconstriction.[162] In an OVA-sensitized mouse model of asthma using sulfated non-anticoagulant low molecular weight heparin (S-NACH), a low molecular weight form of heparin, S-NACH acted as an inhibitor of STAT6. This led to reduced secreted IL-4. Transcription factor protein amounts, such as GATA3, were also reduced.[163]

Reductions in protein expression levels of IL-4 in mouse lungs occur when STAT6 siRNA is used in asthmatic models.[164] Similarly, in models of allergic rhinitis, sneezing and nasal rubbing in STAT6 siRNA-treated mice are decreased as compared to untreated controls. When submandibular lymph node cells were stimulated with OVA in vitro, secreted IL-4 amounts were undetected in cells originating from the STAT6 siRNA-treated mice.[165] These results corroborate with those from similar studies. In chronic fungal asthma mouse models using *Stat6−/−* mice, IL-4 levels increased much slower than in WT mice.[166] The lack of complete abolishment of IL-4 suggests that STAT6 does not prevent all features of asthma from developing.[167] Chemokines such as MCP-1, RANTES, and eotaxin were altered compared to WT animals. This confirmed the results of several *Stat6−/−* in vitro studies.[166,168,169] Studies using STAT6-deficient mice and acute and chronic challenge with OVA, also show decreased amounts of IL-4 in BAL.[167]

Several chemical compounds provide insights into the role of STAT6 and IL-4 in allergic airway disease. In rat allergic airway models using YM-341619 hydrochloride, a compound that exclusively suppresses the differentiation of T cells to T_H2 cells, STAT6 expression was suppressed. When the compound was used on cultured T-cells, IL-4 production was decreased. Airway hyperresponsiveness was similarly decreased when the compound is administered orally in a DNP-*Ascaris* asthmatic-mouse model.[170] This finding mimics the role of Prednisolone, an orally available corticosteroid, in altering IL-4 levels in asthma. In an OVA sensitization and methacholine-challenge model using BALB/c mice, the use of Boswellic acid led to decreased IL-4 secretion, phopho-STAT6, and GATA3. Overall, airway inflammation was attenuated.[171] In a study using tetrahydrocurcumin (THC), a major metabolite of Curcumin, a pigment from the Indian spice turmeric, OVA-induced asthmatic mice displayed decreased asthmatic symptoms, such as nasal rubbing, tissue eosinophilia, macrophage infiltration, and mucus production. Cytokine levels in BAL fluid such as IL-4, IL-5, and IL-13 were reduced compared to OVA-challenge THC-untreated mice. In examining protein secretion and phosphorylation levels, the data suggested that these decreases were due to decreased expression of IL4 receptor, STAT6, and GATA3, further suggesting a role for these proteins in asthma.[172]

Natural anti-inflammatory compounds such as ursolic acid have been used in lung inflammation treatments. In one study, ursolic acid prevented the development of airway eosinophilia in OVA-challenged mice. Levels of several cytokines such as IL-13 and IL-17 were also decreased as compared to wild type controls. Interestingly, when examining the levels of transcription factor expression, both STAT6 and GATA3 were decreased compared to controls.[173]

4.3.6.3 STAT6 and IL-13

Eotaxin is of great interest in studies of asthma and atopy. Elevated levels of eotaxin are routinely present in asthma patients.[174] Several cell types secrete eotaxin in response to IL-13 treatment. For example, airway epithelial cells induced by IL-13 secrete eotaxin.[175] In human skin cells, when treated with STAT6 siRNA and stimulated with IL-13, eotaxin levels are decreased.[164] When human bronchial epithelial cells are treated with IL-13, STAT6 is activated, suggesting a link between IL-13, STAT6, and eotaxin, and potentially explaining the mechanism for increased levels of eosinophils in some forms of asthma.[176] In in vitro studies using human airway cell lines, such as A549, eotaxin family levels (CCL11, CCL24, CCL26) are upregulated in response to IL-4 and IL-13. In this same study, STAT6 siRNA blocked the upregulation of these eotaxins. When secreted protein levels were examined, only CCL26 was detectable while the others were not. In this same study, eotaxin gene expression was decreased when primary human bronchial smooth muscle cells and primary epithelial cells were treated with IL-13 and STAT6 siRNA.[177]

In a study where transgenic mice that expressed STAT6 under control of a CC10 promoter, a protein that is specific to airway epithelial cells, were crossed to STAT6 deficient mice, they selectively expressed STAT6 only in lung epithelium cells. These mice were crossed further with transgenic IL-13 mice and examined for the requirement of STAT6 in IL-13 induced asthma. While researchers found that STAT6-deficient mice were protected from airway hyperresponsiveness, they saw that mice that only expressed STAT6 in the airways still developed airway hyperresponsiveness.[178]

4.3.5.4 STAT6 and IL-13/IL-17

Like IL-4 and IL-13, IL-17A is increased in asthmatics. IL-17A induces bronchial cells to secrete cytokines such as IL.-6[179] IL-17 can induce the secretion of IL.-13[180] In studies that transferred T_H17 cells into both WT and STAT6-deficient mice, IL-13 protein expression was assessed after OVA challenge. Harvested lungs from both groups showed higher IL-13 protein expression when compared to PBS controls. When comparing both the OVA-treated groups, the STAT6-deficient mice displayed higher levels of IL-13 protein expression. Overall, STAT6 negatively regulated IL-13 protein expression.[181]

The role of IL-17 on receptor-mediated signaling has been explored. The IL-4Rα chain can link with the IL-13Rα2 receptor to form the type II IL-4/IL-13 receptor. When this happens, IL-17 can induce IL-13 secretion. Thus, studies have attempted to identify the role of IL-17 on the IL-13 receptor complex.[182] Recent studies focusing on IL-13rα2, which has traditionally not been linked to the development of asthma, have shown that it makes a modest contribution to the development of airway hyperresponsiveness.[183] When IL-13rα2-deficient mice were given IL-17, IL-13 driven airway hyperresponsiveness was enhanced.[180] While IL-13rα1 clearly plays a role in STAT6-mediated signaling, it appears that IL-13rα2 does not. When non-transgenic mice are compared to IL-13rα2 deficient mice, and STAT6 levels are examined, differences in STAT6 activation are not observed.[183]

4.3.5.5 STAT6 and Mucus Gene Overproduction

Since one of the hallmark features of asthma is mucus overproduction, studies have defined ways that STAT6 may, directly or indirectly, play a role in the regulation of mucus genes.[127,132,184]

Twelve mucin-related genes are expressed in humans and several of these, such as MUC2 and MUC4, are sometimes increased in asthmatics when compared to healthy controls.[185,186] One of the more heavily studied genes, MUC5AC, is increased up to 60% higher than normal in asthmatic patients.[185]

Intratracheal IL-13 induces MUC5AC in wild type but not STAT6-deficient mice. Consistent with this, when cells were isolated and then treated with IL-13 for 24 hours, MUC5AC expression was only slightly increased in STAT6 deficient animals compared to untreated cells, and to wild type controls.[132] Moreover, another gene involved in goblet cell metaplasia, Gob-5, was unresponsive in cells deficient in STAT6 as compared to wild type controls, suggesting a role for STAT6 in controlling both genes.[132] In a study using STAT6 siRNA to knockdown STAT6 levels, *Muc5ac* promoter activation was strongly decreased.[187] In another study using STAT6 siRNA, repression of STAT6 in the presence of IL-13 led to decreased MUC5AC expression as well as SAM domain-containing prostate-derived Ets factor (SPDEF), which is a transcription factor that plays a role in goblet cell hyperplasia and mucus secretion.[188] Reductions in IL-13-mediated expression of the transmembrane protein 16A, a calcium-activated chloride channel that is a key regulator of mucus overproduction in airway epithelial cells, were shown to be a result of STAT6 inhibition. This reduction corresponded to decreased MUC5AC expression, further highlighting the role of STAT6 in mucin expression.[189] In a study exploring the effects of using a Chinese herb extract, *Glycyrrhiza uralensis*, on asthma, the flavonoid 7′4′-dihydroxyflavone was shown to possess anti-inflammatory properties. When this compound was used on human cell cultures stimulated with PMA, Muc5AC secretion was decreased compared to untreated controls. This decrease was positively correlated with declining phosphorylated STAT6.[190]

The role of STAT6 in controlling mucin genes has additionally been explored in Lyn kinase-deficient mice. Lyn kinase negatively regulates the progression of asthma.[191] In an in vitro study using human bronchial epithelial cells treated with either IL-4 or IL-13, when Lyn siRNA was used, STAT6 expression and phosphorylation increased in both conditions, as compared with untreated cells. When Lyn was overexpressed in their system, STAT6 expression was decreased. Taken together, the results suggest that Lyn regulates MUC5AC expression via the STAT6 pathway.[187]

4.3.6 STAT Inhibitors

Theoretically, like JAKinibs, inhibitors of STAT proteins could be promising in immune-related disorders (see Table 4.2). Efforts in studying STAT inhibitors have been challenged by a lack of bioavailability and low efficacy and specificity. Some groups have tried to circumvent issues of specificity by employing oligonucleotide-based STAT inhibitors. One example is the STAT1 decoy oligonucleotide which, in mouse models, has been shown to reduce lung infiltration and antigen-specific AHR.[192]

A STAT3 inhibitor that targets the SH2 domain (C1889) has been used to treat animals intranasally during allergic asthma model. This treatment significantly reduced T_H2 and T_H17 development and allergic asthma.[193] Currently, some small molecule inhibitors of STAT3 are in clinical trials to be tested for indications other than asthma.[194,195]

In mouse models of allergic bronchial asthma, IL-13 administered intranasally leads to induced phosphorylation of STAT6 as well as the upregulation of the RhoA protein, a protein involved in smooth muscle contraction. In this same study, when human bronchial smooth muscle cells were used and treated with a STAT6 inhibitor, leflunomide, STAT6 phosphorylation and the RhoA upregulation induced by IL13 were abolished.[196] In another study, when the STAT6 inhibitor, AS15174999, was used concurrent with a 1-hour exposure of human bronchial smooth muscle cells to IL-13, STAT6 phosphorylation was inhibited in a dose-dependent manner. Similarly, RhoA expression was decreased in these cells. In the same study, when STAT6 siRNA was used, RhoA expression was partially inhibited.[197]

Another approach taken to find inhibitors of STAT proteins is the use of clinically available drugs that, through unknown mechanisms, reduce STAT activation. For example, the psychotrophic drug pimozide has been shown to reduce the phosphorylation of STAT5[198] This drug was recently used in an ex-vivo study to show that the steroid resistance seen in the ILC2 cells of asthmatic humans was dependent on TSLP and STAT5 activation.[199] Additionally, JAK inhibitors are used to assess the role of STAT proteins in vivo. In particular, the JAK1/3 inhibitor (R256) can prevent STAT5 activation and T_H2 development in culture. Treatment of animals with R256 can reduce AHR, eosinophilia, and mucus production.[200]

4.4 Conclusion

Despite some of the challenges, JAK/STAT pathway constituents remain attractive targets for pharmaceutical intervention. Several intrinsic properties of JAK/STAT signaling highlight the complexity of this pathway. STAT proteins can be expressed in a cell- and tissue-specific manner. One cytokine can activate multiple JAK and STAT proteins to varying degrees. STAT proteins can create homodimers and heterodimers which allow for differential DNA binding.[7,201] STAT proteins may also be able to compete with or compensate for one another and some evidence does exist supporting this claim.[7,202] It has therefore been a challenge to study the effects of inhibiting STAT proteins in multicellular and multi-organ systems and existing studies that tackle this question should be interpreted cautiously.

The JAK/STAT pathway is clearly central in the development of asthma. STAT3, STAT5, and STAT6 are required in multiple cell types for myriad cytokine responses. While the role of other STAT proteins is less clear, there are likely some situations where they are required in the spectrum of asthma endotypes that span disease pathologies from purely T_H2 to more mixed T_H cell phenotypes. As the use of JAKinibs and more specific STAT inhibitors move more broadly into the clinic, it will be interesting to see how they affect various patients with asthma and how personalized medicine can help direct the use of drugs to inhibit these critical pathways.

REFERENCES

1. McKinley, L. *et al.* TH17 cells mediate steroid-resistant airway inflammation and airway hyperresponsiveness in mice. *J. Immunol. Baltim. Md 1950.* **181**, 4089–4097 (2008).
2. Lai, T. *et al.* Interleukin-31 expression and relation to disease severity in human asthma. *Sci. Rep.* **6**, 22835 (2016).
3. Stott, B. *et al.* Human IL-31 is induced by IL-4 and promotes TH2-driven inflammation. *J. Allergy Clin. Immunol.* **132**, 446–454.e5 (2013).
4. Chen, M. *et al.* Complex effects of naturally occurring mutations in the JAK3 pseudokinase domain: Evidence for Interactions between the kinase and pseudokinase domains. *Mol. Cell. Biol.* **20**, 947–956 (2000).
5. Lindauer, K., Loerting, T., Liedl, K. R. & Kroemer, R. T. Prediction of the structure of human Janus kinase 2 (JAK2) comprising the two carboxy-terminal domains reveals a mechanism for autoregulation. *Protein Eng.* **14**, 27–37 (2001).
6. Williams, N. K. *et al.* Dissecting specificity in the Janus kinases: The structures of JAK-specific inhibitors complexed to the JAK1 and JAK2 protein tyrosine kinase domains. *J. Mol. Biol.* **387**, 219–232 (2009).
7. Villarino, A. V., Kanno, Y., & O'Shea, J. J. Mechanisms and consequences of Jak–STAT signaling in the immune system. *Nat. Immunol.* **18**, 374–384 (2017).
8. Zouein, F. A., Duhé, R. J., & Booz, G. W. JAKs go nuclear: Emerging role of nuclear JAK1 and JAK2 in gene expression and cell growth. *Growth Factors Chur Switz.* **29**, 245–252 (2011).
9. Frucht, D. M. *et al.* Unexpected and variable phenotypes in a family with JAK3 deficiency. *Genes Immun.* **2**, 422–432 (2001).

10. Karaghiosoff, M. *et al.* Partial impairment of cytokine responses in Tyk2-deficient mice. *Immunity.* **13**, 549–560 (2000).
11. Minegishi, Y. *et al.* Human tyrosine kinase 2 deficiency reveals its requisite roles in multiple cytokine signals involved in innate and acquired immunity. *Immunity.* **25**, 745–755 (2006).
12. Leonard, W. J. & O'Shea, J. J. Jaks and STATs: Biological implications. *Annu. Rev. Immunol.* **16**, 293–322 (1998).
13. Howell, M. D., Fitzsimons, C., & Smith, P. A. JAK/STAT inhibitors and other small molecule cytokine antagonists for the treatment of allergic disease. *Ann. Allergy Asthma Immunol. Off. Publ. Am. Coll. Allergy Asthma Immunol.* **120**, 367–375 (2018).
14. Malaviya, R. *et al.* Treatment of allergic asthma by targeting janus kinase 3-dependent leukotriene synthesis in mast cells with 4-(3', 5'-dibromo-4'-hydroxyphenyl)amino-6,7-dimethoxyquinazoline (WHI-P97). *J. Pharmacol. Exp. Ther.* **295**, 912–926 (2000).
15. Simeone-Penney, M. C. *et al.* Airway epithelial STAT3 is required for allergic inflammation in a murine model of asthma. *J. Immunol.* **178**, 6191–6199 (2007).
16. Wiegman, C. H. *et al.* The selective pan-Janus kinase (JAK) inhibitor VR588 demonstrates potent anti-inflammatory activity in a murine chronic house dust mite (HDM) model of asthma. in *C34. New Basic Science in Asthma: Allergic Inflammation I.* A6435–A6435. (American Thoracic Society, 2015). doi:10.1164/ajrccm-conference.2015.191.1_MeetingAbstracts.A6435.
17. O'Shea, J. J. *et al.* The JAK-STAT pathway: Impact on human disease and therapeutic intervention. *Annu. Rev. Med.* **66**, 311–328 (2015).
18. Bergauer, A. *et al.* IFN-α/IFN-λ responses to respiratory viruses in paediatric asthma. *Eur. Respir. J.* **49**, 1600969 (2017).
19. Jordan, W. J. *et al.* Human interferon lambda-1 (IFN-λ1/IL-29) modulates the Th1/Th2 response. *Genes Immun.* **8**, 254–261 (2007).
20. Koltsida, O. *et al.* IL-28A (IFN-λ2) modulates lung DC function to promote Th1 immune skewing and suppress allergic airway disease. *EMBO Mol. Med.* **3**, 348–361 (2011).
21. Srinivas, S. *et al.* Interferon-λ1 (interleukin-29) preferentially down-regulates interleukin-13 over other T helper type 2 cytokine responses in vitro. *Immunology.* **125**, 492–502 (2008).
22. Shi, Z. O.-Q., Fischer, M. J., Sanctis, G. T. D., Schuyler, M. R., & Tesfaigzi, Y. IFN-γ, but not fas, mediates reduction of allergen-induced mucous cell metaplasia by inducing apoptosis. *J. Immunol.* **168**, 4764–4771 (2002).
23. Fulkerson, P. C., Zimmermann, N., Hassman, L. M., Finkelman, F. D., & Rothenberg, M. E. Pulmonary chemokine expression is coordinately regulated by STAT1, STAT6, and IFN-γ. *J. Immunol.* **173**, 7565–7574 (2004).
24. Sampath, D., Castro, M., Look, D. C., & Holtzman, M. J. Constitutive activation of an epithelial signal transducer and activator of transcription (STAT) pathway in asthma. *J. Clin. Invest.* **103**, 1353–1361 (1999).
25. Wjst, M., Lichtner, P., Meitinger, T., & Grimbacher, B. STAT3 single-nucleotide polymorphisms and STAT3 mutations associated with hyper-IgE syndrome are not responsible for increased serum IgE serum levels in asthma families. *Eur. J. Hum. Genet.* **17**, 352–356 (2009).
26. Zafra, M. P. *et al.* SOCS3 silencing attenuates eosinophil functions in asthma patients. *Int. J. Mol. Sci.* **16**, 5434–5451 (2015).
27. Stritesky, G. L. *et al.* The transcription factor STAT3 is required for T helper 2 cell development. *Immunity.* **34**, 39–49 (2011).
28. Ulrich, B. J., Verdan, F. F., McKenzie, A. N. J., Kaplan, M. H., & Olson, M. R. STAT3 activation impairs the stability of Th9 cells. *J. Immunol. Baltim. Md 1950.* **198**, 2302–2309 (2017).
29. Egwuagu, C. E. STAT3 in CD4+ T helper cell differentiation and inflammatory diseases. *Cytokine.* **47**, 149–156 (2009).
30. Yamamoto, K. *et al.* Stat4, a novel gamma interferon activation site-binding protein expressed in early myeloid differentiation. *Mol. Cell. Biol.* **14**, 4342–4349 (1994).
31. Nakayamada, S. *et al.* Early Th1 cell differentiation is marked by a Tfh cell-like transition. *Immunity.* **35**, 919–931 (2011).
32. Schmitt, N. *et al.* The cytokine TGF-β co-opts signaling via STAT3-STAT4 to promote the differentiation of human TFH cells. *Nat. Immunol.* **15**, 856–865 (2014).

33. Jacobson, N. G. *et al.* Interleukin 12 signaling in T helper type 1 (Th1) cells involves tyrosine phosphorylation of signal transducer and activator of transcription (Stat)3 and Stat4. *J. Exp. Med.* **181**, 1755–1762 (1995).
34. Kaplan, M. H., Sun, Y. L., Hoey, T., & Grusby, M. J. Impaired IL-12 responses and enhanced development of Th2 cells in Stat4-deficient mice. *Nature.* **382**, 174–177 (1996).
35. Kaplan, M. H. & Grusby, M. J. Regulation of T helper cell differentiation by STAT molecules. *J. Leukoc. Biol.* **64**, 2–5 (1998).
36. Grant, L. *et al.* Stat4-dependent, T-bet-independent regulation of IL-10 in NK cells. *Genes Immun.* **9**, 316–327 (2008).
37. Glosson-Byers, N. L., Sehra, S., & Kaplan, M. H. STAT4 is required for IL-23 responsiveness in Th17 memory cells and NKT cells. *JAK-STAT* **3**, e955393 (2014).
38. Xu, J. *et al.* Stat4 is critical for the balance between Th17 cells and regulatory T cells in colitis. *J. Immunol. Baltim. Md 1950.* **186**, 6597–6606 (2011).
39. Iida, K. *et al.* STAT4 is required for IFN-β-induced MCP-1 mRNA expression in murine mast cells. *Int. Arch. Allergy Immunol.* **155 Suppl 1**, 71–76 (2011).
40. Di Stefano, A. *et al.* STAT4 activation in smokers and patients with chronic obstructive pulmonary disease. *Eur. Respir. J.* **24**, 78–85 (2004).
41. Zhang, Y., Li, J., Wang, C., & Zhang, L. Association between the interaction of key genes involved in effector T-cell pathways and susceptibility to developallergic rhinitis: A population-based case-control association study. *PLoS One.* **10**, e0131248 (2015).
42. Li, X. *et al.* Genome-wide association study identifies TH1 pathway genes associated with lung function in asthmatic patients. *J. Allergy Clin. Immunol.* **132**, 313–320.e15 (2013).
43. Nakamura, Y. *et al.* Therapeutic implication of genetic variants of IL13 and STAT4 in airway remodelling with bronchial asthma. *Clin. Exp. Allergy J. Br. Soc. Allergy Clin. Immunol.* **46**, 1152–1161 (2016).
44. Nguyen, K. B. *et al.* Critical role for STAT4 activation by type 1 interferons in the interferon-gamma response to viral infection. *Science.* **297**, 2063–2066 (2002).
45. Wurtz, O., Bajénoff, M., & Guerder, S. IL-4-mediated inhibition of IFN-gamma production by CD4+ T cells proceeds by several developmentally regulated mechanisms. *Int. Immunol.* **16**, 501–508 (2004).
46. Kaplan, M. H., Schindler, U., Smiley, S. T., & Grusby, M. J. Stat6 is required for mediating responses to IL-4 and for development of Th2 cells. *Immunity.* **4**, 313–319 (1996).
47. Ferber, I. A. *et al.* GATA-3 significantly downregulates IFN-gamma production from developing Th1 cells in addition to inducing IL-4 and IL-5 levels. *Clin. Immunol. Orlando Fla.* **91**, 134–144 (1999).
48. Ouyang, W. *et al.* Inhibition of Th1 development mediated by GATA-3 through an IL-4-independent mechanism. *Immunity.* **9**, 745–755 (1998).
49. Chen, Y. *et al.* IFN-γ-expressing Th17 cells are required for development of severe ocular surface autoimmunity. *J. Immunol. Baltim. Md 1950.* **199**, 1163–1169 (2017).
50. Annunziato, F. *et al.* Phenotypic and functional features of human Th17 cells. *J. Exp. Med.* **204**, 1849–1861 (2007).
51. Kotake, S., Yago, T., Kobashigawa, T., & Nanke, Y. The plasticity of Th17 cells in the pathogenesis of rheumatoid arthritis. *J. Clin. Med.* **6**, E67 (2017).
52. Glosson-Byers, N. L. *et al.* Th17 cells demonstrate stable cytokine production in a proallergic environment. *J. Immunol. Baltim. Md 1950.* **193**, 2631–2640 (2014).
53. Mathur, A. N. *et al.* T-bet is a critical determinant in the instability of the IL-17-secreting T-helper phenotype. *Blood.* **108**, 1595–1601 (2006).
54. Lee, Y. K. *et al.* Late developmental plasticity in the T helper 17 lineage. *Immunity.* **30**, 92–107 (2009).
55. Gavett, S. H. *et al.* Interleukin 12 inhibits antigen-induced airway hyperresponsiveness, inflammation, and Th2 cytokine expression in mice. *J. Exp. Med.* **182**, 1527–1536 (1995).
56. Schwarze, J. *et al.* Local treatment with IL-12 is an effective inhibitor of airway hyperresponsiveness and lung eosinophilia after airway challenge in sensitized mice. *J. Allergy Clin. Immunol.* **102**, 86–93 (1998).

57. Hofstra, C. L. *et al.* Prevention of Th2-like cell responses by coadministration of IL-12 and IL-18 is associated with inhibition of antigen-induced airway hyperresponsiveness, eosinophilia, and serum IgE levels. *J. Immunol. Baltim. Md 1950.* **161**, 5054–5060 (1998).
58. Iwamoto, I., Kumano, K., Kasai, M., Kurasawa, K., & Nakao, A. Interleukin-12 prevents antigen-induced eosinophil recruitment into mouse airways. *Am. J. Respir. Crit. Care Med.* **154**, 1257–1260 (1996).
59. Matsubara, S. *et al.* IL-2 and IL-18 attenuation of airway hyperresponsiveness requires STAT4, IFN-γ, and natural killer cells. *Am. J. Respir. Cell Mol. Biol.* **36**, 324–332 (2007).
60. Meuronen, A. *et al.* Attenuated expression of tenascin-c in ovalbumin-challenged STAT4−/− mice. *Respir. Res.* **12**, 2 (2011).
61. Raman, K., Kaplan, M. H., Hogaboam, C. M., Berlin, A., & Lukacs, N. W. STAT4 signal pathways regulate inflammation and airway physiology changes in allergic airway inflammation locally via alteration of chemokines. *J. Immunol.* **170**, 3859–3865 (2003).
62. Thibodeaux, D. K. *et al.* Autocrine regulation of IL-12 receptor expression is independent of secondary IFN-γ secretion and not restricted to T and NK cells. *J. Immunol.* **163**, 5257–5264 (1999).
63. Lawless, V. A. *et al.* Stat4 regulates multiple components of IFN-gamma-inducing signaling pathways. *J. Immunol. Baltim. Md 1950.* **165**, 6803–6808 (2000).
64. Sur, S. *et al.* Long term prevention of allergic lung inflammation in a mouse model of asthma by CpG oligodeoxynucleotides. *J. Immunol.* **162**, 6284–6293 (1999).
65. Kim, Y.-K. *et al.* Airway exposure levels of lipopolysaccharide determine type 1 versus type 2 experimental asthma. *J. Immunol. Baltim. Md 1950.* **178**, 5375–5382 (2007).
66. Kim, Y.-S. *et al.* IL-12-STAT4-IFN-γ axis is a key downstream pathway in the development of IL-13-mediated asthma phenotypes in a Th2 type asthma model. *Exp. Mol. Med.* **42**, 533–546 (2010).
67. Zhang, Y.-L. *et al.* Peripheral blood MDSCs, IL-10 and IL-12 in children with asthma and their importance in asthma development. *PLoS One.* **8**, e63775 (2013).
68. Chen, T. *et al.* Association of single nucleotide polymorphisms in interleukin 12 (IL-12A and -B) with asthma in a Chinese population. *Hum. Immunol.* **72**, 603–606 (2011).
69. Cui, A.-H., Zhao, J., Liu, S.-X., & Hao, Y.-S. Associations of IL-4, IL-6, and IL-12 levels in peripheral blood with lung function, cellular immune function, and quality of life in children with moderate-to-severe asthma. *Medicine (Baltimore).* **96**, e6265 (2017).
70. Schwantes, E. A. *et al.* Interferon gene expression in sputum cells correlates with the Asthma Index Score during virus-induced exacerbations. *Clin. Exp. Allergy J. Br. Soc. Allergy Clin. Immunol.* **44**, 813–821 (2014).
71. Brooks, G. D., Buchta, K. A., Swenson, C. A., Gern, J. E., & Busse, W. W. Rhinovirus-induced interferon-gamma and airway responsiveness in asthma. *Am. J. Respir. Crit. Care Med.* **168**, 1091–1094 (2003).
72. Smart, J. M., Horak, E., Kemp, A. S., Robertson, C. F., & Tang, M. L. K. Polyclonal and allergen-induced cytokine responses in adults with asthma: Resolution of asthma is associated with normalization of IFN-gamma responses. *J. Allergy Clin. Immunol.* **110**, 450–456 (2002).
73. Noma, T. *et al.* Pattern of cytokine production by T cells from adolescents with asthma in remission, after stimulation with dermatophagoides farinae antigen. *Pediatr. Res.* **38**, 187–193 (1995).
74. Leonard, C., Tormey, V., Burke, C., & Poulter, L. W. Allergen-induced cytokine production in atopic disease and its relationship to disease severity. *Am. J. Respir. Cell Mol. Biol.* **17**, 368–375 (1997).
75. Smith, N. L. D. & Denning, D. W. Clinical implications of interferon-γ genetic and epigenetic variants. *Immunology.* **143**, 499–511 (2014).
76. Litonjua, A. A. *et al.* Serum interferon-gamma is associated with longitudinal decline in lung function among asthmatic patients: The Normative Aging Study. *Ann. Allergy Asthma Immunol. Off. Publ. Am. Coll. Allergy Asthma Immunol.* **90**, 422–428 (2003).
77. Runyon, R. S. *et al.* Asthma discordance in twins is linked to epigenetic modifications of T cells. *PLoS One.* **7**, e48796 (2012).
78. Ouyang, B. *et al.* Interferon-γ promoter is hypermethylated in blood DNA from workers with confirmed diisocyanate asthma. *Toxicol. Sci.* **133**, 218–224 (2013).
79. Pykäläinen, M. *et al.* Association analysis of common variants of STAT6, GATA3, and STAT4 to asthma and high serum IgE phenotypes. *J. Allergy Clin. Immunol.* **115**, 80–87 (2005).

80. Singh, D. P., Bagam, P., Sahoo, M. K., & Batra, S. Immune-related gene polymorphisms in pulmonary diseases. *Toxicology.* **383**, 24–39 (2017).
81. Park, B. L. *et al.* Association analysis of signal transducer and activator of transcription 4 (*STAT4*) polymorphisms with asthma. *J. Hum. Genet.* **50**, 133–138 (2005).
82. Loisel, D. A. *et al.* Genetic associations with viral respiratory illnesses and asthma control in children. *Clin. Exp. Allergy J. Br. Soc. Allergy Clin. Immunol.* **46**, 112–124 (2016).
83. Li, Y., Wu, B., Xiong, H., Zhu, C., & Zhang, L. Polymorphisms of STAT-6, STAT-4 and IFN-gamma genes and the risk of asthma in Chinese population. *Respir. Med.* **101**, 1977–1981 (2007).
84. Caslin, H. L. *et al.* Controlling mast cell activation and homeostasis: Work influenced by Bill Paul that continues today. *Front. Immunol.* **9**, 868 (2018).
85. Kanai, T., Jenks, J., & Nadeau, K. C. The STAT5b pathway defect and autoimmunity. *Front. Immunol.* **3**, 234 (2012).
86. Saeedfar, K., Behmanesh, M., Mortaz, E., & Masjedi, M. R. The expression of STAT3 and STAT5A genes in severe refractory asthma. *Tanaffos.* **16**, 1–8 (2017).
87. Barnstein, B. O. *et al.* Stat5 expression is required for IgE-mediated mast cell function. *J. Immunol.* **177**, 3421–3426 (2006).
88. Schmitt, E. *et al.* IL-9 production of naive CD4+ T cells depends on IL-2, is synergistically enhanced by a combination of TGF-beta and IL-4, and is inhibited by IFN-gamma. *J. Immunol.* **153**, 3989–3996 (1994).
89. Papathanassiou, E., Loukides, S., & Bakakos, P. Severe asthma: Anti-IgE or anti-IL-5? *Eur. Clin. Respir. J.* **3**, 31813 (2016).
90. Liao, W. *et al.* Opposing actions of IL-2 and IL-21 on Th9 differentiation correlate with their differential regulation of BCL6 expression. *Proc. Natl. Acad. Sci.* **111**, 3508–3513 (2014).
91. Olson, M. R., Verdan, F. F., Hufford, M. M., Dent, A. L., & Kaplan, M. H. STAT3 impairs STAT5 activation in the development of IL-9–secreting T cells. *J. Immunol.* **196**, 3297–3304 (2016).
92. Li, G., Liu, Z., Ran, P., Qiu, J., & Zhong, N. Activation of signal transducer and activator of transcription 5 (STAT5) in splenocyte proliferation of asthma mice induced by ovalbumin. *Cell. Mol. Immunol.* **1**, 471–474 (2004).
93. Hamid, Q. *et al.* Expression of mRNA for interleukin-5 in mucosal bronchial biopsies from asthma. *J. Clin. Invest.* **87**, 1541–1546 (1991).
94. Haldar, P. *et al.* Mepolizumab and exacerbations of refractory eosinophilic asthma. *N. Engl. J. Med.* **360**, 973–984 (2009).
95. Ortega, H. G. *et al.* Mepolizumab treatment in patients with severe eosinophilic asthma. *N. Engl. J. Med.* **371**, 1198–1207 (2014).
96. Bel, E. H. *et al.* Oral glucocorticoid-sparing effect of mepolizumab in eosinophilic asthma. *N. Engl. J. Med.* **371**, 1189–1197 (2014).
97. Kopf, M. *et al.* IL-5-deficient mice have a developmental defect in CD5+ B-1 cells and lack eosinophilia but have normal antibody and cytotoxic T cell responses. *Immunity.* **4**, 15–24 (1996).
98. Foster, P. S., Hogan, S. P., Ramsay, A. J., Matthaei, K. I., & Young, I. G. Interleukin 5 deficiency abolishes eosinophilia, airways hyperreactivity, and lung damage in a mouse asthma model. *J. Exp. Med.* **183**, 195–201 (1996).
99. Kagami, S. *et al.* Both Stat5a and Stat5b are required for antigen-induced eosinophil and T-cell recruitment into the tissue. *Blood.* **95**, 1370–1377 (2000).
100. Kagami, S. *et al.* Stat5a regulates T helper cell differentiation by several distinct mechanisms. *Blood.* **97**, 2358–2365 (2001).
101. Zhu, J., Cote-Sierra, J., Guo, L., & Paul, W. E. Stat5 activation plays a critical role in Th2 differentiation. *Immunity.* **19**, 739–748 (2003).
102. Takatori, H. *et al.* Indispensable role of Stat5a in Stat6-independent Th2 cell differentiation and allergic airway inflammation. *J. Immunol.* **174**, 3734–3740 (2005).
103. Rochman, Y. *et al.* Thymic stromal lymphopoietin-mediated STAT5 phosphorylation via kinases JAK1 and JAK2 reveals a key difference from IL-7–induced signaling. *Proc. Natl. Acad. Sci. U. S. A.* **107**, 19455–19460 (2010).
104. Kashyap, M., Rochman, Y., Spolski, R., Samsel, L., & Leonard, W. J. Thymic stromal lymphopoietin is produced by dendritic cells. *J. Immunol. Baltim. Md 1950.* **187**, 1207–1211 (2011).

105. Zhou, B. *et al.* Thymic stromal lymphopoietin as a key initiator of allergic airway inflammation in mice. *Nat. Immunol.* **6**, 1047–1053 (2005).
106. Al-Shami, A., Spolski, R., Kelly, J., Keane-Myers, A., & Leonard, W. J. A role for TSLP in the development of inflammation in an asthma model. *J. Exp. Med.* **202**, 829–839 (2005).
107. Nguyen, K. D., Vanichsarn, C., & Nadeau, K. C. TSLP directly impairs pulmonary Treg function: Association with aberrant tolerogenic immunity in asthmatic airway. *Allergy Asthma Clin. Immunol. Off. J. Can. Soc. Allergy Clin. Immunol.* **6**, 4 (2010).
108. Lei, L., Zhang, Y., Yao, W., Kaplan, M. H., & Zhou, B. TSLP interferes with airway tolerance by suppressing the generation of antigen-specific regulatory T cells. *J. Immunol. Baltim. Md 1950.* **186**, 2254–2261 (2011).
109. Quelle, F. W. *et al.* Cloning of murine Stat6 and human Stat6, Stat proteins that are tyrosine phosphorylated in responses to IL-4 and IL-3 but are not required for mitogenesis. *Mol. Cell. Biol.* **15**, 3336–3343 (1995).
110. Hou, J. *et al.* An interleukin-4-induced transcription factor: IL-4 Stat. *Science.* **265**, 1701–1706 (1994).
111. Takeda, K., Kamanaka, M., Tanaka, T., Kishimoto, T., & Akira, S. Impaired IL-13-mediated functions of macrophages in STAT6-deficient mice. *J. Immunol. Baltim. Md 1950.* **157**, 3220–3222 (1996).
112. Hershey, G. K. K. IL-13 receptors and signaling pathways: An evolving web. *J. Allergy Clin. Immunol.* **111**, 677–690; quiz 691 (2003).
113. Masuda, A. *et al.* Interleukin-15 induces rapid tyrosine phosphorylation of STAT6 and the expression of interleukin-4 in mouse mast cells. *J. Biol. Chem.* **275**, 29331–29337 (2000).
114. Masuda, A., Matsuguchi, T., Yamaki, K., Hayakawa, T., & Yoshikai, Y. Interleukin-15 prevents mouse mast cell apoptosis through STAT6-mediated Bcl-xL expression. *J. Biol. Chem.* **276**, 26107–26113 (2001).
115. Hebenstreit, D., Wirnsberger, G., Horejs-Hoeck, J., & Duschl, A. Signaling mechanisms, interaction partners, and target genes of STAT6. *Cytokine Growth Factor Rev.* **17**, 173–188 (2006).
116. Maier, E., Duschl, A., & Horejs-Hoeck, J. STAT6-dependent and -independent mechanisms in Th2 polarization. *Eur. J. Immunol.* **42**, 2827–2833 (2012).
117. Asnagli, H., Afkarian, M., & Murphy, K. M. Cutting edge: Identification of an alternative GATA-3 promoter directing tissue-specific gene expression in mouse and human. *J. Immunol. Baltim. Md 1950.* **168**, 4268–4271 (2002).
118. Stütz, A. M. *et al.* The Th2 cell cytokines IL-4 and IL-13 regulate found in inflammatory zone 1/resistin-like molecule alpha gene expression by a STAT6 and CCAAT/enhancer-binding protein-dependent mechanism. *J. Immunol. Baltim. Md 1950.* **170**, 1789–1796 (2003).
119. Hoeck, J. & Woisetschläger, M. Activation of eotaxin-3/CCLl26 gene expression in human dermal fibroblasts is mediated by STAT6. *J. Immunol. Baltim. Md 1950.* **167**, 3216–3222 (2001).
120. Matsukura, S. *et al.* Activation of eotaxin gene transcription by NF-kappa B and STAT6 in human airway epithelial cells. *J. Immunol. Baltim. Md 1950.* **163**, 6876–6883 (1999).
121. Wirnsberger, G., Hebenstreit, D., Posselt, G., Horejs-Hoeck, J., & Duschl, A. IL-4 induces expression of TARC/CCL17 via two STAT6 binding sites. *Eur. J. Immunol.* **36**, 1882–1891 (2006).
122. Hoeck, J. & Woisetschläger, M. STAT6 mediates eotaxin-1 expression in IL-4 or TNF-alpha-induced fibroblasts. *J. Immunol. Baltim. Md 1950.* **166**, 4507–4515 (2001).
123. Maier, E., Wirnsberger, G., Horejs-Hoeck, J., Duschl, A., & Hebenstreit, D. Identification of a distal tandem STAT6 element within the CCL17 locus. *Hum. Immunol.* **68**, 986–992 (2007).
124. Goenka, S. & Kaplan, M. H. Transcriptional regulation by STAT6. *Immunol. Res.* **50**, 87–96 (2011).
125. Nguyen, V. T. & Benveniste, E. N. IL-4-activated STAT-6 inhibits IFN-gamma-induced CD40 gene expression in macrophages/microglia. *J. Immunol. Baltim. Md 1950.* **165**, 6235–6243 (2000).
126. Bennett, B. L., Cruz, R., Lacson, R. G., & Manning, A. M. Interleukin-4 suppression of tumor necrosis factor α-stimulated E-selectin gene transcription is mediated by STAT6 antagonism of NF-κB. *J. Biol. Chem.* **272**, 10212–10219 (1997).
127. Holgate, S. T. Pathogenesis of asthma. *Clin. Exp. Allergy J. Br. Soc. Allergy Clin. Immunol.* **38**, 872–897 (2008).
128. Umetsu, D. T. & DeKruyff, R. H. A role for natural killer T cells in asthma. *Nat. Rev. Immunol.* **6**, 953–958 (2006).

129. Akbari, O. *et al.* CD4+ invariant T-cell-receptor+ natural killer T cells in bronchial asthma. *N. Engl. J. Med.* **354**, 1117–1129 (2006).
130. Bradding, P., Walls, A. F., & Holgate, S. T. The role of the mast cell in the pathophysiology of asthma. *J. Allergy Clin. Immunol.* **117**, 1277–1284 (2006).
131. Brightling, C. E. *et al.* Interleukin-4 and -13 expression is co-localized to mast cells within the airway smooth muscle in asthma. *Clin. Exp. Allergy J. Br. Soc. Allergy Clin. Immunol.* **33**, 1711–1716 (2003).
132. Thai, P., Chen, Y., Dolganov, G., & Wu, R. Differential regulation of MUC5AC/Muc5ac and hCLCA-1/mGob-5 expression in airway epithelium. *Am. J. Respir. Cell Mol. Biol.* **33**, 523–530 (2005).
133. Venkayya, R. *et al.* The Th2 lymphocyte products IL-4 and IL-13 rapidly induce airway hyperresponsiveness through direct effects on resident airway cells. *Am. J. Respir. Cell Mol. Biol.* **26**, 202–208 (2002).
134. Huo, Y. & Zhang, H.-Y. Genetic mechanisms of asthma and the implications for drug repositioning. *Genes* **9**, E237 (2018).
135. Goswami, R. *et al.* STAT6-dependent regulation of Th9 development. *J. Immunol. Baltim. Md 1950.* **188**, 968–975 (2012).
136. Zhu, J., Guo, L., Watson, C. J., Hu-Li, J., & Paul, W. E. Stat6 is necessary and sufficient for IL-4's role in Th2 differentiation and cell expansion. *J. Immunol.* **166**, 7276–7281 (2001).
137. Summers, S. A. *et al.* Signal transducer and activation of transcription 6 (STAT6) regulates T helper type 1 (Th1) and Th17 nephritogenic immunity in experimental crescentic glomerulonephritis. *Clin. Exp. Immunol.* **166**, 227–234 (2011).
138. Doherty, T. A. *et al.* Alternaria induces STAT6-dependent acute airway eosinophilia and epithelial FIZZ1 expression that promotes airway fibrosis and epithelial thickness. *J. Immunol. Baltim. Md 1950.* **188**, 2622–2629 (2012).
139. Krishnamurthy, P. & Kaplan, M. H. STAT6 and PARP family members in the development of T cell-dependent allergic inflammation. *Immune Netw.* **16**, 201–210 (2016).
140. Antczak, A. *et al.* Analysis of changes in expression of IL-4/IL-13/STAT6 pathway and correlation with the selected clinical parameters in patients with atopic asthma. *Int. J. Immunopathol. Pharmacol.* **29**, 195–204 (2016).
141. Gao, P. S. *et al.* Variants of STAT6 (signal transducer and activator of transcription 6) in atopic asthma. *J. Med. Genet.* **37**, 380–382 (2000).
142. Al-Muhsen, S. *et al.* Association of the STAT-6 rs324011 (C2892T) variant but not rs324015 (G2964A), with atopic asthma in a Saudi Arabian population. *Hum. Immunol.* **75**, 791–795 (2014).
143. Nagarkatti, R., B-Rao, C., Vijayan, V., Sharma, S. K., & Ghosh, B. Signal transducer and activator of transcription 6 haplotypes and asthma in the Indian population. *Am. J. Respir. Cell Mol. Biol.* **31**, 317–321 (2004).
144. Godava, M., Vrtel, R., & Vodicka, R. STAT6 - polymorphisms, haplotypes and epistasis in relation to atopy and asthma. *Biomed. Pap. Med. Fac. Univ. Palacky Olomouc Czechoslov.* **157**, 172–180 (2013).
145. Eifan, A. O. *et al.* Reduced T-bet in addition to enhanced STAT6 and GATA3 expressing T cells contribute to human allergen-induced late responses. *Clin. Exp. Allergy J. Br. Soc. Allergy Clin. Immunol.* **42**, 891–900 (2012).
146. Yu, Q. N. *et al.* ILC2 frequency and activity are inhibited by glucocorticoid treatment via STAT pathway in patients with asthma. *Allergy.* **73**, 1860–1870 (2018).
147. Taha, R., Hamid, Q., Cameron, L., & Olivenstein, R. T helper type 2 cytokine receptors and associated transcription factors GATA-3, c-MAF, and signal transducer and activator of transcription factor-6 in induced sputum of atopic asthmatic patients. *Chest.* **123**, 2074–2082 (2003).
148. Stephenson, L., Johns, M. H., Woodward, E., Mora, A. L., & Boothby, M. An IL-4R alpha allelic variant, I50, acts as a gain-of-function variant relative to V50 for Stat6, but not Th2 differentiation. *J. Immunol. Baltim. Md 1950.* **173**, 4523–4528 (2004).
149. Risma, K. A. *et al.* V75R576 IL-4 receptor α is associated with allergic asthma and enhanced IL-4 receptor function. *J. Immunol.* **169**, 1604–1610 (2002).
150. Khoo, S.-K. *et al.* Associations of a novel IL4RA polymorphism, Ala57Thr, in Greenlander Inuit. *J. Allergy Clin. Immunol.* **118**, 627–634 (2006).

196. Chiba, Y. *et al.* Interleukin-13 augments bronchial smooth muscle contractility with an up-regulation of RhoA protein. *Am. J. Respir. Cell Mol. Biol.* **40**, 159–167 (2009).
197. Goto, K. *et al.* The proximal STAT6 and NF-kappaB sites are responsible for IL-13- and TNF-alpha-induced RhoA transcriptions in human bronchial smooth muscle cells. *Pharmacol. Res.* **61**, 466–472 (2010).
198. Nelson, E. A. *et al.* The STAT5 inhibitor pimozide displays efficacy in models of acute myelogenous leukemia driven by FLT3 mutations. *Genes Cancer.* **3**, 503–511 (2012).
199. Liu, S. *et al.* Steroid resistance of airway type 2 innate lymphoid cells from patients with severe asthma: The role of thymic stromal lymphopoietin. *J. Allergy Clin. Immunol.* **141**, 257–268.e6 (2018).
200. Ashino, S. *et al.* JAK1/3 signaling pathways are key initiators of TH2 differentiation and lung allergic responses. *J. Allergy Clin. Immunol.* **133**, 1162–1174.e4 (2014).
201. Delgoffe, G. M. & Vignali, D. A. A. STAT heterodimers in immunity. *JAK-STAT* **2**, e23060 (2013).
202. Yang, X.-P. *et al.* Opposing regulation of the locus encoding IL-17 through direct, reciprocal actions of STAT3 and STAT5. *Nat. Immunol.* **12**, 247–254 (2011).

5

Role of JAK-STAT Signaling in Atopic Dermatitis

Radomir M. Slominski
Department of Dermatology
Indiana University School of Medicine
Indianapolis, Indiana

Matthew J. Turner
Department of Dermatology
Indiana University School of Medicine
Indianapolis, Indiana

Richard L. Roudebush VA Medical Center
Indianapolis, Indiana

5.1 Introduction

Atopic dermatitis (AD) is one of the most common inflammatory diseases known to humankind. Severe itch, lost sleep, and secondary skin infections are major contributors to the morbidity of this disease (Holm et al. 2006; Ong and Leung 2010). Allergic skin inflammation and associated impaired skin barrier function are key drivers of AD pathogenesis, and Th2 (and other) cytokines are implicated in initiation and preservation of skin inflammation in AD. Targeted therapeutics have recently entered the market for treatment of AD, including monoclonal anti-human IL-4Rα antibody, dupilumab, which was FDA approved in 2017 for treatment of moderate-to-severe AD (Beck et al. 2014). Several other types of targeted therapeutics are either FDA approved or under investigation; the latter group includes oral and topical forms of Janus kinase (JAK) inhibitors. This chapter will discuss clinical and histopathologic aspects of AD, and mechanisms of AD immunopathogenesis, including the role of JAK/STAT signaling in AD, and conclude with a review of available data regarding use of JAK inhibitors in human and canine AD.

5.2 Clinical Characteristics of AD

Based on estimates of up to 17% of children in the United States developing atopic dermatitis (AD) and 2009 U.S. census data, more than 13 million children in the United States have AD (Laughter et al. 2000). The worldwide prevalence of AD in adults is 1–3%, equating to millions of additional cases (Katsarou and Armenaka 2011). There are two different phases of AD lesions, acute and chronic (Rajka 1986). Acute lesions are red (erythematous), scaly, minimally elevated plaques with poorly defined borders. In contrast to acute lesions, chronic lesions are more well defined, often less scaly, and are more elevated appearing with accentuated skin lines (i.e. lichenified). Chronic lesions develop from acute lesions, in part as a result of repetitive rubbing and scratching of the skin (Rajka 1986). The areas of the body affected by AD (i.e. the distribution) change with age. In infancy, the flexor aspects of the extremities and face are more prominently involved. In contrast, the involvement of the extensor aspects of the extremities is more common later in childhood and adulthood. As implied above, the extent of skin affected by AD patients can be limited to specific body regions; however, disease can be much more extensive involving the majority of a patient's skin. In addition to AD lesions themselves, AD patients often have other associated clinical findings including dry skin (xerosis), ichthyosis vulgaris (fish-like scaling of the skin caused by autosomal semidominant mutations in the *Filaggrin* gene) and keratosis pilaris as well as comorbid allergic diseases such as asthma, allergic rhinitis and food allergies (Rajka 1986; Smith et al. 2006).

As noted in the introduction, severe itch (pruritus), poor sleep and secondary skin infections are major sources of morbidity in AD. Regardless of the extent of skin involvement, itch is one of the defining features of AD (Rajka 1986). Patients often respond to itch by scratching the skin, which can have several effects including: minor bleeding, inoculation of injured skin with microbial pathogens, promotion of acute to chronic lesion conversion, and development of scarring and/or skin dyspigmentation. In addition, itching and scratching are important contributors to sleep disruption in AD (Stores et al. 1998). With respect to secondary skin infections, *Staphylococcus aureus*, which also frequently colonizes skin in AD patients without causing disease, is probably the most common pathogen (Ong and Leung 2016; Totte et al. 2016). Typically, *S. aureus* infections manifest as impetigo, a superficial skin infection that may resolve even without antibiotic therapy if a patient's AD is treated with anti-inflammatory agents (Travers et al. 2012). In contrast, herpes simplex virus infections of AD skin lesions (eczema herpeticum) benefits from prompt treatment with systemic antiviral therapies, particularly in the case of those that put the patient at risk of developing eye infections, which can result in permanent vision loss (Young et al. 2010). As a result of the aforementioned morbidities, the quality of life in AD can be dramatically impaired. For example, children with generalized AD reported greater impairment in quality of life than patients with asthma, diabetes, and epilepsy (Beattie and Lewis-Jones 2006). Healthcare costs for AD are also significant (Mancini et al. 2008). In light of the above and the historical lack of widely effective treatments for moderate-to-severe disease, patients and the medical community recognize the need to develop more effective therapeutic agents for AD.

5.3 Histologic Aspects of AD

This section begins with a review of the histologic architecture of healthy skin and the histopathologic changes seen in AD. These findings are summarized in Figure 5.1. Although the architecture varies between body regions, the skin can generally be thought of as being composed of three layers, the upper layer (epidermis), a middle layer (dermis) and the underlying subcutaneous fat. The outermost layer of the skin, the epidermis, is also composed of multiple layers as detailed below. While this simplistic three-layer concept of the skin will be discussed below for understanding AD pathogenesis, it is worth noting that the skin is a highly specialized and complex organ system composed of many resident cell types and structures including nerves, blood vessels, hair follicles, eccrine glands, apocrine glands, and others. The role of all of these structures and cell types in AD is beyond the scope of this discussion.

The lowest (basal) layer of the epidermis is composed of stem cells and undifferentiated keratinocytes. These undifferentiated keratinocytes leave the basal layer and transit upward through the epidermis, undergoing progressive and eventually terminal differentiation, at which point their nucleus is lost and they become protein-rich devitalized structures called corneocytes that are embedded in a lipid-rich matrix. This composite of corneocytes and lipid matrix is the uppermost layer of skin called the stratum corneum, a key barrier structure that prevents harmful irritants, allergens, and pathogens from entering the skin and promotes water retention in the skin (Matsui and Amagai 2015).

Histopathologic changes in AD include increased thickness of the stratum corneum (ortho- and parakeratosis), increased epidermal thickness (acanthosis) due to epidermal hyperplasia, spongiosis due to intercellular edema in the epidermis, and lymphocytic and histiocytic inflammatory cell infiltration into the dermis and epidermis. In comparing acute and chronic AD lesions, acanthosis and inflammatory infiltrates are more prominent and spongiosis is often less prominent in chronic

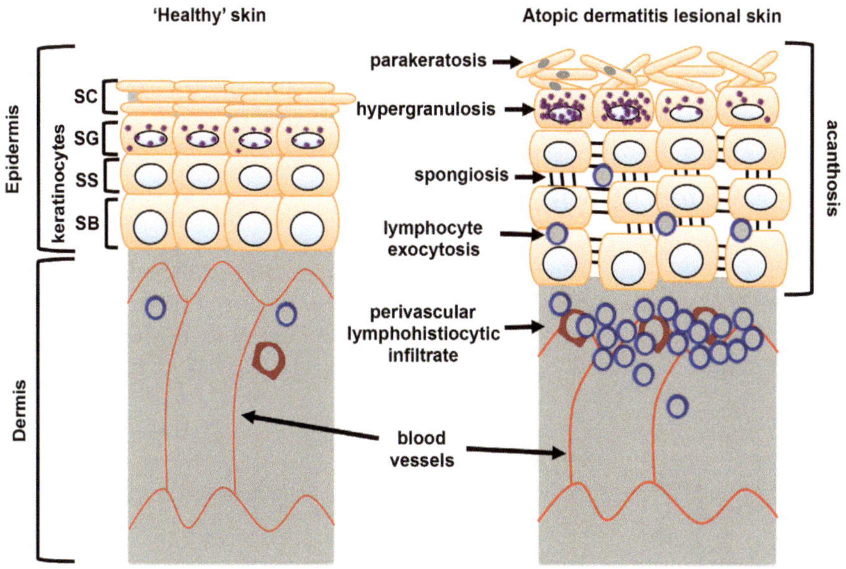

FIGURE 5.1 Basic histologic features of healthy human skin and atopic dermatitis lesional skin. The epidermis is predominately composed of keratinocytes. Other cells in the epidermis include melanocytes and Langerhans cells (not shown). The four layers of the epidermis include the stratum basale (SB), stratum spinosum (SS), stratum granulosum (SG), and the stratum corneum (SC). The SB is the proliferative and least differentiated layer. Keratinocytes in the SG have keratohyalin granules (purple dots), which contain filaggrin. The SC is the devitalized terminally differentiated layer of the epidermis and is composed of corneocytes embedded in a lipid matrix. The SC and SG are major contributors to epidermal barrier function. Spongiosis, which is caused by intercellular edema leading to keratinocyte separation, is a common histologic feature of atopic dermatitis; however, spongiosis is not always observed in chronic atopic dermatitis lesions. A perivascular lymphohistiocytic infiltrate is also seen in atopic dermatitis lesional skin; the density and composition of this infiltrate varies between acute and chronic atopic dermatitis lesional skin and different atopic dermatitis endotypes. Other histologic features that can be seen in atopic dermatitis lesional skin include parakeratosis (thickened SC composed of corneocytes with retained nuclei), hypergranulosis (increased numbers of keratohyalin granules in the SG), and lymphocyte exocytosis (lymphocytes in epidermis). Of note, the histologic features of atopic dermatitis are not distinguishable from many other types of eczematous dermatitis.

lesions. Immunohistochemical and other analyses demonstrate that T cells are abundant in the epidermal and dermal inflammatory cell infiltrates (Zachary et al. 1985).

5.4 Pathogenesis of AD

The pathogenesis of AD is mediated by a complex interplay between skin barrier dysfunction and immune system dysregulation (Werfel et al. 2016). Atopic dermatitis is a complex genetic disease; *Filaggrin* gene mutations are perhaps the best-characterized genetic defect in AD (Palmer et al. 2006; Weidinger et al. 2006). There is also a great deal of genetic, clinical, histologic, and immunologic variation between subjects with AD. For example, most AD patients do not harbor *Filaggrin* gene mutations, and there is variation in the type and frequency of Filaggrin mutations between different populations (Akiyama 2010). The presence (extrinsic AD) or absence (intrinsic AD) of elevated total serum IgE levels provide another was to subdivide AD (Novak and Bieber 2003). There are also differences in clinical, histologic and immunologic characteristics of AD in certain Asian populations who can manifest with a more psoriasiform phenotype compared to classical AD features commonly seen in those of European descent (Noda et al. 2015). Taken together, these observations highlight the heterogeneity of AD and support the concept that there are subtypes (endotypes) of disease (Werfel et al. 2016). However, there are characteristic clinical, histologic, cellular, and molecular aspects of disease that help explain AD pathogenesis. As this chapter is focused on JAK-STAT signaling and JAK inhibition as a potential treatment of AD, much of what follows in this section examines the immunologic aspects of AD pathogenesis, some of which are summarized in Figure 5.2.

5.4.1 Epidermal Barrier Dysfunction

Development of AD involves two key and interrelated processes, epidermal barrier dysfunction and T helper type 2 (Th2) cell-polarized immune responses (Bieber 2008; Sehra et al. 2008b; De Benedetto et al. 2009). The stratum corneum is the outermost part of the epidermal barrier and is composed of devitalized keratinocytes (i.e. corneocytes) embedded in a lipid matrix (Elias and Schmuth 2009; Kypriotou et al. 2012). Cornified envelope proteins including filaggrin, loricrin and

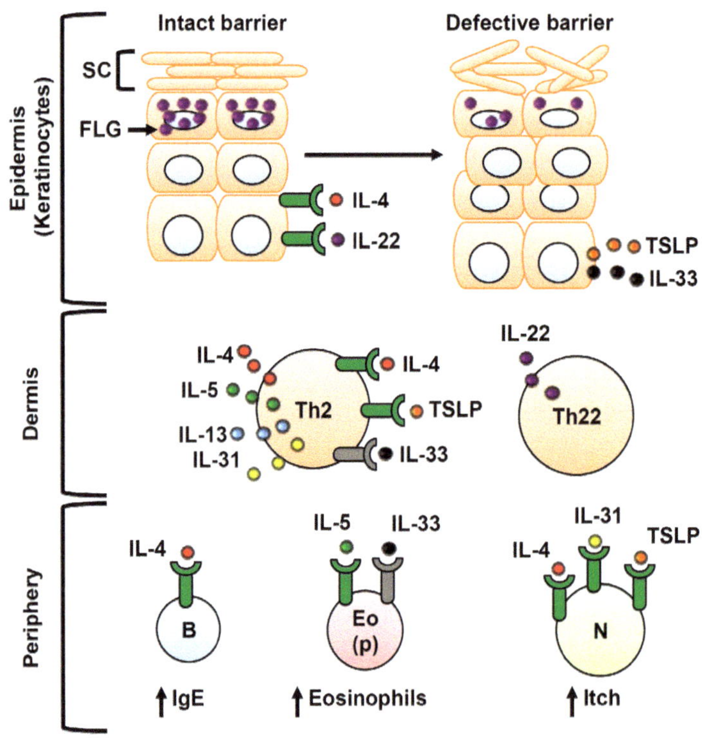

FIGURE 5.2 Immunopathogenic mechanisms of atopic dermatitis that involve JAK-STAT signaling. In AD, a variety of cytokines are produced by keratinocytes (TSLP, IL-33), Th2 cells (IL-4, IL-5, IL-13 and IL-31), and Th22 cells (IL-22). With the exception of IL-33, these cytokines activate receptors (green) that employ JAK-STAT pathways. Top panel: the upper layer of the skin (epidermis) is composed of keratinocytes that undergo progressive differentiation. As keratinocytes differentiate and progress upward through the epidermis, they acquire the ability to produce the epidermal barrier protein, filaggrin (FLG), which is stored in intracellular granules (purple circles). Keratinocytes then lose their nucleus and become devitalized corneocytes embedded in a lipid matrix; this layer is called the stratum corneum (SC) and forms a key part of the skin barrier. Stimulation of keratinocytes with IL-4 reduces keratinocyte maturation, filaggrin expression, and proper stratum corneum formation. In contrast, IL-22 stimulation of keratinocytes promotes epidermal hyperplasia. Middle panel: Stimulation of Th2 cells with IL-4, IL-33, and TSLP promotes Th2 cell differentiation and/or cytokine production. Bottom panel: Stimulation of B cells with IL-4 promotes IgE class switching and production. Peripheral eosinophilia is stimulated by IL-5 and IL-33. Cytokines that act directly on neurons to stimulate itch include IL-4, IL-31, and TSLP.

involucrin are key structural and functional proteins in corneocytes and are critical for proper epidermal barrier function (Elias and Schmuth 2009; Kypriotou et al. 2012). These proteins are encoded within the epidermal differentiation complex (EDC) gene cluster on human chromosome 1q21 (Elias and Schmuth 2009; Kypriotou et al. 2012). Barrier dysfunction in AD has been classically attributed to inherent defects (i.e. genetic) in the keratinocyte; the most well-established example of such a defect is the presence of mutations in the Filaggrin gene (Palmer et al. 2006; Weidinger et al. 2006).

Filaggrin is a cornified envelope protein that normally aggregates keratin filaments in corneocytes facilitating the ordered formation of a 'bricks and mortar' type structure of the stratum corneum and an effective barrier that helps impede water loss from the skin and entry of certain microbes and environmental agents into the skin (Elias and Schmuth 2009; Scharschmidt et al. 2009; Kypriotou et al. 2012). Filaggrin mutations are detected in a significant proportion of patients with AD and result in insufficient quantities of functional protein thereby leading to defective bundling of keratin filaments and disarray in the organization of the stratum corneum (Manabe et al. 1991; Brown and McLean 2012). Filaggrin breakdown products are important components of the so-called natural moisturizing factor (NMF) that acts as a humectant helping to retain water in the skin. In patients with *Filaggrin* gene mutations and AD, decreased levels of filaggrin and as a result decreased levels of NMF, are thought to contribute to skin dryness (Rawlings and Harding 2004; Kezic et al. 2008). As noted earlier, *Filaggrin* gene mutations also cause ichthyosis vulgaris, thus explaining the association between this scaling skin condition and AD.

The preceding discussion on Filaggrin shows that intrinsic defects in keratinocytes can and often do occur in AD, resulting in defective epidermal barrier function. These primary defects in keratinocyte function leading to impaired skin barrier function can in turn promote cutaneous inflammation in AD (Kim and Leung 2018). Interestingly, the majority of patients with AD do not harbor *Filaggrin* gene mutations (Margolis et al. 2012). This suggests the impaired keratinocyte differentiation and function in AD do not necessarily require intrinsic defects in keratinocytes. One explanation for this observation is that in AD, the Th2 cytokine IL-4 signals through its cognate receptor on keratinocytes leading to impaired keratinocyte differentiation and epidermal barrier dysfunction (Howell et al. 2007; Sehra et al. 2010). This suggests that inflammation is sufficient to induce keratinocyte and skin barrier dysfunction in AD. Finally, environmental insults can induce epidermal barrier disruption and resultant production of keratinocyte-derived cytokines (e.g. TSLP, IL-33), which can in turn promote Th2-polarized inflammation in AD (Oyoshi et al. 2010). Thus, extrinsic and intrinsic pathways can promote keratinocyte/epidermal barrier dysfunction and immune dysregulation. Furthermore, there is a bidirectional positive feedback loop between keratinocyte/epidermal barrier dysfunction and Th2 cytokine-mediated immune dysregulation in AD pathogenesis. These data suggest a central role for IL-4-mediated signaling in AD and suggest JAK inhibition could counteract the ability of IL-4 to impair keratinocyte function in this disease.

5.4.2 T Cell Subsets and Cytokines

Histologic analyses of nonlesional skin from patients with AD reveal sparse perivascular T cell-rich inflammatory infiltrates, while acute lesional skin reveals denser T cell-rich inflammatory infiltrates of CD4+ and less abundant CD8+ T cells (Zachary et al. 1985). In acute lesions, Th2 cells and Th2 cytokines such as IL-4, IL-5 and IL-13 are predominant. In chronic lesions, Th2 cells and cytokines persist, but the T cell infiltrate becomes more diverse to include Th1 cells (producing IFNγ) and other subsets (Hamid et al. 1994, 1996; Leung and Soter 2001). Clinical investigations using atopy patch testing to mimic AD lesion formation revealed similar patterns of T helper cell infiltrates with an early influx of Th2 cells and a later accumulation of Th1 cells (Grewe et al. 1998). Th2 cytokines contribute to AD pathogenesis through multiple mechanisms including promotion of Th2 cell differentiation, class switching and IgE production by B cells,

with no serious or severe events, including death, occurring during the study. The most common adverse event noted in the JTE-052 treatment group was nasopharyngitis (3.4%).

5.5.2 JAK Inhibition for Canine AD

While no JAK inhibitors have been FDA approved for treatment of AD in humans, oclacitinib, marketed as Apoquel, was FDA approved in 2013 for treatment of AD in dogs (canine AD). Canine AD is a common dermatologic disease; one study demonstrated 8.7% of dogs were reported to have AD (Lund et al. 1999). Canine AD shares similarities to human AD with respect to some clinical, histologic, and microbiologic features and treatment modalities used (Marsella and Girolomoni 2009; Cosgrove et al. 2013; Gedon and Mueller 2018). Environmental and food allergens also appear to have a more prominent role in contributing to skin inflammation in canine AD (Marsella and Girolomoni 2009). Although the pathogenic mechanisms of canine AD differ in some respects to human disease, Th2 cytokines are important drivers of allergic skin inflammation in both human and canine AD (Gedon and Mueller 2018).

Cosgrove et al. performed a randomized double-blinded placebo-controlled trial of oral oclacitinib for canine AD (Cosgrove et al. 2013). The study used 299 dogs (12 months of age or older) with AD and was divided into placebo, 0.4 mg/kg and 0.6 mg/kg oclacitinib treatment arms. Dogs first received placebo or oclacitinib by mouth twice daily for 14 days, followed by once daily dosing thereafter for 98 additional days (total treatment duration 112 days). Response to therapy was determined periodically using the canine AD extent and severity index (CADESI-02) determined by the clinician and the dog owner's assessment of their dog's itching (owner pruritus visual analogue scale). Disease severity and itch were both significantly reduced in oclacitinib-treated dogs compared to placebo-treated controls. For example, a 48.4% reduction in CADESI-02 score was noted on days 14 and 28 in oclacitinib-treated dogs compared to 1.7% and 3.6% reductions in placebo-treated dogs on days 14 and 28. With respect to itching, pruritus scores on days 1, 2, 7, 14 and 28 improved by 29.5–66.7% in oclacitinib-treated dogs compared to 3.9–10.4% improvement in the placebo group. Data analysis after day 28 was limited by the fact that most (>86%) placebo-treated dogs experienced worsening itch and/or AD and were therefore transitioned to an open-label arm in which they received oclacitinib. Cosgrove et al. also performed a long-term study of oclacitinib therapy in canine AD in which dogs were followed for up to 683 days of treatment with oclacitinib; in this study, dog owners reported 91% of dogs exhibited an improved quality of life (Cosgrove et al. 2015).

In the 2013 study by Cosgrove et al., the majority of placebo-treated dogs were switched to oclacitinib by day 16 due to worsening AD. This switch limited the ability of the investigators to compare adverse events between oclacitinib and placebo-treated groups beyond day 16. Adverse events between days 0 and 16 in the placebo and oclacitinib groups included diarrhea, vomiting, anorexia and lethargy; these events affected less than 5% of dogs with no clear difference between study groups and resolved in the majority of dogs despite continuation of treatment. Of the 299 dogs used in this study, 283 received at least one dose of oclacitinib. Adverse events affecting at least 5% of oclacitinib-treated dogs included pyoderma (12%), nonspecified dermal nodules (12%), otitis, (9.9%), vomiting (9.2%) and diarrhea (6%). Leukopenia due to neutropenia occurred in some dogs but the percentage of dogs affected was not reported. Although some variance in hematologic parameters was seen, cell counts remained in the normal range, as did serum chemistry values and urinalysis parameters. In a long-term investigation of oclacitinib therapy for canine AD, Cosgrove et al. studied dogs for up to 683 days of treatment with oclacitinib (Cosgrove et al. 2015). Adverse events and laboratory values in oclacitinib-treated dogs were similar to those reported in the author's earlier short-term study.

5.6 Conclusions

After decades of studying atopic dermatitis, some key aspects of pathogenesis are clear. Skin inflammation, skin barrier disruption, and itch are key manifestations and targets for treatment of AD. Cellular mediators of these manifestations include epithelial cells (keratinocytes), Th2 and other immune cells and neurons. Each of these cell types utilize JAK/STAT signaling to respond to one or more cytokines implicated in AD. In light of this, there is a strong rationale for the realized and potential utility of JAK inhibitors for treatment of this disease. In canine AD, systemic therapy with the FDA-approved JAK inhibitor, oclacitinib, has helped revolutionize treatment of AD in dogs. As clinical studies progress, it will be interesting to see how JAK inhibitors impact treatment of AD in humans.

ACKNOWLEDGMENTS

MJT was supported by awards from the Dermatology Foundation (Physician Scientist CDA), the Veterans Administration (VA CDA2; IK2CX001019) and the Showalter Trust as well as Indiana University School of Medicine. MJT also thanks Dr. Stephen Shideler for his kind support. RMS was supported by the Indiana Cutaneous Biology Training Program NIH T32 fellowship award (T32AR062495).

REFERENCES

Akiyama, M. 2010. FLG mutations in ichthyosis vulgaris and atopic eczema: spectrum of mutations and population genetics. *Br J Dermatol* 162 (3):472–477.

Amano, W., S. Nakajima, H. Kunugi et al. 2015. The Janus kinase inhibitor JTE-052 improves skin barrier function through suppressing signal transducer and activator of transcription 3 signaling. *J Allergy Clin Immunol* 136 (3):667–677.e7.

Angelova-Fischer, I., I. M. Fernandez, M. H. Donnadieu et al. 2010. Injury to the stratum corneum induces in vivo expression of human thymic stromal lymphopoietin in the epidermis. *J Invest Dermatol* 130 (10):2505–2507.

Beattie, P. E., and M. S. Lewis-Jones. 2006. A comparative study of impairment of quality of life in children with skin disease and children with other chronic childhood diseases. *Br J Dermatol* 155 (1):145–151.

Beck, L. A., D. Thaci, J. D. Hamilton et al. 2014. Dupilumab treatment in adults with moderate-to-severe atopic dermatitis. *N Engl J Med* 371 (2):130–139.

Bieber, T. 2008. Atopic dermatitis. *N Engl J Med* 358 (14):1483–1494.

Bilsborough, J., D. Y. Leung, M. Maurer et al. 2006. IL-31 is associated with cutaneous lymphocyte antigen-positive skin homing T cells in patients with atopic dermatitis. *J Allergy Clin Immunol* 117 (2):418–425.

Bissonnette, R., K. A. Papp, Y. Poulin et al. 2016. Topical tofacitinib for atopic dermatitis: a phase IIa randomized trial. *Br J Dermatol* 175 (5):902–911.

Briot, A., C. Deraison, M. Lacroix et al. 2009. Kallikrein 5 induces atopic dermatitis-like lesions through PAR2-mediated thymic stromal lymphopoietin expression in Netherton syndrome. *J Exp Med* 206 (5):1135–1147.

Brown, S. J., and W. H. McLean. 2012. One remarkable molecule: filaggrin. *J Invest Dermatol* 132 (3 Pt 2):751–762.

Bruns, H. A., U. Schindler, and M. H. Kaplan. 2003. Expression of a constitutively active Stat6 in vivo alters lymphocyte homeostasis with distinct effects in T and B cells. *J Immunol* 170 (7):3478–3487.

Carmi-Levy, I., B. Homey, and V. Soumelis. 2011. A modular view of cytokine networks in atopic dermatitis. *Clin Rev Allergy Immunol* 41 (3):245–253.

Cosgrove, S. B., D. M. Cleaver, V. L. King et al. 2015. Long-term compassionate use of oclacitinib in dogs with atopic and allergic skin disease: safety, efficacy and quality of life. *Vet Dermatol* 26 (3):171–179. e35.

Cosgrove, S. B., J. A. Wren, D. M. Cleaver et al. 2013. Efficacy and safety of oclacitinib for the control of pruritus and associated skin lesions in dogs with canine allergic dermatitis. *Vet Dermatol* 24 (5):479–e114.

DaSilva-Arnold, S. C., A. Thyagarajan, L. J. Seymour et al. 2018. Phenotyping acute and chronic atopic dermatitis-like lesions in Stat6VT mice identifies a role for IL-33 in disease pathogenesis. *Arch Dermatol Res* 310 (3):197–207.

De Benedetto, A., R. Agnihothri, L. Y. McGirt, L. G. Bankova, and L. A. Beck. 2009. Atopic dermatitis: a disease caused by innate immune defects? *J Invest Dermatol* 129 (1):14–30.

Elias, P. M., and M. Schmuth. 2009. Abnormal skin barrier in the etiopathogenesis of atopic dermatitis. *Curr Allergy Asthma Rep* 9 (4):265–272.

Gedon, N. K. Y., and R. S. Mueller. 2018. Atopic dermatitis in cats and dogs: a difficult disease for animals and owners. *Clin Transl Allergy* 8:41.

Grewe, M., C. A. Bruijnzeel-Koomen, E. Schopf et al. 1998. A role for Th1 and Th2 cells in the immunopathogenesis of atopic dermatitis. *Immunol Today* 19 (8):359–361.

Guttman-Yassky, E., J. I. Silverberg, O. Nemoto et al. 2019. Baricitinib in adult patients with moderate-to-severe atopic dermatitis: a phase 2 parallel, double-blinded, randomized placebo-controlled multiple-dose study. *J Am Acad Dermatol* 80 (4):913–921.

Halabi-Tawil, M., F. M. Ruemmele, S. Fraitag et al. 2009. Cutaneous manifestations of immune dysregulation, polyendocrinopathy, enteropathy, X-linked (IPEX) syndrome. *Br J Dermatol* 160 (3):645–651.

Hamid, Q., M. Boguniewicz, and D. Y. Leung. 1994. Differential in situ cytokine gene expression in acute versus chronic atopic dermatitis. *J Clin Invest* 94 (2):870–876.

Hamid, Q., T. Naseer, E. M. Minshall, Y. L. Song, M. Boguniewicz, and D. Y. Leung. 1996. In vivo expression of IL-12 and IL-13 in atopic dermatitis. *J Allergy Clin Immunol* 98 (1):225–231.

Hammad, H., and B. N. Lambrecht. 2015. Barrier Epithelial Cells and the Control of Type 2 Immunity. *Immunity* 43 (1):29–40.

Hanifin, J. M., M. Thurston, M. Omoto, R. Cherill, S. J. Tofte, and M. Graeber. 2001. The eczema area and severity index (EASI): assessment of reliability in atopic dermatitis. EASI Evaluator Group. *Exp Dermatol* 10 (1):11–18.

Holm, E. A., H. C. Wulf, H. Stegmann, and G. B. Jemec. 2006. Life quality assessment among patients with atopic eczema. *Br J Dermatol* 154 (4):719–725.

Howell, M. D., B. E. Kim, P. Gao et al. 2007. Cytokine modulation of atopic dermatitis filaggrin skin expression. *J Allergy Clin Immunol* 120 (1):150–155.

Imai, Y., K. Yasuda, Y. Sakaguchi et al. 2013. Skin-specific expression of IL-33 activates group 2 innate lymphoid cells and elicits atopic dermatitis-like inflammation in mice. *Proc Natl Acad Sci U S A* 110 (34):13921–13926.

Katsarou, A., and M. Armenaka. 2011. Atopic dermatitis in older patients: particular points. *J Eur Acad Dermatol Venereol* 25 (1):12–18.

Kezic, S., P. M. Kemperman, E. S. Koster et al. 2008. Loss-of-function mutations in the filaggrin gene lead to reduced level of natural moisturizing factor in the stratum corneum. *J Invest Dermatol* 128 (8):2117–2119.

Kim, B. E., D. Y. Leung, M. Boguniewicz, and M. D. Howell. 2008. Loricrin and involucrin expression is down-regulated by Th2 cytokines through STAT-6. *Clin Immunol* 126 (3):332–337.

Kim, B. E., and D. Y. M. Leung. 2018. Significance of skin barrier dysfunction in atopic dermatitis. *Allergy Asthma Immunol Res* 10 (3):207–215.

Kim, B. S., M. C. Siracusa, S. A. Saenz et al. 2013. TSLP elicits IL-33-independent innate lymphoid cell responses to promote skin inflammation. *Sci Transl Med* 5 (170):170ra16.

Kunz, B., A. P. Oranje, L. Labreze, J. F. Stalder, J. Ring, and A. Taieb. 1997. Clinical validation and guidelines for the SCORAD index: consensus report of the European Task Force on atopic dermatitis. *Dermatology* 195 (1):10–19.

Kypriotou, M., M. Huber, and D. Hohl. 2012. The human epidermal differentiation complex: cornified envelope precursors, S100 proteins and the 'fused genes' family. *Exp Dermatol* 21 (9):643–649.

Laughter, D., J. A. Istvan, S. J. Tofte, and J. M. Hanifin. 2000. The prevalence of atopic dermatitis in Oregon schoolchildren. *J Am Acad Dermatol* 43 (4):649–655.

Leung, D. Y., and N. A. Soter. 2001. Cellular and immunologic mechanisms in atopic dermatitis. *J Am Acad Dermatol* 44 (1 Suppl):S1–S12.

Levy, L. L., J. Urban, and B. A. King. 2015. Treatment of recalcitrant atopic dermatitis with the oral Janus kinase inhibitor tofacitinib citrate. *J Am Acad Dermatol* 73 (3):395–399.

Lund, E. M., P. J. Armstrong, C. A. Kirk, L. M. Kolar, and J. S. Klausner. 1999. Health status and population characteristics of dogs and cats examined at private veterinary practices in the United States. *J Am Vet Med Assoc* 214 (9):1336–1341.

Manabe, M., M. Sanchez, T. T. Sun, and B. A. Dale. 1991. Interaction of filaggrin with keratin filaments during advanced stages of normal human epidermal differentiation and in ichthyosis vulgaris. *Differentiation* 48 (1):43–50.

Mancini, A. J., K. Kaulback, and S. L. Chamlin. 2008. The socioeconomic impact of atopic dermatitis in the United States: a systematic review. *Pediatr Dermatol* 25 (1):1–6.

Margolis, D. J., A. J. Apter, J. Gupta et al. 2012. The persistence of atopic dermatitis and filaggrin (FLG) mutations in a US longitudinal cohort. *J Allergy Clin Immunol* 130 (4):912–917.

Marsella, R., and G. Girolomoni. 2009. Canine models of atopic dermatitis: a useful tool with untapped potential. *J Invest Dermatol* 129 (10):2351–2357.

Matsui, T., and M. Amagai. 2015. Dissecting the formation, structure and barrier function of the stratum corneum. *Int Immunol* 27 (6):269–280.

Meephansan, J., H. Tsuda, M. Komine, S. Tominaga, and M. Ohtsuki. 2012. Regulation of IL-33 expression by IFN-gamma and tumor necrosis factor-alpha in normal human epidermal keratinocytes. *J Invest Dermatol* 132 (11):2593–2600.

Milovanovic, M., V. Volarevic, G. Radosavljevic et al. 2012. IL-33/ST2 axis in inflammation and immunopathology. *Immunol Res* 52 (1–2):89–99.

Nakagawa, H., O. Nemoto, A. Igarashi, and T. Nagata. 2018. Efficacy and safety of topical JTE-052, a Janus kinase inhibitor, in Japanese adult patients with moderate-to-severe atopic dermatitis: a phase II, multicentre, randomized, vehicle-controlled clinical study. *Br J Dermatol* 178 (2):424–432.

Noda, S., M. Suarez-Farinas, B. Ungar et al. 2015. The Asian atopic dermatitis phenotype combines features of atopic dermatitis and psoriasis with increased TH17 polarization. *J Allergy Clin Immunol* 136 (5):1254–1264.

Nograles, K. E., L. C. Zaba, A. Shemer et al. 2009. IL-22-producing "T22" T cells account for upregulated IL-22 in atopic dermatitis despite reduced IL-17-producing TH17 T cells. *J Allergy Clin Immunol* 123 (6):1244–52.e2.

Novak, N., and T. Bieber. 2003. Allergic and nonallergic forms of atopic diseases. *J Allergy Clin Immunol* 112 (2):252–262.

Ong, P. Y., and D. Y. Leung. 2010. The infectious aspects of atopic dermatitis. *Immunol Allergy Clin North Am* 30 (3):309–321.

Ong, P. Y., and D. Y. Leung. 2016. Bacterial and viral infections in atopic dermatitis: a comprehensive review. *Clin Rev Allergy Immunol* 51 (3):329–337.

Oyoshi, M. K., R. P. Larson, S. F. Ziegler, and R. S. Geha. 2010. Mechanical injury polarizes skin dendritic cells to elicit a T(H)2 response by inducing cutaneous thymic stromal lymphopoietin expression. *J Allergy Clin Immunol* 126 (5):976–984. 984.e1–5.

Palmer, C. N., A. D. Irvine, A. Terron-Kwiatkowski et al. 2006. Common loss-of-function variants of the epidermal barrier protein filaggrin are a major predisposing factor for atopic dermatitis. *Nat Genet* 38 (4):441–446.

Rajka, G. 1986. Natural history and clinical manifestations of atopic dermatitis. *Clin Rev Allergy* 4 (1):3–26.

Rawlings, A. V., and C. R. Harding. 2004. Moisturization and skin barrier function. *Dermatol Ther* 17 (Suppl 1):43–48.

Saenz, S. A., B. C. Taylor, and D. Artis. 2008. Welcome to the neighborhood: epithelial cell-derived cytokines license innate and adaptive immune responses at mucosal sites. *Immunol Rev* 226:172–190.

Savinko, T., S. Matikainen, U. Saarialho-Kere et al. 2012. IL-33 and ST2 in atopic dermatitis: expression profiles and modulation by triggering factors. *J Invest Dermatol* 132 (5):1392–1400.

Scharschmidt, T. C., M. Q. Man, Y. Hatano et al. 2009. Filaggrin deficiency confers a paracellular barrier abnormality that reduces inflammatory thresholds to irritants and haptens. *J Allergy Clin Immunol* 124 (3):496–506. 506.e1–6.

Schlapbach, C., A. Gehad, C. Yang et al. 2014. Human TH9 cells are skin-tropic and have autocrine and paracrine proinflammatory capacity. *Sci Transl Med* 6 (219):219ra8.

Schmitz, J., A. Owyang, E. Oldham et al. 2005. IL-33, an interleukin-1-like cytokine that signals via the IL-1 receptor-related protein ST2 and induces T helper type 2-associated cytokines. *Immunity* 23 (5):479–490.

Sehra, S., H. A. Bruns, A. N. Ahyi et al. 2008a. IL-4 is a critical determinant in the generation of allergic inflammation initiated by a constitutively active Stat6. *J Immunol* 180 (5):3551–3559.

Sehra, S., F. M. Tuana, M. Holbreich et al. 2008b. Scratching the surface: towards understanding the pathogenesis of atopic dermatitis. *Crit Rev Immunol* 28 (1):15–43.
Sehra, S., Y. Yao, M. D. Howell et al. 2010. IL-4 regulates skin homeostasis and the predisposition toward allergic skin inflammation. *J Immunol* 184 (6):3186–3190.
Shreberk-Hassidim, R., Y. Ramot, and A. Zlotogorski. 2017. Janus kinase inhibitors in dermatology: a systematic review. *J Am Acad Dermatol* 76 (4):745–753.e19.
Smith, F. J., A. D. Irvine, A. Terron-Kwiatkowski et al. 2006. Loss-of-function mutations in the gene encoding filaggrin cause ichthyosis vulgaris. *Nat Genet* 38 (3):337–342.
Sonkoly, E., A. Muller, A. I. Lauerma et al. 2006. IL-31: a new link between T cells and pruritus in atopic skin inflammation. *J Allergy Clin Immunol* 117 (2):411–417.
Soumelis, V., P. A. Reche, H. Kanzler et al. 2002. Human epithelial cells trigger dendritic cell mediated allergic inflammation by producing TSLP. *Nat Immunol* 3 (7):673–680.
Stores, G., A. Burrows, and C. Crawford. 1998. Physiological sleep disturbance in children with atopic dermatitis: a case control study. *Pediatr Dermatol* 15 (4):264–268.
Stritesky, G. L., R. Muthukrishnan, S. Sehra et al. 2011. The transcription factor STAT3 is required for T helper 2 cell development. *Immunity* 34 (1):39–49.
Takaoka, A., I. Arai, M. Sugimoto et al. 2006. Involvement of IL-31 on scratching behavior in NC/Nga mice with atopic-like dermatitis. *Exp Dermatol* 15 (3):161–167.
Tanimoto, A., Y. Shinozaki, Y. Yamamoto et al. 2018. A novel JAK inhibitor JTE-052 reduces skin inflammation and ameliorates chronic dermatitis in rodent models: comparison with conventional therapeutic agents. *Exp Dermatol* 27 (1):22–29.
Totte, J. E., W. T. van der Feltz, M. Hennekam, A. van Belkum, E. J. van Zuuren, and S. G. Pasmans. 2016. Prevalence and odds of Staphylococcus aureus carriage in atopic dermatitis: a systematic review and meta-analysis. *Br J Dermatol* 175 (4):687–695.
Travers, J. B., A. Kozman, Y. Yao et al. 2012. Treatment outcomes of secondarily impetiginized pediatric atopic dermatitis lesions and the role of oral antibiotics. *Pediatr Dermatol* 29 (3):289–296.
Turner, M. J., S. DaSilva-Arnold, N. Luo et al. 2014. STAT6-mediated keratitis and blepharitis: a novel murine model of ocular atopic dermatitis. *Invest Ophthalmol Vis Sci* 55 (6):3803–3808.
Turner, M. J., and B. Zhou. 2014. A new itch to scratch for TSLP. *Trends Immunol* 35 (2):49–50.
Vu, A. T., T. Baba, X. Chen et al. 2010. Staphylococcus aureus membrane and diacylated lipopeptide induce thymic stromal lymphopoietin in keratinocytes through the Toll-like receptor 2-Toll-like receptor 6 pathway. *J Allergy Clin Immunol* 126 (5):985–993. 993.e1–3.
Weidinger, S., T. Illig, H. Baurecht et al. 2006. Loss-of-function variations within the filaggrin gene predispose for atopic dermatitis with allergic sensitizations. *J Allergy Clin Immunol* 118 (1):214–219.
Werfel, T., J. P. Allam, T. Biedermann et al. 2016. Cellular and molecular immunologic mechanisms in patients with atopic dermatitis. *J Allergy Clin Immunol* 138 (2):336–349.
Wilson, S. R., L. The, L. M. Batia et al. 2013. The epithelial cell-derived atopic dermatitis cytokine TSLP activates neurons to induce itch. *Cell* 155 (2):285–295.
Young, R. C., D. O. Hodge, T. J. Liesegang, and K. H. Baratz. 2010. Incidence, recurrence, and outcomes of herpes simplex virus eye disease in Olmsted County, Minnesota, 1976–2007: the effect of oral antiviral prophylaxis. *Arch Ophthalmol* 128 (9):1178–1183.
Zachary, C. B., M. H. Allen, and D. M. MacDonald. 1985. In situ quantification of T-lymphocyte subsets and Langerhans cells in the inflammatory infiltrate of atopic eczema. *Br J Dermatol* 112 (2):149–156.
Ziegler, S. F. 2012. Thymic stromal lymphopoietin and allergic disease. *J Allergy Clin Immunol* 130 (4):845–852.

6

JAK-STAT Signaling Pathway and Gliosis in Neuroinflammatory Diseases

Han-Chung Lee
Laboratory Centre
Xiamen University Malaysia
Sepang, Malaysia

Kai-Leng Tan

Institute of Biomedical and Pharmaceutical Sciences
Guangdong University of Technology
Guangzhou, China

Pike-See Cheah

Department of Human Anatomy, Faculty of Medicine and Health Sciences
Universiti Putra Malaysia
Seri Kembangan, Malaysia

King-Hwa Ling

Department of Biomedical Sciences, Faculty of Medicine and Health Sciences
Universiti Putra Malaysia
Seri Kembangan, Malaysia

Department of Genetics
Harvard Medical School
Boston, Massachusetts United States

6.1 Introduction

It is commonly perceived that the brain composes of glia and neuron cells (GNR) in the ratio of 10:1. However, a recent review suggested that the GNR could be only about 1:1 (von Bartheld, Bahney, and Herculano-Houzel 2016). In the human cerebral cortex, there are about 10–20 billion neuronal cells, while the number of glial cells could be about 15–30 billion. There are three types of glial cells in the human cerebral cortex with 45–75% of them are oligodendrocytes, 19–40% are astrocytes, and 10% or less are microglia (von Bartheld, Bahney, and Herculano-Houzel 2016).

Each type of glial cell has a different role; for example, oligodendrocytes are the myelinating cells that function to produce the insulating sheath of axons. On the other hand, microglia act as immune cells in the central nervous system. In contrast, astrocyte has broader functions, which include the maintenance of water and ion homeostasis, the blood–brain barrier, and modulation of synaptic activity (Jakel and Dimou 2017; Dossi, Vasile, and Rouach 2018). However, reactive microglia and astrocytes could have a detrimental effect on neuropathological conditions, for example, when they are stimulated by factors that are toxic to neurons such as β-amyloid and α-synuclein, which are the critical molecule in the pathogenesis of Alzheimer's disease (Hardy and Higgins 1992) and Parkinson's disease (Zhang et al. 2005), respectively.

Neuroinflammation is pathological inflammatory responses of the central nervous system (CNS), which can lead to further damage/insult to the brain and spinal cord. The pathological conditions in the brain include neurodegenerative diseases such as Alzheimer's disease, Parkinson's disease, and autoimmune diseases. These conditions induce both astrocytes and microglia, which are the most significant and smallest neuroglial cells, respectively, to become reactive in concert. Shreds of evidence now support that the reactive state of astrocytes and microglia are deleterious and neurotoxic to the brain cells.

The Janus kinase–signal transducers and activators of transcription (JAK-STAT) signaling pathway is an intracellular signaling pathway that involves the activation of two families of proteins: The Janus kinases (JAK) and the signal transducer and activator of transcriptions (STAT). JAK is a class of four cytoplasmic protein tyrosine kinases that includes JAK1, JAK2, JAK3, and TYK2. The STAT family contains seven transcription factors: STAT1, STAT2, STAT3, STAT4, STAT5A, STAT5B, and STAT6 (Kisseleva et al. 2002). JAK-STAT signaling pathway is essential for the initiation of innate immunity, subsequently coordinating the adaptive immune mechanism and ultimately constraining inflammatory responses (O'Shea and Plenge 2012). Specifically, the JAK-STAT signaling pathway is critical for T-cell proliferation, myeloid and lymphoid cell differentiation, B cells, and macrophage activation. Under normal condition, the protein expression level of JAK-STAT signaling pathway candidates such as JAK2, STAT1, and STAT3 in the brain is low, which probably is due to the low cytokine secretion as the number of immune cell present is small, and the glial cells are not in a reactive state (Na et al. 2007; Yan et al. 2018).

Neurons, astrocytes, microglia, and immune cells are the essential sources for the cytokine production in the brain (Galic, Riazi, and Pittman 2012). In general, cytokines are secreted in response to infection or injury, and whether they are pro-inflammatory or anti-inflammatory cytokines, the outcome will be different. Pro-inflammatory cytokines have been shown elevated in many neurodegenerative diseases, which have a detrimental effect—such as damage to CNS tissue—and are toxic to neurons and glial cells (Wang et al. 2015). Various cytokines have been reported to activate the JAK-STAT signaling pathway, and different cytokines have the propensity to activate a specific JAK and STAT. These include the interferons family such as IFN-α, IFN-β, IFN-γ, IL-10, IL-19, IL-20, and IL-22; the gp130 family including IL-6, IL-11, oncostatin M, leukaemia inhibitory factor, cardiotrophin-1, granulocyte colony-stimulating factor, IL-12, leptin, ciliary neurotrophic factor, and cytokines under γ-chain family, such as IL-2, IL-4, IL-7, IL-9, IL-15, and IL-21 (Rawlings, Rosler, and Harrison 2004; Schindler, Levy, and Decker 2007).

Aberrant activation of the JAK-STAT pathway is becoming evident in contributing to neuroinflammatory diseases, with the involvement of activated microglia and astrocytes. Importantly, the downregulation of JAK2–STAT3 signaling pathway was reported to cause memory impairment in Alzheimer's disease (Chiba et al. 2009), and potently promote the neurogenic-to-gliogenic shift in Down syndrome brain (Lee et al. 2016). In this review, we aim to review the role of JAK-STAT signaling pathway in neuroinflammation such as neurodegenerative disorders (e.g., Alzheimer's disease, Parkinson's disease, Huntington's disease), autoimmune disease (e.g., multiple sclerosis), and developmental disorder due to chromosomal abnormalities (e.g., Down syndrome).

6.1.1 JAK-STAT and Reactive Astrogliosis and Microgliosis

Reactive gliosis is a pathological term referred to as the accumulation of astrocytes and microglia at the injury site in the CNS (Streit, Mrak, and Griffin 2004). Advances in the neurobiology field have pinpointed that glial cells, particularly astrocytes and microglia, are the essential players in neuroinflammation (Heneka et al. 2015). Microglia are the resident phagocytes within the CNS

(Heneka et al. 2015) and make up about 5–12% of the glial cell population (Gomez-Nicola and Perry 2015). Generally, microglia exist in two distinct states, which are surveillance/resting state versus activated state. Microglia in the resting state usually is elongated in shape with long processes to facilitate its surveillance function. They are highly mobile, actively scouting their niche by extending and retracting ramified processes while remaining stationary (Hristovska and Pascual 2016). Upon detection of immunogen or pathogenic debris, microgliosis occurs and initiates an immune response (Heneka et al. 2015).

When activated, they alter the cell morphology (Streit and Graeber 1993) and transform into amoeboid shape with extensive cell processes toward the lesion site. It is the first-line defense that protects the CNS after traumatic injury or infection (Gaudet et al. 2018). However, this supposed beneficial mechanism when it is prolonged, becomes detrimental to neurons, with or without the chronic persistence of the injury or immunogen (Cherry, Olschowka, and Kerry O'Banion 2014). Activated microglia are further categorized into two different states: M1 and M2 states. In neuroinflammation, the dynamic between M1 and M2 polarization is crucial to determine the possible beneficial or detrimental outcomes (Cherry, Olschowka, and Kerry O'Banion 2014). M1 microglia are responsive to harmful stimuli and secrete pro-inflammatory cytokines such as TNF-α, IL-6, IL-1β, IFN-γ, and IL-23 that are known to kick-start the inflammation cascades. In contrast, M2 microglia express inflammation inhibitory cytokines, clear cellular debris through phagocytosis, and are capable of restoring homeostasis (Hu et al. 2015).

JAK-STAT signaling pathway is the canonical pathway in immune and inflammatory responses. It plays a vital role in microglial M1/M2 polarization (Hu et al. 2015). Extracellular signals such as IFN-γ, IL-2, IL-6, and TNF-α trigger M1 polarization in microglia, while IL-4, IL-10, and Galectin-1 stimulate M2 phenotype polarization. Members of the STAT family, including STAT1, STAT3, and STAT6, are the intracellular switches in microglia M1/M2 phenotype interchange. IFN-γ is found to induce M1 polarization via STAT1, while IL-13 and IL-4 stimulation polarize microglia to M2 phenotype via STAT6 (Wang et al. 2015). Upon IFN-γ binds to its receptor, JAK1/2 was activated, and STAT1 was phosphorylated, which subsequently trigger the expression of inflammatory genes such as iNOS and IL-12. In contrast, JAK1/JAK3 or JAK1/Tyk2 was activated by IL-4 or IL-13, respectively, which leads to phosphorylation of STAT6 and the release of anti-inflammatory cytokine including IL-10 and IL-1RA (Tugal, Liao, and Jain 2013; Orihuela, McPherson, and Harry 2016).

These microglial-secreted factors are acting through autocrine and paracrine manner to neurons and astrocyte. The interplay between these cells can either further damage or repair the injury site. Furthermore, cytokines (such as IFN-β and TNF-α) released from activated microglia are known to stimulate astrocytes to secrete more inflammatory mediators including CCL2, CXCL1, CXCL2, CXCL10, GM-CSF, and IL-6, as a secondary immune response (Pugazhenthi et al. 2013; Yan et al. 2018).

In response to pathological brain conditions, astrocytes exhibit cardinal features by becoming hypertrophied and expressed a high level of GFAP, vimentin, and S100β (Zamanian et al. 2012). This reactive state of astrocyte can be detrimental or beneficial to neuronal function, which depends on the reactive state of the astrocyte, A1 or A2 type. Neuroinflammation-induced astrocyte into A1 type led to downregulated genes such as *Gpc6*, *Sparcl1*, *Mertk*, and *Megf10* that are critical for promoting synapse formation and phagocytic capacity (Liddelow et al. 2017). A2 reactive astrocyte induced by ischemia upregulated many neurotrophic factors such as CLCF1, LIF, IL6, and thrombospondins, which promote survival and growth of neurons, and synapse repair (Zamanian et al. 2012; Liddelow et al. 2017).

Interestingly, activation of the JAK-STAT3 pathway has been reported as the central player in the induction of reactive astrocytes, suggesting the pathway as a common mediator of astrocyte reactivity was found highly conserved between disease states, species, and brain regions (Ben Haim and Ceyzeriat 2015). Mounting evidence suggests that the JAK2–STAT3 pathway serves as the critical regulator for astrocyte reactivity. See Figure 6.1 for a schematic illustration of microglia and astrocyte activation in response to pathological brain conditions.

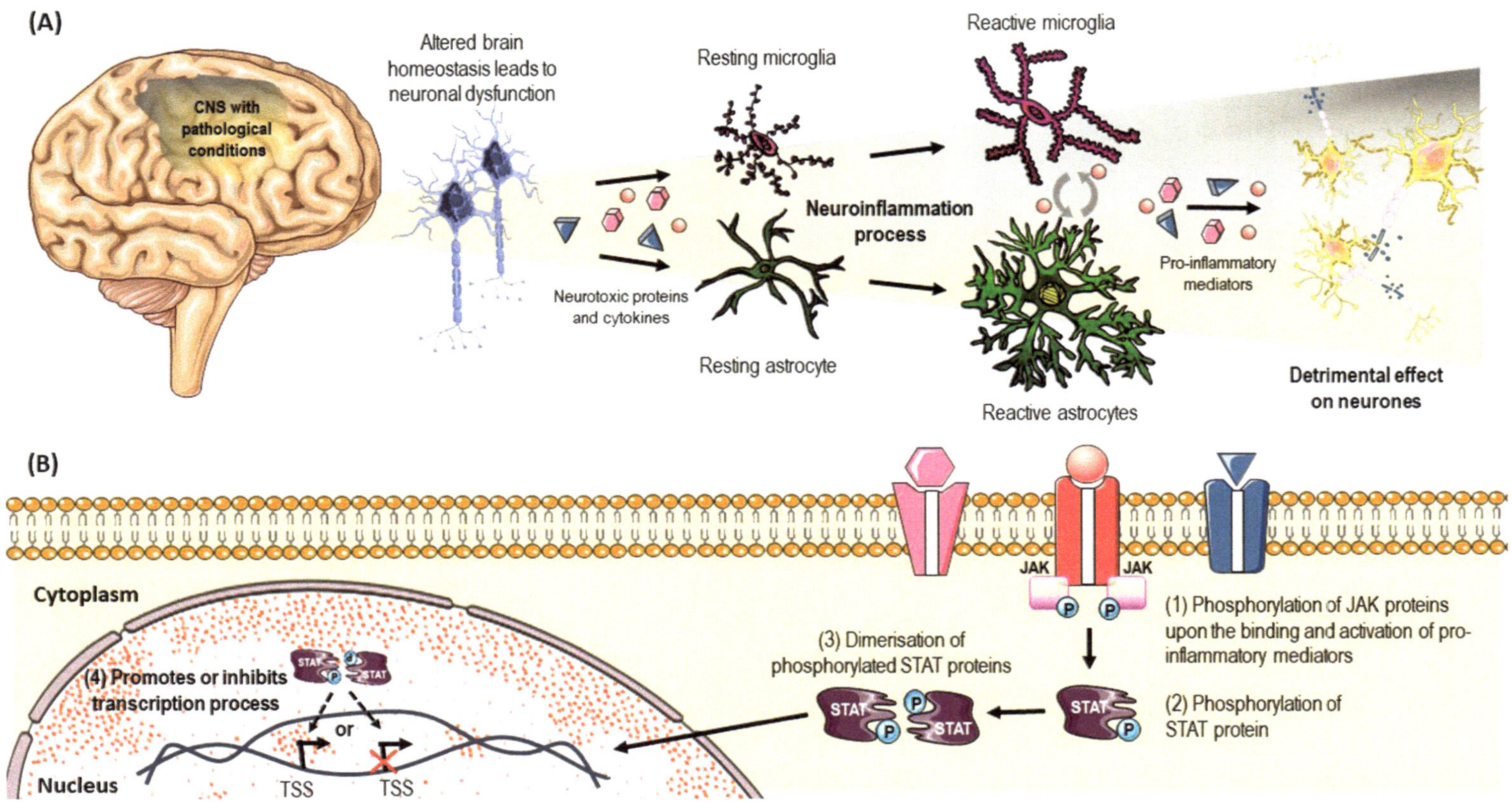

FIGURE 6.1 Microglia and astrocyte activation in response to pathological brain conditions. (A) Neuroinflammation process involving the activation of both resting microglia and astrocytes into M1/M2 and A1/A2 states, respectively. (B) Multistep cellular activation of a JAK-STAT signaling pathway in microglia, astrocytes, or neurons upon the activation by pro-inflammatory mediators. The activation of the JAK-STAT signaling pathway leads to the upregulation or downregulation of the transcription of targeted genes. TSS: transcription start site.

6.1.2 JAK-STAT Signaling Pathway in Alzheimer's Disease

Alzheimer's disease (AD) is a progressive neurodegenerative disorder, which has been associated with the extracellular deposition of neurotoxic β-amyloid (Aβ) plaques and intracellular neurofibrillary tangles (composed of hyperphosphorylated tau protein). There is ample evidence to show that the deposition of Aβ neuritic plaques activates both the astrocytes and microglia to secrete pro-inflammatory cytokines that are capable of inducing neuroinflammation (Lyman et al. 2014). Pro-inflammatory cytokines such as IL-1β, IL-6, IFN-γ, TNF-α, and TNF-β have been reported increased in human AD brain (Su, Bai, and Zhang 2016) and transgenic mouse models of AD (Benzing et al. 1999; Apelt and Schliebs 2001). These cytokines exert an inflammatory effect through activation of the JAK-STAT signaling pathway (Pugazhenthi et al. 2013; Rothaug, Becker-Pauly, and Rose-John 2016), particularly the JAK2–STAT3 pathway. It was reported that the activation of the JAK2–STAT3 pathway increased astrocyte reactivity, which in turn increased β-amyloid deposition (Ceyzeriat et al. 2018). The mechanism is still largely unknown. STAT3, however, has been associated with the β site APP cleaving enzyme 1 (BACE1), which is responsible for Aβ generation (Wen et al. 2008). Besides, the upregulation of STAT3 promotes inflammation genes expression such as iNOS and COX-2 and thus promotes neuroinflammation (Lee et al. 2017).

Cytokines of the interleukins family activate the JAK-STAT signaling pathway. Upon cytokine binding, the kinase JAK is activated and leads to STAT3 phosphorylation, which dimerizes, translocated to the nucleus, and, subsequently, regulates the transcription of its target genes. STAT3 activation in neurons is triggered in response to the increased release of TNF-α from surrounding activated glial cells (Wan et al. 2010; Carret-Rebillat et al. 2015). It is noteworthy that the upregulation of Tyk2 increases the phosphorylation of STAT3 and contributes to neuronal apoptosis (Wan et al. 2010). The JAK-STAT signaling pathway was found activated in microglial even before the first plaque deposition (Boza-Serrano et al. 2018). When amyloidogenesis continues and elevates Aβ plaque deposition, microglia and astrocyte can be activated via toll-like receptor (TLR) and receptor for advanced glycoxidation end products (RAGE)-dependent pathways. Subsequently, the activation of caspases and signal-dependent transcription factors, such as NF-kB and activator protein 1 (AP-1) in microglia and astrocytes, leading to the release of pro-inflammatory cytokines (Glass et al. 2010).

Furthermore, a pro-inflammatory cytokine, such as TNF-α, induced phosphorylation of STAT3, which potently bound to 14-3-3 epsilon gene promoter and increased the expression that could lead to more microglia activation (Xu et al. 2007; Tanabe et al. 2010; Zhang et al. 2012; Eufemi et al. 2015). Activated microglia also secrete several factors, IL-1α, TNFα, and complement component 1, q subcomponent (C1q), which can induce more astrocytes into the reactive state (Figure 6.2) (Liddelow et al. 2017). Indeed, activation of STAT3 promotes more microglia and astrocyte into a reactive state. Reactive astrocytes are observed close to amyloid depositions in both patients with AD (Probst, Ulrich and Heitz 1982) and mouse models of AD (Itagaki et al. 1989). Similar to microglia, upon activated, reactive astrocytes reorganized arborization with increased number and length of GFAP-positive processes (Wilhelmsson et al. 2006), polarize toward the site of amyloid plaques aggregation (Nagele et al. 2003). It is unknown whether inhibition of STAT3 reduces β-amyloid (Aβ) plaques aggregates. However, a study showed that suppression of STAT3 reduced the expression of Aβ-induced gene transcription, such as *iNOS*, which markedly attenuates neuronal cell death (Wan et al. 2010).

6.1.3 JAK-STAT Signaling Pathway in Parkinson's Disease

Parkinson's disease (PD) is an age-related and ranked as second most frequent chronic neurodegenerative disease worldwide (Darnell, Kerr, and Stark 1994; Troncoso-Escudero et al. 2018), typically presented with decreased motor activity. Loss of dopaminergic neurons (DA) in the substantia nigra pars compacta (SNc) and their projecting fibers in the striatum are observable

model, compared to the baseline level in naïve rats. Most importantly, the stereological analysis of tyrosine hydroxylase stained DA showed that AZD1480 attenuated the ~50% DA degeneration in the SNc of AAV2-α-syn-transduced rat model after 3 months (Qin et al. 2016). RNAseq analysis from the same sample also highlighted IFN-inducible genes, particularly *IFNAR, STAT1, IFNB1, IRF7*, and *TRIM24* that are regulated by the JAK-STAT signaling pathway in α-syn transduced group. This proposes that repressing microglia activation via the JAK-STAT signaling pathway might be beneficial to PD. It has been shown that *in vivo* LPS-induced microglia activation led to PD-like symptoms, including the aggregation of α-syn and it was IL-1 dependent (Sakai et al. 1995; Tanaka et al. 2013). Collectively, the JAK-STAT signaling pathway in neuroinflammation is a possible target to halt PD progression.

6.1.4 JAK-STAT Signaling Pathway in Huntington's Disease

Huntington's disease (HD) is a neurodegenerative disorder. It is characterized by progressive motor dysfunction, psychiatric disorders, and cognitive deficits (Ross and Tabrizi 2011). The disease is caused by a mutation in the huntingtin (*HTT*) gene (Träger et al. 2013; Crotti and Glass 2015). The CAG repeat expansion in the *HTT* gene encodes an expanded polyglutamine tract in the huntingtin protein (HTT), resulting in a mutant protein. Mutant HTT protein with 40 or more CAG repeats causes HD with full penetrance (MacDonald et al. 1993). While HD brain quintessentially shows selective loss of neurons in the striatum and cortex, neuroinflammation has been one of the leading clues to understand the pathogenesis of HD neurodegeneration. Notably, the expression of mutant HTT increased microglial activation (Kraft et al. 2012). Immunostaining of microglial marker, ionized calcium-binding adaptor molecule 1 (IBA1) in HD transgenic mice striatum showed an increase of reactive microglia compared to wild-type mice (Simmons et al. 2007). Furthermore, mutant HTT expressing microglia promotes not only cell-autonomous pro-inflammatory gene expression such as *IL-6*, but also triggers neighboring cells to secrete pro-inflammatory cytokine (Crotti et al. 2014). The upregulation of several pro-inflammatory cytokines, such as IL-6 and IL-8, was observed in the cerebrospinal fluid of HD patients (Bjorkqvist et al. 2008). In addition, IL-6 upregulation was also reported in the cortex, cerebellum, and striatum of HD patients (Silvestroni et al. 2009).

Increasing pro-inflammatory cytokine IL-6 potentially activates JAK-STAT3 signaling, which in turn promotes astrocytic activation (Ben Haim and Ceyzeriat 2015; Luo and Zheng 2016). The reactive astrocyte was observed in the striatum and cortex of patients with HD and HD-transgenic mouse models. Besides, the increase of mutant HTT expression also contributes to astrogliosis. A study showed that astrocytes were more hypertrophic and had increased STAT3 expression when the mutant HTT was delivered into the brain cells derived from mice and monkeys (Ben Haim and Ceyzeriat 2015). Bioinformatics analysis identified STAT1 and STAT2 as potentially involved in the progression of HD. Protein–protein interaction network analysis based on the differentially expressed genes (DEG) in the mutant huntingtin mouse cell line, STHdh111/111, has identified 18 hub genes, including STAT1 and STAT2, due to the high connectivity to the network; however, the study did not further investigate the role of these genes in HD (Dong and Cong 2018). Besides, the analysis also found a novel regulatory pathway (*miR-124*-STAT3-RNASE4) that could be a possible target for HD treatment (Dong and Cong 2018). RNASE4 has been reported to have a neuroprotective effect, and it was significantly upregulated in HD. MiRNA and transcription factor prediction showed that RNASE4 is regulated by *miR-124*-STAT3 complex (Li et al. 2013; Dong and Cong 2018). Taken together, the data showed that JAK-STAT signaling pathway is associated with the pathogenesis of HD; however, to what extent that the pathway involves in HD neuroinflammation remains unknown.

Immunostaining analysis on post-mortem brain tissue from patients with HD showed that many astrocytes in the brain were complement component 3 (C3)-positive, which is the specific marker for A1 astrocytes (Liddelow et al. 2017). A1-reactive astrocytes lost the normal function and secreted many cytokines and chemokines, such as IL-6, chemokine CXCL-1, and IL-8 (Liu et al. 2018). In consequence, this may attract circulating leukocytes to pass through the blood–brain barrier and contribute to the chronic inflammatory progression in the brain (Liebner et al. 2018). On the other hand, the elevation of pSTAT5 level in HD monocytes has been reported previously, which enhances the DNA-binding activity of NFκB leading to IL-6 production (Kawashima et al. 2001; Träger et al. 2013). STAT5 does not directly activate the IL-6; it binds to NFκB promoter and transactivates the promoter of IL-6, which has the NFκB-binding site (Kawashima et al. 2001). Notably, the increased level of plasma IL-6 has been detected in HD patients (Chang et al. 2015). Moreover, gene set enrichment analysis (GSEA) also found differential expression of the JAK-STAT signaling pathway in HD monocyte (Miller et al. 2016). Taken together, it is postulated that the elevation of circulating pro-inflammatory cytokine released by HD monocyte infiltration into the brain and leading to chronic inflammation. In brief, the existence of mutant HTT in the glial cells potentially induce neuroinflammation via JAK-STAT signaling pathway in HD brains, both locally and systemically.

6.1.5 JAK-STAT Signaling Pathway in Multiple Sclerosis and Experimental Autoimmune Encephalomyelitis

Multiple sclerosis (MS) is a demyelinating disease of the CNS with chronic neuroinflammation and neurodegeneration (Constantinescu et al. 2011). MS exhibits four pathological features, which are inflammation, demyelination, axonal loss, and gliosis. The pathological hallmark of MS is demyelination, which is the destruction of the myelin sheath or the oligodendrocyte cell body by the inflammatory process (Dendrou, Fugger, and Friese 2015). The etiology of MS remains unknown, but it is widely accepted that neuroinflammation, specifically auto-reactive lymphocytes, $CD4^+$ T cells target CNS components and initiate the disease (Goverman 2009). T helper type 1 (Thl) and T helper type 17 (Th17) lymphocytes produce IFN-γ and IL-17, respectively (Constantinescu et al. 2011). In MS patients, T cells are peripherally primed by the antigens, which possess molecular similarity with some CNS antigen (Sospedra and Martin 2005). Later they differentiate into Thl and Th17 cells that are capable of crossing the blood–brain barrier (BBB) and causing a series of myelin-targeted autoimmune neuroinflammation. Other immune cell types $CD8^+$ T cells, B cells, and myeloid cells are also involved in promoting lesion formation and neuronal damage (Yan et al. 2018).

Experimental autoimmune encephalomyelitis (EAE) is the animal model of MS. Auto-antigens, such as myelin basic protein, proteolipid protein, and preferably myelin oligodendrocyte glycoprotein (MOG), has been used to induce EAE in mice. Infiltrating macrophages and activated microglia are the common populations found on the lesion site (Yan et al. 2018), and the EAE model showed autoimmune response and MS-like clinical symptoms (Robinson et al. 2014). The M1/M2 polarization of microglia and infiltrating macrophage is determined by the environmental cytokine produced by Thl cells and Th17 cells in EAE. The summary of how JAK-STAT interplays between auto-reactive T cells and microglia polarization is shown in Figure 6.4. On the infiltrating macrophage, IFN-γ secreted by Thl cells promotes M1 polarization by activating STAT1 in the JAK-STAT signaling pathway, while GM-CSF produced by Th17 cells promotes M1 polarization via JAK2–STAT5 activation (Hamilton 2008). M1 macrophage polarization is a disastrous event for MS and EAE progression. This has been proven by a study that showed mice with GM-CSF receptor-deficient-$CCR2^+$ monocytes are completely unresponsive from EAE induction (Croxford et al. 2015). Subsequently, M1 macrophages produce IL-12 that polarize T cells into Thl cells via JAK2–TYK2 and STAT4 activation (Natarajan and Bright 2002; Krausgruber et al. 2011). Besides, IL-23 and IL-6 from M1 macrophages facilitate Th17 polarization by activating JAK1/2 and STAT3 in T cells (Samoilova et al. 1998; Cua et al. 2003). Thus, M1 macrophages are the kick starter of the vicious cycle in MS and EAE by activating JAK-STAT signaling pathway within. From a different perspective, it is also shown that macrophage and microglia are the most feasible candidates to halt the neuroinflammation in MS or EAE. Studies showed that external stimuli, such as IL-4 (Francos-Quijorna et al.

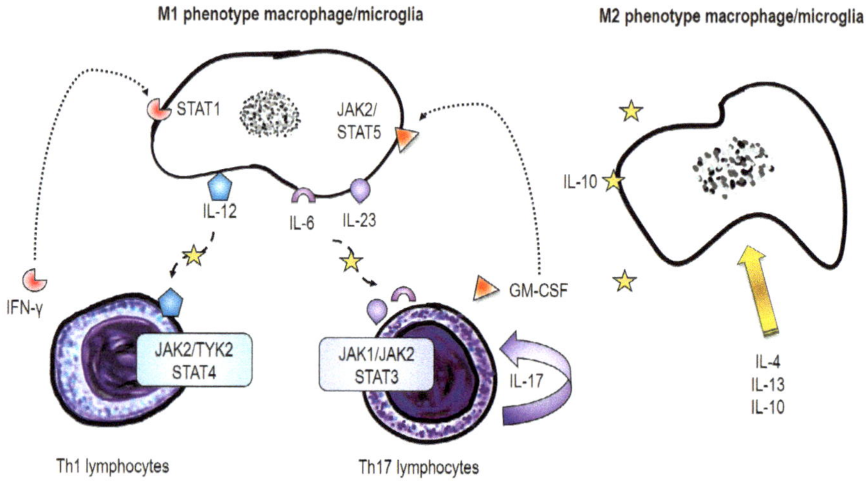

FIGURE 6.4 Summary of JAK-STAT expression in macrophage or microglia M1/M2 phenotype polarization and released factors in MS and EAE models.

T lymphocytes transformed into autoreactive T helper type I (Th1) and Th17 lymphocytes in MS. Th1 and Th17 release IFN-γ and GM-CSF, respectively. These two factors induced M1 phenotype via STAT1 and JAK2–STAT5 activation. Pro-inflammatory cytokines, such as IL-12, IL6, and IL-23, further activate T cells into Th1 and Th17 cells. IL-17 is Th17 autocrine factor. IL-4, IL-13, and IL-10 drive M2 phenotype and secrete IL-10 that are capable of inactivating Th1 and Th17 auto-activation.

2016; Liu et al. 2016), IL-13 (Rutschman et al. 2001; Wynn, Chawla, and Pollard 2013), and IL-10 (Lang et al. 2002; Gordon 2003), could drive them into M2 polarization. M2 macrophages and microglia exhibit strong anti-inflammatory reaction via an elevated level of IL-10 production (Moore et al. 2001; Lobo-Silva et al. 2016) reduced production of IL-23 (Correa et al. 2011) and IL-6 (Sanchez-Ventura et al. 2019). Furthermore, M2 macrophage-secreted IL-10 demonstrated a strong suppression on Th1 and Th17 phenotypes via STAT3 activation, which then alleviate EAE symptoms (Qin et al. 2012; Liu et al. 2013; Li et al. 2018; Lotfi et al. 2019). These showed that M2 macrophage and microglia are the pivots in relieving clinical symptoms of MS and EAE. In brief, the JAK-STAT signaling pathway and the responded cytokines or interleukins are utilized by both auto-reactive T cells and macrophages. It can be either deleterious or beneficial to the MS and EAE neuroinflammation, by tweaking the M1/M2 phenotype polarization and the environmental cytokines via the JAK-STAT signaling pathway.

Astrocytes are also responsive to environmental cytokines. They respond to IFN-β and secrete suppressors of cytokine signaling (SOCS) via activation of the JAK-STAT signaling pathway to negatively regulate immune cell infiltration, such as monocytes and $CD4^+$ T cells, into the CNS (Qin et al. 2008). IFN-β is a successful MS treatment as it reduces exacerbation rates and delays MS progression (Weinstock-Guttman et al. 1995; Dhib-Jalbut 1997; Hohlfeld and Wekerle 2004). IFN-β induced SOCS protein expression via STATs activation in primary murine astrocyte (Qin et al. 2008). The induction of SOCS-1 and SOCS-3 expression are dependent on STAT1α and STAT3, respectively. In the same study, SOCS-1 and SOCS-3 silenced-astrocytes show greater chemoattraction to bone marrow-derived primary macrophages and CD4+ T cells under IFN-β stimulation (Qin et al. 2008). This strengthens the importance of the JAK-STAT signaling pathway in ameliorating neuroinflammation in MS. The depletion of IFN-γ receptors on astrocytes has been shown to reduce the severity of EAE, which suggest that by ignoring IFN-γ detected in the niche may lower the chance to trigger neuroinflammation (Ding et al. 2015). Similar to M1 macrophages, the secretion of IL-6 by astrocytes in response to endoplasmic reticulum stress

(Meares et al. 2014) has been reported to be a determining factor of EAE neuroinflammation (Quintana et al. 2009; Giralt et al. 2013). IL-6 deficient mouse is completely resistant to EAE induced by MOG (Giralt et al. 2013). In the same study, the mouse with the only astrocyte produced IL-6 demonstrated demyelination, leukocytes infiltration, and astrogliosis, restricted within the cerebellar region, upon MOG stimulation. Like Th17 cells, it also produced GM-CSF during the EAE process (Mayo et al. 2014) that will polarize macrophage into M1 phenotype via JAK2–STAT5 (Hamilton 2008). Upregulated expression of lactosylceramide (LacCer) has been reported in the CNS of EAE mice, and it regulates the astrocytes transcriptional program (Mayo et al. 2014). In the same study, LacCer in astrocytes regulates the chemokine CCL2 and GM-CSF production in a non-cell autonomous manner, which modulates the recruitment of activated microglia and CNS-infiltrating monocytes. JAK-STAT in astrocytes is undeniably the central control system in switching between beneficial and detrimental functions during neuroinflammation. Thus, the JAK-STAT signaling pathway might be the candidate to manipulate astrocyte to counteract the progression of EAE and MS.

6.1.6 JAK-STAT Signaling Pathway in Down Syndrome

Very little is known about the neuroinflammation in the Down syndrome (DS) brain. However, inflammatory receptor genes are found overexpressed in both human and mouse models for DS (Ferrando-Miguel et al. 2003; Amano et al. 2004; Wilcock and Griffin 2013; Ling et al. 2014). Notably, pro-inflammatory receptor genes located on chromosome 21, including *IFNAR1*, *IFNAR2*, and *IFNGR2* are triplicated in DS samples (Wilcock and Griffin 2013), and they can activate JAK-STAT signaling pathway when interferons bind to the receptor. Once activated, the JAK-STAT signaling pathway promotes astrocytes differentiation (Bonni 1997). Besides, the analysis of protein expression level demonstrated that there was an approximately 2-fold increment of IFNAR2 expression in human fetal DS brain at 19–21 weeks of gestational age (Ferrando-Miguel et al. 2003). The increased expression levels of *Ifnar1*, *Ifnar2*, and *Il10rb* genes were also seen in a DS mouse model, Ts1Cje. These trisomic genes were overexpressed about 1.5-fold in Ts1Cje mouse whole brain (Amano et al. 2004). In addition to *Ifn* receptor, the downstream target, *STAT1*, was also found upregulated in the P84 Ts1Cje cerebellum and cerebral cortex (Ling et al. 2014). Furthermore, western blot analysis on fibroblast cell lines from individuals with trisomy 21 confirmed the upregulation of IFN receptor proteins (Sullivan et al. 2016).

The number of astrocytes is increased in human fetuses with DS. Immunohistochemical staining on the frontal lobe of DS human fetuses showed increased GFAP positive cells as compared to the age-matched controls (Zdaniuk et al. 2011). The finding was also consistent with previous studies on both human fetuses and Ts65Dn mice brain (another mouse model for DS) (Contestabile et al. 2007; Guidi et al. 2008). Astrocytes in DS are not only more proliferative and abundant, but they also display altered processes (Dossi, Vasile, and Rouach 2018).

In addition, pro-inflammatory cytokines, such as IL-6, TNF-α, and TGF-β, are well-known activators of the JAK-STAT signaling pathway. These cytokines were at least 2-fold higher in autopsied human DS brain tissues (<40 years old) when compared to age-matched controls. In contrast, it showed at least a 3-fold increase in Down syndrome with AD (>40 years old) (Wilcock et al. 2015). Besides, the level of IFN-γ was significantly increased in the whole brain of the trisomy 16 mouse (Hallam et al. 2000). The presence of the extra copy of IFN receptor and the elevation of IFN level have been postulated to sensitize the cells to interferon interaction and lead to activation of the JAK-STAT signaling pathway (Ling et al. 2014; Lee et al. 2016).

Furthermore, it has been indicated that TNF-α and IFN-γ bind to receptors on microglia and astrocytes (Benveniste and Benos 1995), which can activate both cell type and potently induces JAK-STAT signaling pathway in microglia and astrocytes. Activation of glial cells includes triggering of the JAK-STAT signaling pathway that could lead to neuroinflammation via the release of nitric oxide (NO) (Figure 6.5) (Kim et al. 2002; Lively and Schlichter 2018). A previous study reported that the gene expression level of inducible nitric oxide synthase (*iNOS*) that stimulates NO generation was higher in DS astroglia than in control astroglia (Chen et al. 2014).

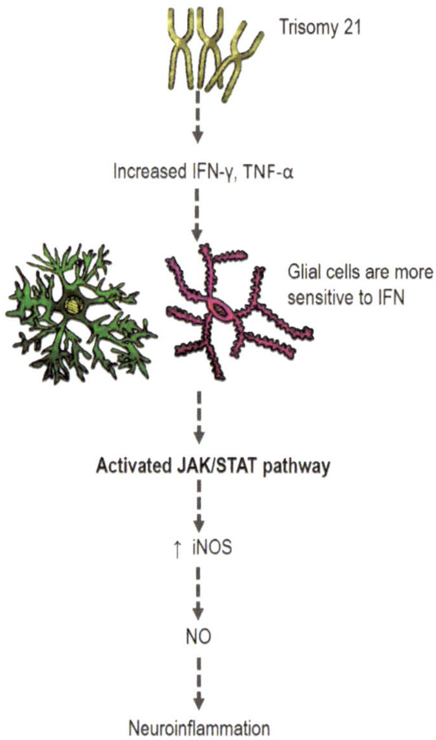

FIGURE 6.5 Neuroinflammation in Down Syndrome.

When TNF-α and IFN-γ bind to receptors on microglia and astrocytes, it is potently to activate JAK-STAT signaling pathway. Besides, an extra copy of IFN receptors on glial cells sensitizes the cells to interferon interaction, which further triggers the JAK-STAT signaling pathway. In consequence, the expression level of *iNOS* increased in and released NO, which leads to neuroinflammation.

In addition, increased of reactive oxygen species (ROS) production was observed in primary cultures of hippocampal neurons and astrocytes derived from Ts1Cje, a mouse model for DS (Shukkur et al. 2006). The expression of STAT1 and STAT3 were then highly induced in response to ROS (Simon et al. 1998). In consequence, chronic expression of STAT1 and STAT3 stimulate an inflammatory environment in the brain and lead to neuroinflammation. This could be through upregulation of downstream transcriptional targets of STATs, such as *Jmjd3*, *Ccl5*, *Ezr*, *Ifih1*, *Irf7*, *Uba7*, and *Pim1*, which are the genes that code for inflammatory proteins (Przanowski et al. 2014). JAK inhibitors or JAKinibs are currently a type of pharmacological approach that aids in inhibiting the activity of one or more enzymes of the JAK family, thereby interfering with the JAK-STAT signaling pathway. The application of JAKinibs, targeting astroglial and microglial activation states in Down syndrome, offers excellent therapeutic potential and warrant further investigation.

6.2 Concluding Remarks

The JAK-STAT signaling pathway in response to a wide range of cytokines is now known as one of the critical factors to promote neuroinflammation in neurodegenerative disease. However, it is conditional by whether it is pro-inflammatory cytokines or anti-inflammatory cytokines that trigger the JAK-STAT signaling pathway. In neurodegenerative disease, M1-reactive microglial or A1-reactive astrocyte secretes pro-inflammatory cytokines that create an environment conducive to inflammation. Cytokines act on the cells that secrete them, or nearby cells, and stimulate JAK-STAT signaling pathways. Activated JAK-STAT signaling leads to more microglia change into the reactive state as well as increase more pro-inflammatory cytokines and upregulated inflammatory gene expression. JAKinib is the inhibitor to one or more of the JAK family, thereby suppressing JAK-STAT signaling pathway that could reduce the releasing pro-inflammatory cytokines. However, inhibition of JAKs may potentially reduce anti-inflammatory response as the JAK-STAT signaling pathway also triggers by anti-inflammatory cytokines. The cytokines released in the microenvironment conditions the detrimental or beneficial effect; therefore, modulation of the balance between pro-inflammatory and anti-inflammatory cytokines, which would affect the function of the JAK-STAT signaling pathway, could be another way to reduce inflammation.

ACKNOWLEDGEMENTS

This work was supported by Malaysian Ministry of Science, Technology and Innovation Science-fund (Project ID: 02-01-04-SF2336) awarded to K.-H.L. and National Natural Science Foundation of China (Project ID: 81850410549) awarded to K.-L.T.

REFERENCES

Amano, K., H. Sago, C. Uchikawa, et al. 2004. Dosage-dependent over-expression of genes in the trisomic region of Ts1Cje mouse model for Down syndrome. *Hum Mol Genet* 13(13):1333–40.

Apelt, J., and R. Schliebs. 2001. Beta-amyloid-induced glial expression of both pro- and anti-inflammatory cytokines in cerebral cortex of aged transgenic Tg2576 mice with Alzheimer plaque pathology. *Brain Res* 894(1):21–30.

Ben Haim, L., and K. Ceyzeriat. 2015. The JAK–STAT3 pathway is a common inducer of astrocyte reactivity in Alzheimer's and Huntington's diseases. *J Neurosci* 35(6):2817–29.

Benzing, W. C., J. R. Wujek, E. K. Ward, et al. 1999. Evidence for glial-mediated inflammation in aged APP(SW) transgenic mice. *Neurobiol Aging* 20(6):581–89.

Benveniste, E. N., and D. J. Benos. 1995. TNF-alpha- and IFN-gamma-mediated signal transduction pathways: effects on glial cell gene expression and function. *Faseb J* 9(15):1577–84.

Bjorkqvist, M., E. J. Wild, J. Thiele, et al. 2008. A novel pathogenic pathway of immune activation detectable before clinical onset in Huntington's disease. *J Exp Med* 205(8):1869–77.

Bonni, A. 1997. Regulation of gliogenesis in the central nervous system by the JAK–STAT signaling pathway. *Science* 278(5337):477–83.

Boza-Serrano, A., Y. Yang, A. Paulus, and T. Deierborg. 2018. Innate immune alterations are elicited in microglial cells before plaque deposition in the Alzheimer's disease mouse model 5xFAD. *Sci Rep* 8 (1):1550.

Carret-Rebillat, A. S., C. Pace, S. Gourmaud, et al. 2015. Neuroinflammation and Abeta accumulation linked to systemic inflammation are decreased by genetic PKR down-regulation. *Sci Rep* 5:8489.

Ceyzeriat, K., L. Ben Haim, A. Denizot, et al. 2018. Modulation of astrocyte reactivity improves functional deficits in mouse models of Alzheimer's disease. *Acta Neuropathol Comm* 6(1):104.

Chang, K. H., Y. R. Wu, Y. C. Chen, and C. M. Chen. 2015. Plasma inflammatory biomarkers for Huntington's disease patients and mouse model. *Brain Behav Immun* 44:121–27.

Chen, C., P. Jiang, H. Xue, et al. 2014. Role of astroglia in Down's syndrome revealed by patient-derived human-induced pluripotent stem cells. *Nat Commun* 5:4430.

Chen, H., E. J. O'Reilly, M. A. Schwarzschild, and A. Ascherio. 2008. Peripheral inflammatory biomarkers and risk of Parkinson's disease. *Am J Epidemiol* 167(1):90–95.

Cherry, J. D., J. A. Olschowka, and M. Kerry O'Banion. 2014. Neuroinflammation and M2 microglia: the good, the bad, and the inflamed. *J Neuroinflammation* 11(1):98.

Chiba, T., M. Yamada, J. Sasabe, et al. 2009. Amyloid-beta causes memory impairment by disturbing the JAK2/STAT3 axis in hippocampal neurons. *Mol Psychiatry* 14(2):206–22.

Constantinescu, C. S., N. Farooqi, K. O'Brien, and B. Gran. 2011. Experimental autoimmune encephalomyelitis (EAE) as a model for multiple sclerosis (MS). *British J Pharmacol* 164 (4):1079–106.

Contestabile, A., T. Fila, C. Ceccarelli, et al. 2007. Cell cycle alteration and decreased cell proliferation in the hippocampal dentate gyrus and in the neocortical germinal matrix of fetuses with Down syndrome and in Ts65Dn mice. *Hippocampus* 17(8):665–78.

Correa, F., M. Hernangomez-Herrero, L. Mestre, F. Loria, F. Docagne, and C. Guaza. 2011. The endocannabinoid anandamide downregulates IL-23 and IL-12 subunits in a viral model of multiple sclerosis: evidence for a cross-talk between IL-12p70/IL-23 axis and IL-10 in microglial cells. *Brain Behav Immun* 25(4):736–49.

Crotti, A., C. Benner, B. E. Kerman, et al. 2014. Mutant Huntingtin promotes autonomous microglia activation via myeloid lineage-determining factors. *Nat Neurosci* 17(4):513–21.

Crotti, A., and C. K. Glass. 2015. The choreography of neuroinflammation in Huntington's disease. *Trends Immunol* 36(6):364–73.

Croxford, A. L., M. Lanzinger, F. J. Hartmann, et al. 2015. The cytokine GM-CSF drives the inflammatory signature of CCR2+ monocytes and licenses autoimmunity. *Immunity* 43(3):502–14.

Cua, D. J., J. Sherlock, Y. Chen, et al. 2003. Interleukin-23 rather than interleukin-12 is the critical cytokine for autoimmune inflammation of the brain. *Nature* 421(6924):744–48.

Darnell, J. E., Jr., I. M. Kerr, and G. R. Stark. 1994. Jak-STAT pathways and transcriptional activation in response to IFNs and other extracellular signaling proteins. *Science* 264(5164):1415–21.

Dendrou, C. A., L. Fugger, and M. A. Friese. 2015. Immunopathology of multiple sclerosis. *Nat Rev Immunol* 15:545.

Dhib-Jalbut, S. 1997. Mechanisms of interferon beta action in multiple sclerosis. *Mult Scler* 3(6):397–401.

Ding, X., Y. Yan, X. Li, et al. 2015. Silencing IFN-gamma binding/signaling in astrocytes versus microglia leads to opposite effects on central nervous system autoimmunity. *J Immunol* 194 (9):4251–64.

Dong, X., and S. Cong. 2018. Identification of differentially expressed genes and regulatory relationships in Huntington's disease by bioinformatics analysis. *Mol Med Rep* 17(3):4317–26.

Dossi, E., F. Vasile, and N. Rouach. 2018. Human astrocytes in the diseased brain. *Brain Res Bull* 136:139–56.

Eufemi, M., R. Cocchiola, D. Romaniello, et al. 2015. Acetylation and phosphorylation of STAT3 are involved in the responsiveness of microglia to beta amyloid. *Neurochem Int* 81:48–56.

Fellner, L., R. Irschick, K. Schanda, et al. 2013. Toll-like receptor 4 is required for α-synuclein dependent activation of microglia and astroglia. *Glia* 61(3):349–60.

Ferrando-Miguel, R., K. S. Shim, M. S. Cheon, M. Gimona, M. Furuse, and G. Lubec. 2003. Over-expression of Interferon α/β receptor β chain in fetal Down syndrome brain. *Neuroembryol Aging* 2 (4):147–55.

Francos-Quijorna, I., J. Amo-Aparicio, A. Martinez-Muriana, and R. Lopez-Vales. 2016. IL-4 drives microglia and macrophages toward a phenotype conducive for tissue repair and functional recovery after spinal cord injury. *Glia* 64(12):2079–92.

Galic, M. A., K. Riazi, and Q. J. Pittman. 2012. Cytokines and brain excitability. *Front Neuroendocrinol* 33(1):116–25.

Gao, H. M., Zhang, F., Zhou, H., Kam, W., Wilson, B., Hong, J. S. 2011. Neuroinflammation and α-synuclein dysfunction potentiate each other, driving chronic progression of neurodegeneration in a mouse model of Parkinson's disease. *Environ Health Perspect* 119(6):807–814.

Gaudet, A. D., L. K. Fonken, L. R. Watkins, R. J. Nelson, and P. G. Popovich. 2018. MicroRNAs: roles in regulating neuroinflammation. *The Neuroscientist* 24(3):221–45.

Giralt, M., R. Ramos, A. Quintana, et al. 2013. Induction of atypical EAE mediated by transgenic production of IL-6 in astrocytes in the absence of systemic IL-6. *Glia* 61(4):587–600.

Glass, C. K., K. Saijo, B. Winner, M. C. Marchetto, and F. H. Gage. 2010. Mechanisms underlying inflammation in neurodegeneration. *Cell* 140(6):918–34.

Gomez-Nicola, D., and V. H. Perry. 2015. Microglial dynamics and role in the healthy and diseased brain: a paradigm of functional plasticity. *Neuroscientist* 21(2):169–84.

Gordon, S. 2003. Alternative activation of macrophages. *Nat Rev Immunol* 3(1):23–35.

Gough, D. J., D. E. Levy, R. W. Johnstone, and C. J. Clarke. 2008. IFNgamma signaling-does it mean JAK–STAT? *Cytokine Growth Factor Rev* 19(5–6):383–94.
Goverman, J. 2009. Autoimmune T cell responses in the central nervous system. *Nat Rev Immunol* 9 (6):393–407.
Guidi, S., P. Bonasoni, C. Ceccarelli, et al. 2008. Neurogenesis impairment and increased cell death reduce total neuron number in the hippocampal region of fetuses with Down syndrome. *Brain Pathol* 18(2):180–97.
Hallam, D. M., N. L. Capps, A. L. Travelstead, G. J. Brewer, and L. E. Maroun. 2000. Evidence for an interferon-related inflammatory reaction in the trisomy 16 mouse brain leading to caspase-1-mediated neuronal apoptosis. *J Neuroimmunol* 110(1–2):66–75.
Hamilton, J. A. 2008. Colony-stimulating factors in inflammation and autoimmunity. *Nat Rev Immunol* 8 (7):533–44.
Hardy, J. A., and G. A. Higgins. 1992. Alzheimer's disease: the amyloid cascade hypothesis. *Science* 256 (5054):184–85.
Heneka, M. T., M. J. Carson, J. E. Khoury, et al. 2015. Neuroinflammation in Alzheimer's disease. *The Lancet Neurol* 14(4):388–405.
Hoenen, C., Gustin, A., Birck, C., Kirchmeyer, M., Beaume, N., Felten, P., Grandbarbe, L., Heuschling, P., Heurtaux, T. 2016. Alpha-Synuclein Proteins Promote Pro-Inflammatory Cascades in Microglia: Stronger Effects of the A53T Mutant. *PLoS One* 11(9):e0162717.
Hohlfeld, R., and H. Wekerle. 2004. Autoimmune concepts of multiple sclerosis as a basis for selective immunotherapy: from pipe dreams to (therapeutic) pipelines. *Proc Natl Acad Sci U S A* 101(Suppl 2): 14599–606.
Hristovska, I., and O. Pascual. 2016. Deciphering resting microglial morphology and process motility from a synaptic prospect. *Front Integrat Neurosci* 9:73–73.
Hu, X., R. K. Leak, Y. Shi, et al. 2015. Microglial and macrophage polarization-new prospects for brain repair. *Nat Rev Neurol* 11(1):56–64.
Itagaki, S., P. L. McGeer, H. Akiyama, S. Zhu, and D. Selkoe. 1989. Relationship of microglia and astrocytes to amyloid deposits of Alzheimer disease. *J Neuroimmunol* 24(3):173–82.
Jakel, S., and L. Dimou. 2017. Glial cells and their function in the adult brain: a journey through the history of their ablation. *Front Cell Neurosci* 11:24.
Kawanokuchi, J., T. Mizuno, H. Takeuchi, et al. 2006. Production of interferon-gamma by microglia. *Mult Scler* 12(5):558–64.
Kawashima, T., K. Murata, S. Akira, et al. 2001. STAT5 induces macrophage differentiation of M1 leukemia cells through activation of IL-6 production mediated by NF-kappaB p65. *J Immunol* 167 (7):3652–60.
Kim, O. S., E. J. Park, E. H. Joe, and I. Jou. 2002. JAK–STAT signaling mediates gangliosides-induced inflammatory responses in brain microglial cells. *J Biol Chem* 277(43):40594–601.
Kisseleva, T., S. Bhattacharya, J. Braunstein, and C. W. Schindler. 2002. Signaling through the JAK–STAT pathway, recent advances and future challenges. *Gene* 285(1–2):1–24.
Kraft, A. D., L. S. Kaltenbach, D. C. Lo, and G. J. Harry. 2012. Activated microglia proliferate at neurites of mutant huntingtin-expressing neurons. *Neurobiol Aging* 33(3):621.e17–33.
Krausgruber, T., K. Blazek, T. Smallie, et al. 2011. IRF5 promotes inflammatory macrophage polarization and TH1-TH17 responses. *Nat Immunol* 12(3):231–38.
Lang, R., D. Patel, J. J. Morris, R. L. Rutschman, and P. J. Murray. 2002. Shaping gene expression in activated and resting primary macrophages by IL-10. *J Immunol* 169(5):2253–63.
Laurence, A., M. Pesu, O. Silvennoinen, and J. O'Shea. 2012. JAK kinases in health and disease: an update. *Open Rheumatol J* 6:232–44.
Lee, D. Y., C. J. Hwang, J. Y. Choi, et al. 2017. KRICT-9 inhibits neuroinflammation, amyloidogenesis and memory loss in Alzheimer's disease models. *Oncotarget* 8(40):68654–67.
Lee, H. C., K. L. Tan, P. S. Cheah, and K. H. Ling. 2016. Potential role of JAK–STAT signaling pathway in the neurogenic-to-gliogenic shift in Down syndrome brain. *Neural Plast* 2016:7434191.
Li, S., J. Sheng, J. K. Hu, et al. 2013. Ribonuclease 4 protects neuron degeneration by promoting angiogenesis, neurogenesis, and neuronal survival under stress. *Angiogenesis* 16(2):387–404.
Li, X., L. Zhao, J.-J. Han, et al. 2018. Carnosol modulates Th17 cell differentiation and microglial switch in experimental autoimmune encephalomyelitis. *Front Immunol* 9:1807.

Liddelow, S. A., K. A. Guttenplan, L. E. Clarke, et al. 2017. Neurotoxic reactive astrocytes are induced by activated microglia. *Nature* 541(7638):481–87.

Liebner, S., R. M. Dijkhuizen, Y. Reiss, K. H. Plate, D. Agalliu, and G. Constantin. 2018. Functional morphology of the blood-brain barrier in health and disease. *Acta Neuropathol* 135(3):311–36.

Ling, K. H., C. A. Hewitt, K. L. Tan, et al. 2014. Functional transcriptome analysis of the postnatal brain of the Ts1Cje mouse model for Down syndrome reveals global disruption of interferon-related molecular networks. *BMC Genomics* 15:624.

Liu, C. Y., Y. Yang, W. N. Ju, X. Wang, and H. L. Zhang. 2018. Emerging roles of astrocytes in neuro-vascular unit and the tripartite synapse with emphasis on reactive gliosis in the context of Alzheimer's disease. *Front Cell Neurosci* 12:193.

Liu, C., Y. Li, J. Yu, et al. 2013. Targeting the shift from M1 to M2 macrophages in experimental autoimmune encephalomyelitis mice treated with fasudil. *PLOS One* 8(2):e54841.

Liu, X., J. Liu, S. Zhao, et al. 2016. Interleukin-4 is essential for microglia/macrophage M2 polarization and long-term recovery after cerebral ischemia. *Stroke* 47(2):498–504.

Lively, S., and L. C. Schlichter. 2018. Microglia responses to pro-inflammatory stimuli (LPS, IFNgamma+TNFalpha) and reprogramming by resolving cytokines (IL-4, IL-10). *Front Cell Neurosci* 12:215.

Lobo-Silva, D., G. M. Carriche, A. G. Castro, S. Roque, and M. Saraiva. 2016. Balancing the immune response in the brain: IL-10 and its regulation. *J Neuroinflammation* 13(1):297.

Lotfi, N., R. Thome, N. Rezaei, et al. 2019. Roles of GM-CSF in the pathogenesis of autoimmune diseases: an update. *Front Immunol* 10:1265.

Luo, Y., and S. G. Zheng. 2016. Hall of fame among pro-inflammatory cytokines: interleukin-6 gene and its transcriptional regulation mechanisms. *Front Immunol* 7:604.

Lyman, M., D. G. Lloyd, X. Ji, M. P. Vizcaychipi, and D. Ma. 2014. Neuroinflammation: the role and consequences. *Neurosci Res* 79:1–12.

MacDonald, M. E., C. M. Ambrose, M. P. Duyao, et al. 1993. A novel gene containing a trinucleotide repeat that is expanded and unstable on Huntington's disease chromosomes. *Cell* 72(6):971–83.

Maries, E., B. Dass, T. J. Collier, J. H. Kordower, and K. Steece-Collier. 2003. The role of α-synuclein in Parkinson's disease: insights from animal models. *Nat Rev Neurosci* 4:727.

Mayo, L., S. A. Trauger, M. Blain, et al. 2014. Regulation of astrocyte activation by glycolipids drives chronic CNS inflammation. *Nat Med* 20:1147.

Meares, G. P., Y. Liu, R. Rajbhandari, et al. 2014. PERK-dependent activation of JAK1 and STAT3 contributes to endoplasmic reticulum stress-induced inflammation. *Mol Cell Biol* 34(20):3911–25.

Miller, J. R., K. K. Lo, R. Andre, et al. 2016. RNA-Seq of Huntington's disease patient myeloid cells reveals innate transcriptional dysregulation associated with proinflammatory pathway activation. *Hum Mol Genet* 25(14):2893–904.

Moore, K. W., R. de Waal Malefyt, R. L. Coffman, and A. O'Garra. 2001. Interleukin-10 and the interleukin-10 receptor. *Annu Rev Immunol* 19:683–765.

Mount, M. P., A. Lira, D. Grimes, et al. 2007. Involvement of interferon-γ in microglial-mediated loss of dopaminergic neurons. *J Neurosci* 27(12):3328–37.

Na, Y. J., J. K. Jin, J. I. Kim, E. K. Choi, R. I. Carp, and Y. S. Kim. 2007. JAK–STAT signaling pathway mediates astrogliosis in brains of scrapie-infected mice. *J Neurochem* 103(2):637–49.

Nagele, R. G., M. R. D'Andrea, H. Lee, V. Venkataraman, and H. Y. Wang. 2003. Astrocytes accumulate A beta 42 and give rise to astrocytic amyloid plaques in Alzheimer disease brains. *Brain Res* 971 (2):197–209.

Nakagawa, Y., and K. Chiba. 2015. Diversity and plasticity of microglial cells in psychiatric and neurological disorders. *Pharmacol Ther* 154:21–35.

Natarajan, C., and J. J. Bright. 2002. Curcumin inhibits experimental allergic encephalomyelitis by blocking IL-12 signaling through Janus kinase-STAT pathway in T lymphocytes. *J Immunol* 168 (12):6506–13.

Nateghi Rostami, M., M. Douraghi, A. Miramin Mohammadi, and B. Nikmanesh. 2012. Altered serum pro-inflammatory cytokines in children with down's syndrome. *Eur Cytokine Netw* 23 (2):64–67.

O'Shea, J. J., and R. Plenge. 2012. JAK and STAT signaling molecules in immunoregulation and immune-mediated disease. *Immunity* 36(4):542–50.

Olanow, C. W., and W. G. Tatton. 1999. Etiology and pathogenesis of Parkinson's disease. *Annu Rev Neurosci* 22:123–44.

Orihuela, R., C. A. McPherson, and G. J. Harry. 2016. Microglial M1/M2 polarization and metabolic states. *Br J Pharmacol* 173(4):649–65.

Park, H. J., Oh, S. H., Kim, H. N., Jung, Y. J., Lee, P.H. 2016. Mesenchymal stem cells enhance α-synuclein clearance via M2 microglia polarization in experimental and human parkinsonian disorder. *Acta Neuropathol* 132(5):685–701.

Petrucelli, L., C. O'Farrell, P. J. Lockhart, et al. 2002. Parkin protects against the toxicity associated with mutant α-synuclein: proteasome dysfunction selectively affects catecholaminergic neurons. *Neuron* 36(6):1007–19.

Probst, A., J. Ulrich, and P. U. Heitz. 1982. Senile dementia of Alzheimer type: astroglial reaction to extracellular neurofibrillary tangles in the hippocampus. An immunocytochemical and electron-microscopic study. *Acta Neuropathol* 57(1):75–9.

Przanowski, P., M. Dabrowski, A. Ellert-Miklaszewska, et al. 2014. The signal transducers Stat1 and Stat3 and their novel target Jmjd3 drive the expression of inflammatory genes in microglia. *J Mol Med (Berl)* 92(3):239–54.

Pugazhenthi, S., Y. Zhang, R. Bouchard, and G. Mahaffey. 2013. Induction of an inflammatory loop by interleukin-1beta and tumor necrosis factor-alpha involves NF-kB and STAT-1 in differentiated human neuroprogenitor cells. *PLOS One* 8(7):e69585.

Qin, H., J. A. Buckley, L. Xinru, et al. 2016. Inhibition of the JAK–STAT pathway protects against α-synuclein-induced neuroinflammation and dopaminergic neurodegeneration. *J Neurosci* 36 (18):5144–59.

Qin, H., S. A. Niyongere, S. J. Lee, B. J. Baker, and E. N. Benveniste. 2008. Expression and functional significance of SOCS-1 and SOCS-3 in astrocytes. *J Immunol (Baltimore, MD: 1950)* 181(5):3167–76.

Qin, H., W.-I. Yeh, P. De Sarno, et al. 2012. Signal transducer and activator of transcription-3/suppressor of cytokine signaling-3 (STAT3/SOCS3) axis in myeloid cells regulates neuroinflammation. *Proc Natl Acad Sci* 109(13):5004–5009.

Quintana, A., M. Muller, R. F. Frausto, et al. 2009. Site-specific production of IL-6 in the central nervous system retargets and enhances the inflammatory response in experimental autoimmune encephalomyelitis. *J Immunol* 183(3):2079–88.

Rawlings, J. S., K. M. Rosler, and D. A. Harrison. 2004. The JAK–STAT signaling pathway. *J Cell Sci* 117 (Pt 8):1281–83.

Robinson, A. P., C. T. Harp, A. Noronha, and S. D. Miller. 2014. The experimental autoimmune encephalomyelitis (EAE) model of MS: utility for understanding disease pathophysiology and treatment. *Handbook Clin Neurol* 122:173–89.

Ross, C. A., and S. J. Tabrizi. 2011. Huntington's disease: from molecular pathogenesis to clinical treatment. *Lancet Neurol* 10(1):83–98.

Rothaug, M., C. Becker-Pauly, and S. Rose-John. 2016. The role of interleukin-6 signaling in nervous tissue. *Biochim Biophys Acta* 1863(6 Pt A):1218–27.

Rutschman, R., R. Lang, M. Hesse, J. N. Ihle, T. A. Wynn, and P. J. Murray. 2001. Cutting edge: Stat6-dependent substrate depletion regulates nitric oxide production. *J Immunol* 166(4):2173–77.

Saijo, K., B. Winner, C. T. Carson, et al. 2009. A Nurr1/CoREST pathway in microglia and astrocytes protects dopaminergic neurons from inflammation-induced death. *Cell* 137(1):47–59.

Sakai, K., T. Ohta, S. Minoshima, et al. 1995. Human ribosomal RNA gene cluster: identification of the proximal end containing a novel tandem repeat sequence. *Genomics* 26(3):521–26.

Samoilova, E. B., J. L. Horton, B. Hilliard, T. S. Liu, and Y. Chen. 1998. IL-6-deficient mice are resistant to experimental autoimmune encephalomyelitis: roles of IL-6 in the activation and differentiation of autoreactive T cells. *J Immunol* 161(12):6480–86.

Sanchez-Ventura, J., J. Amo-Aparicio, X. Navarro, and C. Penas. 2019. BET protein inhibition regulates cytokine production and promotes neuroprotection after spinal cord injury. *J Neuroinflammation* 16(1):124.

Schindler, C., D. E. Levy, and T. Decker. 2007. JAK–STAT signaling: from interferons to cytokines. *J Biol Chem* 282(28):20059–63.
Schulz-Schaeffer, W. J. 2010. The synaptic pathology of α-synuclein aggregation in dementia with Lewy bodies, Parkinson's disease and Parkinson's disease dementia. *Acta Neuropathologica* 120(2):131–43.
Sherer, T. B. 2011. Biomarkers for Parkinson's disease. *Sci Trans Med* 3(79):79ps14–79ps14.
Shukkur, E. A., A. Shimohata, T. Akagi, et al. 2006. Mitochondrial dysfunction and tau hyperphosphorylation in Ts1Cje, a mouse model for Down syndrome. *Hum Mol Genet* 15(18):2752–62.
Silvestroni, A., R. L. Faull, A. D. Strand, and T. Moller. 2009. Distinct neuroinflammatory profile in post-mortem human Huntington's disease. *Neuroreport* 20(12):1098–103.
Simmons, D. A., M. Casale, B. Alcon, N. Pham, N. Narayan, and G. Lynch. 2007. Ferritin accumulation in dystrophic microglia is an early event in the development of Huntington's disease. *Glia* 55 (10):1074–84.
Simon, A. R., U. Rai, B. L. Fanburg, and B. H. Cochran. 1998. Activation of the JAK–STAT pathway by reactive oxygen species. *Am J Physiol* 275(6 Pt 1):C1640–52.
Sospedra, M., and R. Martin. 2005. Immunology of multiple sclerosis. *Annu Rev Immunol* 23:683–747.
Streit, W. J., and M. B. Graeber. 1993. Heterogeneity of microglial and perivascular cell populations: insights gained from the facial nucleus paradigm. *Glia* 7(1):68–74.
Streit, W. J., R. E. Mrak, and W. S. T. Griffin. 2004. Microglia and neuroinflammation: a pathological perspective. *J Neuroinflammation* 1(1):14.
Su, F., F. Bai, and Z. Zhang. 2016. Inflammatory Cytokines and Alzheimer's disease: a review from the perspective of genetic polymorphisms. *Neurosci Bull* 32(5):469–80.
Subramaniam, S. R., and H. J. Federoff. 2017. Targeting microglial activation states as a therapeutic avenue in Parkinson's disease. *Front Aging Neurosci* 9:176.
Sullivan, K. D., H. C. Lewis, A. A. Hill, et al. 2016. Trisomy 21 consistently activates the interferon response. *Elife* 5:e16220.
Tanabe, K., R. Matsushima-Nishiwaki, S. Yamaguchi, H. Iida, S. Dohi, and O. Kozawa. 2010. Mechanisms of tumor necrosis factor-alpha-induced interleukin-6 synthesis in glioma cells. *J Neuroinflammation* 7:16.
Tanaka, S., A. Ishii, H. Ohtaki, S. Shioda, T. Yoshida, and S. Numazawa. 2013. Activation of microglia induces symptoms of Parkinson's disease in wild-type, but not in IL-1 knockout mice. *J Neuroinflammation* 10:143.
Träger, U., A. Magnusson, N. L. Swales, et al. 2013. JAK–STAT signalling in Huntington's disease immune cells. *PLOS Curr.* 5:ecurrents.hd.5791c897b5c3bebeed93b1d1da0c0648.
Troncoso-Escudero, P., A. Parra, M. Nassif, and R. L. Vidal. 2018. Outside in: unraveling the role of neuroinflammation in the progression of Parkinson's disease. *Front Neurol* 9:860.
Tugal, D., X. Liao, and M. K. Jain. 2013. Transcriptional control of macrophage polarization. *Arterioscler Thromb Vasc Biol* 33(6):1135–44.
von Bartheld, C. S., J. Bahney, and S. Herculano-Houzel. 2016. The search for true numbers of neurons and glial cells in the human brain: a review of 150 years of cell counting. *J Comp Neurol* 524(18):3865–95.
Wan, J., A. K. Fu, F. C. Ip, et al. 2010. Tyk2/STAT3 signaling mediates beta-amyloid-induced neuronal cell death: implications in Alzheimer's disease. *J Neurosci* 30(20):6873–81.
Wang, W. Y., M. S. Tan, J. T. Yu, and L. Tan. 2015. Role of pro-inflammatory cytokines released from microglia in Alzheimer's disease. *Ann Transl Med* 3(10):136.
Weinstock-Guttman, B., R. M. Ransohoff, R. P. Kinkel, and R. A. Rudick. 1995. The interferons: biological effects, mechanisms of action, and use in multiple sclerosis. *Ann Neurol* 37(1):7–15.
Wen, Y., W. H. Yu, B. Maloney, et al. 2008. Transcriptional regulation of beta-secretase by p25/cdk5 leads to enhanced amyloidogenic processing. *Neuron* 57(5):680–90.
Wilcock, D. M., and W. S. Griffin. 2013. Down's syndrome, neuroinflammation, and Alzheimer neuropathogenesis. *J Neuroinflammation* 10:84.
Wilcock, D. M., J. Hurban, A. M. Helman, et al. 2015. Down syndrome individuals with Alzheimer's disease have a distinct neuroinflammatory phenotype compared to sporadic Alzheimer's disease. *Neurobiol Aging* 36(9):2468–74.
Wilhelmsson, U., E. A. Bushong, D. L. Price, et al. 2006. Redefining the concept of reactive astrocytes as cells that remain within their unique domains upon reaction to injury. *Proc Natl Acad Sci U S A* 103 (46):17513–18.

Wynn, T. A., A. Chawla, and J. W. Pollard. 2013. Macrophage biology in development, homeostasis and disease. Nature 496(7446):445–55.

Xing-Long, G., C.-X. Long, L. Sun, C. Xie, X. Lin, and H. Cai. 2010. Astrocytic expression of Parkinson's disease-related A53T α-synuclein causes neurodegeneration in mice. *Molecular Brain* 3(1):12.

Xu, Y., M. Ikegami, Y. Wang, Y. Matsuzaki, and J. A. Whitsett. 2007. Gene expression and biological processes influenced by deletion of Stat3 in pulmonary type II epithelial cells. *BMC genomics* 8:455.

Yan, Z., S. A. Gibson, J. A. Buckley, H. Qin, and E. N. Benveniste. 2018. Role of the JAK–STAT signaling pathway in regulation of innate immunity in neuroinflammatory diseases. *Clin Immunol* 189:4–13.

Yoda, A., Y. Yoda, S. Chiaretti, et al. 2010. Functional screening identifies CRLF2 in precursor B-cell acute lymphoblastic leukemia. *Proc Natl Acad Sci U S A* 107(1):252–57.

Zamanian, J. L., L. Xu, L. C. Foo, et al. 2012. Genomic analysis of reactive astrogliosis. *J Neurosci* 32(18):6391–410.

Zdaniuk, G., T. Wierzba-Bobrowicz, G. M. Szpak, and T. Stepien. 2011. Astroglia disturbances during development of the central nervous system in fetuses with Down's syndrome. *Folia Neuropathol* 49(2):109–14.

Zhang, J., F. Chen, W. Li, et al. 2012. 14-3-3ζ interacts with Stat3 and regulates its constitutive activation in multiple myeloma cells. *PLOS One* 7(1):e29554.

Zhang, W., T. Wang, Z. Pei, et al. 2005. Aggregated alpha-synuclein activates microglia: a process leading to disease progression in Parkinson's disease. *Faseb J* 19(6):533–42.

Zhao, R., S. Xing, Z. Li, et al. 2005. Identification of an acquired JAK2 mutation in polycythemia vera. *J Biol Chem* 280(24):22788–92.

7

JAK-STAT Signaling in Cardiovascular Disease

George W. Booz
Department of Pharmacology and Toxicology
The University of Mississippi Medical Center
Jackson, MS

Raffaele Altara

Institute for Experimental Medical Research
Oslo University Hospital and University of Oslo
Oslo, Norway

KG Jebsen Center for Cardiac Research
Oslo, Norway

Department of Pathology, School of Medicine
University of Mississippi Medical Center
Jackson, MS

Sean P. Didion

Department of Pharmacology and Toxicology
The University of Mississippi Medical Center
Jackson, MS

Department of Neurology
The University of Mississippi Medical Center
Jackson, MS

7.1 Introduction

Janus kinase (JAK) and signal transducer and activator of transcription (STAT) comprise a key signaling pathway for a diverse array of stimuli, including a number of growth factors and cytokines. The JAK family of tyrosine kinases includes JAK1, JAK2, JAK3, and tyrosine kinase 2 (Tyk2). The STAT family of transcription factors upon which the JAKs act include STAT1, STAT2, STAT3, STAT4, STAT5A, STAT5B, and STAT6. Pharmacological inhibitors and genetically modified JAK-STAT mice have yielded important insight into the functional role of JAK-STAT signaling. While much of our knowledge of JAK-STAT signaling has been gained from the role of JAK-STAT in the pathophysiology of immunodeficiency and cancer, the role of JAK-STAT in the cardiovascular system in health and disease has only recently begun to emerge. The goal of this chapter is to highlight the role of JAK-STAT in the cardiovascular system in aging, atherosclerosis, diabetes, heart failure, hypertension, and stroke. We will focus our discussion on the roles of JAK1 and JAK2, and STAT1 and STAT3, which account for the majority of JAK and STAT activities in the cardiovascular system. Discussion will be primarily focused on the role of JAK-STAT signaling in the cardiovascular system, as well as some of the specific ligands and receptors pathways that utilize JAK-STAT and the detrimental effects associated with chronic activation of the JAK-STAT pathway. Areas of future research as well as some potential therapeutic approaches to target JAK-STAT in cardiovascular disease will also be briefly highlighted.

7.2 Canonical and Non-Canonical JAK-STAT Signaling

7.2.1 JAK and STAT Families

The JAK-STAT signaling pathway is an integral component of a number of cytokines and growth factors. JAKs are non-receptor tyrosine kinases that contain a kinase and pseudokinase domain, the latter of which plays an important regulatory role for the kinase domain. For that reason, JAKs are named after Janus, the Roman God of gates and passages, who is symbolized with two faces. Of the four mammalian JAK family members, JAK1 and JAK2, and to a lesser extent Tyk2 are expressed in cardiac and vascular cell types (Booz et al., 2002; Didion, 2017; Dutzmann et al., 2015). JAKs are also characterized by seven highly conserved JAK homology (JH) domains. The C-terminal region contains the JH1 domain with an autophosphorylation KE/DYY motif required for activation, as well as the JH2 domain that regulates kinase activity and is a docking site for STAT proteins. Originally thought to be a pseudokinase domain, JH2 has recently been shown to be a dual-specificity protein kinase that negatively regulates JH1 activity, at least for JAK2 (Hubbard, 2017; Ungureanu et al., 2011). The N-terminal regions are important for JAK-receptor interactions. Once activated, the JAKs phosphorylate receptor tyrosine residues that serve as docking sites to recruit STAT proteins, which are subsequently phosphorylated by the JAK kinases. Tyrosine phosphorylated STATs dimerize via intermolecular Src homology (SH2)-phosphotyrosine interactions, translocate to the nucleus, and initiate or suppress transcription of STAT target genes. JAK-STAT-mediated transcription is tightly regulated by a number of inhibitory processes including cytoplasmic and nuclear protein tyrosine phosphatases, as well as expression of suppressor of cytokine signaling (SOCS) and protein inhibitor of activated STAT (PIAS), both endogenous inhibitors of STAT (Chung et al., 1997; Krebs and Hilton, 2001; Shuai, 2006; Xu and Qu, 2008).

All seven mammalian STAT family members are reported to be expressed in both cardiac and vascular cells: STAT1, STAT2, STAT3, STAT4, STAT5a, STAT5b, and STAT6 (Booz et al., 2002). All STAT family members, except for STAT2 and STAT6 (and STAT3β), contain a serine residue within a conserved PMSP or PSP motif of the transcriptional activation domain (TAD), which serves a regulatory function. For instance, Ser727 in both STAT1 and STAT3 enhances its transcriptional activity, but may also participate in terminating DNA binding by recruiting a tyrosine phosphatase. It has also been reported that serine phosphorylated STAT1 and STAT3 may regulate gene expression independently of tyrosine phosphorylation (Booz et al., 2002). Certain G protein-coupled receptors, such as the angiotensin II type 1 receptor (AT1R), also signal, in part, through JAK-STAT (Booz et al., 2002). Tyrosine kinase and G protein-coupled receptors may also use Src kinase to phosphorylate STAT proteins depending upon cell type and the particular STAT isoform involved.

7.2.2 JAK-STAT Stimuli and Inhibitors

JAK-STAT signaling is activated by a large number of cytokines and growth factors that have important effects on the cardiovascular system, including interferon-α (IFN-α), IFN-γ, IL-3, IL-5, and IL-6 as well as angiotensin II and thrombin, as some respective examples (Darnell et al., 1994; Didion, 2017; Guschin et al., 1995; Kaptein et al., 1996; Madamanchi et al., 2001; Marrero, 2005; Marrero et al., 1995; Okuda et al., 1999) (Figure 7.1). In addition, mechanical stretch has been shown to activate JAK-STAT (Pan et al., 1999). Early studies using the specific JAK2 inhibitor, AG490, provided some of the first evidence for JAK-STAT signaling in the response of a number of cardiovascular cells types to various cytokines. However, AG490 is not particularly specific and is also a potent inhibitor of EGF receptor and ErbB2 tyrosine kinases (Booz et al., 2002). In light of the broad nature of JAK signaling, and difficulties surrounding the development of JAK isoform-specific inhibitors, development of JAK-STAT inhibitors has primarily focused on targeting STAT3. A number of small molecular inhibitors targeting the SH2 domain of STAT3 have been developed, including STATTIC and S31-201 (Schust et al., 2006; Siddiquee et al., 2007). These compounds disrupt STAT3 dimerization and decrease its DNA-binding and transcriptional activity albeit through slightly different mechanisms. One of the first-generation inhibitors of this kind was STATTIC (Schust et al., 2006). S3I-201, which is fairly specific for STAT3, is perhaps the most studied inhibitor of this kind recently. Various analogs of S31-201 of increased potency have been developed and some have entered clinical trials. However, since STAT3 may translocate to the nucleus and bind DNA without Y705 phosphorylation, (Zouein et al., 2015) targeting the SH2 domain has met with limited success due to the inability of current inhibitors to completely inhibit STAT3 activity.

For the IL-6 family of cytokines, signaling is terminated by recruitment of SOCS3 to the receptor subunit close to where the JAK is located, after phosphorylation of the docking residues by the activated JAKs (Kurdi and Booz, 2007) (Figure 7.2). SOCS3 inhibits JAK activity through its kinase inhibitory domain and, unlike SOCS1, shows low affinity for JAK1 or JAK2 in the absence of receptor binding. SOCS3 and SOCS1 are induced by IL-6-type cytokines downstream of STAT3 and/or STAT1, and thus serve as endogenous negative feedback inhibitors. Besides directly inhibiting JAK kinase activity, SOCS1 and SOCS3 may also terminate IL-6-type cytokine signaling by facilitating JAK2 ubiquitination and subsequent proteasomal degradation (Kile and Alexander, 2001). Ablation of cardiac SOCS3 appears to be protective against ischemia-reperfusion injury in the mouse by enhancing STAT3 and other protective signaling mechanisms (Nagata et al., 2015; Oba et al., 2012). On the other hand, SOCS3 deletion in cardiomyocytes was reported to aggravate DOCA-salt-induced hypertrophic

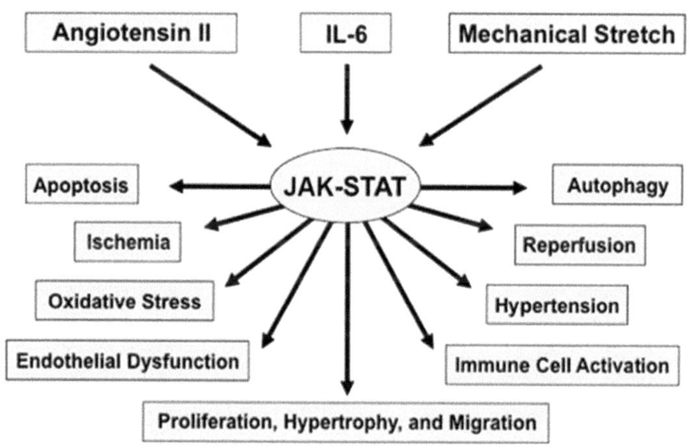

FIGURE 7.1 JAK-STAT signaling in the cardiovascular system is activated by a number of cytokines, such as IL-6, and growth factors, such as angiotensin II. Mechanical stretch, which is a functional component of cardiac and vascular muscle, also involves JAK-STAT signaling and target gene transcription. JAK-STAT signaling has been shown to contribute to cardiovascular disease, in part, by alterations in gene expression that contribute to a number of processes and phenotypes, including that shown above, but most particularly, endothelial dysfunction, immune cell activation, and cardiovascular hypertrophy.

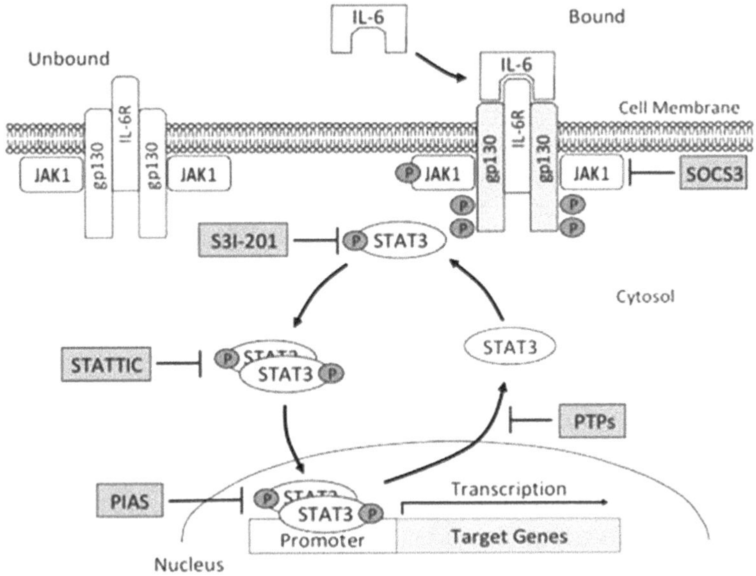

FIGURE 7.2 Interleukin-6 signaling via JAK-STAT in cardiovascular cells involves JAK1 and STAT3. IL-6R lacks intrinsic signaling capability; however, association of the IL-6R with the transmembrane protein, gp130, which contains docking site for the tyrosine kinase JAK1, mediates IL-6 signaling. JAK1 is constitutively associated with gp130 and upon IL-6 ligand/receptor interaction induces autophosphorylation and activation of JAK1. JAK1 then phosphorylates specific tyrosines residues within the cytoplasmic portion of gp130, which function as docking sites for STAT3 and phosphorylation of STAT3 by JAK1. Phosphorylated STAT3 dimerize, then translocates to the nucleus to increase or decrease transcription of STAT3 target genes. STAT3-induced transcription can be inhibited by SOCS3, PIAS, as well as cytosolic or nuclear phosphatases. STAT3-mediated transcription can also be inhibited pharmacologically by the small molecule inhibitors, STATTIC and S3I-201. IL-6, Interleukin-6; IL-6R, Interleukin-6 Receptor; JAK, Janus-associated Kinase; STAT, Signal Transducer and Activator of Transcription; SOCS, Suppressor Of Cytokine Signaling; PIAS, Protein Inhibitor of Signaling; PTPs, protein tyrosine phosphatases.

cardiac remodeling and was associated with age-related cardiomyopathy and ventricular arrhythmias; although a non-specific effect of prolonged cre-recombinase expression cannot be discounted (Liu et al., 2018; Pugach et al., 2015; Yajima et al., 2011). The harmful effects of SOCS3 deletion have been attributed to unbridled inflammatory signaling downstream of the IL-6 type cytokines.

7.2.3 JAK-STAT Transgenic/Knockout Mice

While pharmacological inhibitors of JAK-STAT have provided great insight into the role of JAK-STAT in the transduction of growth factor and cytokine signaling the development and availability of JAK- and STAT-deficient mice has revealed non-redundant roles of specific JAKs and STATs in embryonic development as well as signaling transduction and gene transcription. Global JAK- and STAT-deficient mouse models have been developed for each of the JAK and STAT isoforms (Baird et al., 1998; Durbin et al., 1996; Feldman et al., 1997; Kaplan et al., 1996a, 1996b; Liu et al., 1997; Meraz et al., 1996; Neubauer et al., 1998; Nosaka et al., 1995; Parganas et al., 1998; Park et al., 2000; Rodig et al., 1998; Shimoda et al., 1996; Smit et al., 1997; Takeda et al., 1997; Thomis et al., 1995; Udy et al., 1997). The discussion here, however, will be limited to those JAK and STAT isoforms most relevant to the cardiovascular system, namely JAK1, JAK2, JAK3, STAT1, and STAT3.

Global deficiency of JAK1 is perinatal lethal (Rodig et al., 1998). Despite the lethality associated with global JAK1 deficiency, *JAK1*$^{-/-}$ embryos appear grossly normal despite weighing 40% less than wild-type mice. *JAK1*$^{-/-}$ mice display normal erythrocyte, mononuclear, neutrophil, and platelets numbers (Rodig et al., 1998). *JAK1*$^{-/-}$ mice are, however, characterized by severely reduced B cell and thymocyte cell numbers reflective of primary defects in B cell differentiation and thymocyte production, respectively. Responses to type II cytokines, such as IFN-α and IFN-γ, are completely absent in *JAK1*$^{-/-}$ deficient cells, suggesting a selective role of JAK1 in IFN-α and IFN-γ signaling (Rodig et al., 1998). In contrast, responses to IL-6 in *JAK1*$^{-/-}$ deficient immune cells are diminished, but not completely abolished (Rodig et al., 1998). In order to circumvent some of the issues associated with lethality of global JAK1 deficiency, Cre-recombinase conditional JAK1-deficient mice have been developed (Sakamoto et al., 2016).

Global JAK2 deficiency is associated with lethality around embryonic days 10–13. Global *JAK2*$^{-/-}$ mice are characterized by severe anemia and the absence of red blood cells in the fetal liver, thus, it has been concluded that the lethality associated with loss of JAK2 expression is due to the lack of definitive erythropoiesis (Neubauer et al., 1998; Parganas et al., 1998). Global *JAK2*$^{-/-}$ mice display deficits in cytokine responses to the type II cytokine IFN-γ and no response to cytokines that utilize beta chain receptor signaling, such as thrombopoietin, IL-3, IL-5, and GM-CSF (Neubauer et al., 1998; Parganas et al., 1998). In contrast, cellular responses to IFN-α, IFN-γ, and IL-6 appear normal. Other than the absence of red blood cells, *JAK2*$^{-/-}$ mice display normal immune cell numbers, suggesting that JAK2 is not required for the development of lymphoid lineage progenitors. In order to circumvent the issue of lethality associated with global JAK2-deficiency, a number of conditional and cell-specific (including a cardiac-, endothelial-, and smooth muscle-cell specific) JAK2-deficient mouse models have been developed (Frenzel et al., 2006; Gan et al., 2015; Kirabo et al., 2011a, 2011b; Krempler et al., 2004; Park et al., 2013; Yang et al., 2013). For example, cell–specific-JAK2 knockout models have been used to investigate the role of JAK2 signaling in response to angiotensin II-induced hypertension and endothelial dysfunction (Kirabo et al., 2011a, 2011b).

Unlike JAK1 or JAK2 deficiency, JAK3-deficient mice are viable; however, they are characterized by a severe combined immune deficiency (SCID) phenotype (Nosaka et al., 1995; Thomis et al., 1995). The SCID-like phenotype is due to the fact that the expression of JAK3 is primarily restricted to immune cells, whereas JAK1 and JAK2 are ubiquitously expressed. The SCID-like phenotype of JAK3-deficient mice is primarily due to defects in IL-2, IL-4, and IL-7 signaling both of which utilize the common gamma chain receptor. Examination of T cell development suggests that discrete defects at the CD4-CD8-stage of T cell maturation are responsible for the loss of immature T cells through apoptosis (Baird et al., 1998). The lack of B cell precursors in bone marrow and the spleen and the lack of mature T cells highlight the essential role of JAK3 in the development of lymphoid cells.

Global STAT1 deficient mice are viable and demonstrate no overt developmental abnormalities; however, *Stat1*$^{-/-}$ mice are highly susceptible to pathogenic infection (Durbin et al., 1996; Meraz et al., 1996). *Stat1*$^{-/-}$ mice are completely nonresponsive to IFN-α and IFN-γ indicating that STAT1 is indispensable for interferon signaling. In contrast, *Stat1*$^{-/-}$ mice are characterized by normal, or near normal, responses to leukemia inhibitory factor, IL-6, platelet-derived growth factor, and epidermal growth factor suggesting that there is functional redundancy for these cytokines and growth factors, most likely provided for by STAT3. Interestingly, STAT1 and STAT3 show similar DNA-binding specificity and distinct patterns of binding to various transcriptional enhancer elements. Cardiac cell-specific STAT1 deficiency limits cardiac damage in response to myocardial infarction, in part, by enhancing autophagy (McCormick et al., 2012).

In contrast to STAT1, global STAT3 deficiency is lethal with fetal demise occurring at embryonic days 6–8 (Takeda et al., 1997). Thus, STAT3 expression is essential for early embryonic development. Cell-specific STAT3 deficient mice have been developed to circumvent the issue of embryonic lethality associated with the complete absence of STAT3 (Jacoby et al., 2003; Kano et al., 2003). For example, cardiac specific *Stat3*$^{-/-}$ mice have been found to be more sensitive to doxorubicin-induced heart failure, lipopolysaccharide-induced cardiac cell death, and an increased incidence of heart failure with normal aging, suggesting that STAT3 may play a protective role in the heart (Jacoby et al., 2003). Transgenic expression of a mutant STAT3 gene, SA/SA mice in which Ser727 phosphorylation site has been mutated, is associated with an approximately 50% reduction in STAT3 activity, as phosphorylation of Ser727 is required for maximal STAT3 transcriptional activity (Shen et al., 2004). SA/SA mice have been utilized to understand the role of angiotensin II in cardiac hypertrophy as discussed later on below (Zouein et al., 2013a).

7.3 JAK-STAT in Cardiovascular Aging

Aging is associated with an increased incidence of cardiovascular disease, including atherosclerosis, vascular cognitive impairment, hypertension, and stroke (Buford, 2016; Lo Coco et al., 2016; Wen and Wong, 2018). Blood pressure also increases with age at a rate of approximately 7 mmHg per decade of life over the age of 40 (Wolf-Maier et al., 2003). Interestingly, blood pressure has been found to correlate positively with increases in plasma IL-6 levels in healthy individuals (Chae et al., 2001). Moreover, plasma levels of IL-6 have been found to be significantly higher in aged as compared to young individuals (Maggio et al., 2006). Increases in plasma IL-6 are associated with increases in cardiovascular- and all-cause mortality in elderly populations (Li et al., 2017). In contrast, IL-2 levels have been found to decrease with age (Fulop et al., 2006). Thus, in addition to a global reduction in immune cell responses to pathogens, it appears that aging is also characterized by chronic, low-grade inflammation, and alterations in cytokine levels, particularly alterations in levels of cytokines that signal via JAK-STAT, such as IL-6.

Alterations in immune cell number and activity are altered with aging, such that cytokine-specific responses are reduced in geriatric patients (Longo et al., 2012; Shen-Orr et al., 2016). For example, aging is associated with reductions in T cell responses to IL-6, IFN-α, IFN-γ, and IL-21 and a diminished ability of these cytokines to phosphorylate STAT1, STAT3, and STAT5. (Shen-Orr et al., 2016). While the ability of cytokines to phosphorylate STAT is diminished with aging, this appear to be reflective of the fact that basal levels of phosphorylated STAT are increased in aging, due to a more chronic stimulation of JAK-STAT signaling. Initial reports suggest that inflammatory cytokines, such as IL-6, and STAT1 and STAT3 expression and phosphorylation are increased with age (O'Brown et al., 2015). It is this higher level of phosphorylated STAT that has been suggested to contribute to the greater incidence of atherosclerosis and diastolic dysfunction in aged humans (Shen-Orr et al., 2016). Interestingly, the response to pathogens is greatly diminished with aging, due, in part, to the fact that basal levels of phosphorylated STAT are higher with aging. Thus, the loss of the robustness of the response to cytokines that signal through JAK-STAT may be a major reason why older individuals are more prone to infection.

Despite the fact that JAK-STAT plays a major role in the regulation of immune cells, very little is known regarding the specific alterations that occur in the various immune cell populations with aging and cardiovascular disease. Interestingly, inhibition of JAK signaling has been shown to alleviate cellular senescent phenotypes and frailty commonly associated with aging (Xu et al., 2015). Although pharmacological inhibitors of JAK have provided some insight into the role of JAK-STAT signaling in aging, the role of JAK-STAT signaling in the cardiovascular system with age has not been fully explored and is thus an area that requires further study.

7.4 JAK-STAT in Atherosclerosis

JAK-STAT signaling has been implicated in various aspects of atherosclerosis in a temporal and cellular context. Principally, chronic activation of JAK-STAT signaling contributes to vascular inflammation and atherosclerosis initiation and progression. In endothelial cells from different vascular beds for instance, the IL-6 type cytokines induce via STAT3 the expression of various surface adhesion molecules, chemokines, and cytokines that are important for monocyte/macrophage recruitment (Alturkmani et al., 2012; van Keulen et al., 2018). Induction of SOCS3 in vascular endothelial cells by flavonoids by an ERK1/2-mediated pathway was shown to block IL-6-induced pro-inflammatory signaling (Wiejak et al., 2013). Endothelial cell STAT3 has also been implicated in atherosclerotic lesion development due to the critical importance of STAT3 in vascular cell proliferation and angiogenesis (Li et al., 2018). Lastly, the inflammatory response of platelets to oxidized low-density lipoprotein (ox-LDL) may involve activation of nuclear factor kappa B (NF-κB) and STAT3 (Sun et al., 2018).

STAT3 activation has been implicated as well in the proliferation, hypertrophy, and migration of vascular smooth muscle (Chen et al., 2014; Kirchmer et al., 2014; Wang et al., 2016). The dispersion of intimal smooth muscle cells may contribute to pathologic intimal thickening in human atherosclerosis (Nakagawa and Nakashima, 2018). STAT3 has also been observed to contribute to osteoblastic differentiation of vascular smooth muscle as seen in atherosclerosis, while STAT1 may play a role in vascular smooth muscle de-differentiation (Kakutani et al., 2015; Kirchmer et al., 2014).

Macrophages are key components of atheroma development. The role of STAT3 in macrophage phenotypes is complicated and not fully understood. Evidence that the IL-6 family member cardiotrophin-like cytokine factor 1 (CLCF1) increases macrophage-foam cell transition was recently implicated in this process with ruxolitinib, a JAK inhibitor (Pasquin et al., 2018). Accumulation of fat-laden pro-inflammatory M2 macrophages occurs with plaque development and may lead to a necrotic core that, with plaque rupture, forms a thrombus and precipitates an embolism. STAT3 activation was further implicated in the recruitment of additional macrophages through their formation of chemokine (C-C motif) ligand 2 (CCL2), *aka* monocyte chemoattractant protein 1 (MCP1) (Zhao et al., 2018). Minimally modified LDL-induction of CCL2 was also shown to involve enhancement of STAT3 promoter activity by mTORC1 (Ai et al., 2014). On the other hand, evidence was recently reported to support the conclusion that the anti-atherogenic actions of high-density lipoprotein (HDL) are attributable in part to sphingosine 1-phosphate (S1P)-mediated inhibition of macrophage apoptosis through induction of survivin expression by STAT3 (Feuerborn et al., 2017). Taken together, evidence suggests that JAK-STAT signaling within vascular cells and immune cells are key components of the atherosclerotic processes related to lesion development in the vascular wall.

7.5 JAK-STAT in Diabetes

With type 1 diabetes, levels and activity of STAT3 are reported to be reduced in cardiac and vascular cell types (Das et al., 2015; Drenger et al., 2011; Owais et al., 2015). As a consequence, the effectiveness of protective mechanisms such as ischemic post-conditioning in the heart is less effective (Drenger et al., 2011). In addition, both hyperglycemia and the generation of reactive oxygen species may impact ischemic post-conditioning and STAT3 activation by impairing adiponectin/caveolin-3 signaling in the heart (Li et al., 2016a; Wang et al., 2013). Others have implicated a loss of both PI3K/Akt and JAK2/STAT3 signaling due to increased cardiac PTEN activity, as the basis for ablation of post-conditioning in diabetic hearts (Xue et al., 2016). In diabetic leptin receptor null hearts (a model of type II diabetes), inhibition of mTORC1 with rapamycin, an inhibitor of mTOR, was reported to offer some protection from ischemia-reperfusion injury by restoring activation of STAT3, as well as Akt phosphorylation (Das et al., 2015). The critical importance of STAT3 in the actions of rapamycin was further demonstrated with a cardiomyocyte-specific STAT3 knockout model fed a high fat diet (Das et al., 2015). While reduced STAT3 activation is implicated in compromised protection of the diabetic heart from ischemia-reperfusion injury, a number of preclinical studies have implicated increased STAT3 activation in cardiac fibrosis in both type 1 and 2 diabetes (Chang et al., 2016; Fiaschi et al., 2014; Lo et al., 2017; Wang et al., 2015).

(Kirabo et al., 2011a). These studies highlight the contribution of JAK2 expression and activation within the vasculature to promote hypertension-related end-organ injury.

Arhgef1 is a RhoA-specific guanine exchange factor linked to enhanced rho kinase activity in hypertension. Activated Arhgef1 promotes activation of RhoA and downstream activations of Rho kinase and enhanced vasoconstriction associated with hypertension. Interestingly, JAK2 activation in response to angiotensin II is associated with phosphorylation and activation of Arhgef1, but not Arhgef11 or Arhgef12 (Guilluy et al., 2010). JAK inhibition with AG490, JAK2 siRNA, or a dominant negative JAK2 is associated with reduced JAK2 signaling and reduced angiotensin II-induced phosphorylation of Arhgef1 in cultured smooth muscle (Guilluy et al., 2010). Moreover, smooth muscle-specific Arhgef1 deficiency was associated with loss of Rho kinase activity as well as loss of the pressor response to angiotensin II (Guilluy et al., 2010).

Pharmacological inhibition of STAT3 signaling has been demonstrated to limit hypertension and endothelial dysfunction (Johnson et al., 2013). For example, endothelial dysfunction produced by angiotensin II was substantially reduced by S3I-201 and STATTIC, two small molecule inhibitors of STAT3 with slightly different mechanisms of action (Johnson et al., 2013). S3I-201 has high affinity for the SH2 domain of STAT3, but little affinity for STAT1 or STAT5 SH2 domains, thus selectively preventing STAT3 phosphorylation (Siddiquee et al., 2007). STATTIC directly inhibits STAT3 dimerization thereby directly inhibiting STAT3-mediated gene transcription (Schust et al., 2006). Interestingly, S3I-201 also prevented angiotensin II-induced increases in NADPH oxidase-derived superoxide (Johnson et al., 2013).

Like JAK2-deficient mice, homozygous STAT3 deficiency is associated with a lethal phenotype (Takeda et al., 1997). Thus, genetic insight into the role of STAT3 in hypertension has been through the use of STAT3-mutant and heterozygous STAT3-deficient mice (Johnson et al., 2013; Zouein et al., 2013a). Mutant SA/SA mice have reduced, but not complete absence of STAT3 protein or STAT3 activity (Shen et al., 2004). Basal blood pressure is similar in SA/SA mice as compared to wild-type mice, suggesting that phosphorylation of STAT3 at Ser727 does not contribute to the regulation of blood pressure (Zouein et al., 2013a). Similarly, the development of hypertension in response to angiotensin II is not blunted in SA/SA mice (Zouein et al., 2013a). Despite the general lack of effect of SA/SA on blood pressure, SA/SA appeared to be associated with a greater degree of myocardial fibrosis and myocyte cell death. These data are supportive of the concept that STAT3 plays more of a protective role in the heart.

As homozygous SOCS3 deficiency is also lethal, information regarding the role of SOCS3 deficiency has been garnered from studies involving pharmacological inhibition of SOCS3 or alternatively through the use of mice that are heterozygous deficient for SOCS3 (Li et al., 2016b). Baseline blood pressure under normal conditions is not altered by loss of a single SOCS3 gene (Li et al., 2016b). Surprisingly, heterozygous SOCS3 deficiency has no effect on the hypertension produced by angiotensin II infusion. However, relaxation to acetylcholine in the carotid artery was impaired in angiotensin II-, but not vehicle-, infused *SOCS3* -deficient mice (Li et al., 2016b). These findings suggest that loss of SOCS3 results in lower inhibition of SOCS3 on STAT3 signaling as revealed by enhanced vascular effects of angiotensin II in the presence of a single SOCS3 gene. The lack of effect of heterozygous SOCS3 deficiency on angiotensin-induced hypertension suggests that the protective effects of SOCS3 on STAT3 expression and activity are complex as reflected by the observed differences within the vascular wall and on arterial pressure (Li et al., 2016b).

7.8 JAK-STAT in Stroke

Vascular disease, such as carotid artery disease, is a major risk factor for ischemic stroke (Boehme et al., 2017; Gupta et al., 2013). Although much is known regarding the role of JAK-STAT in heart failure and hypertension, as discussed above, the role of JAK-STAT in stroke and functional recovery of neurons post-stroke is limited. Initial studies have found that STAT1 is activated in response to reactive oxygen species in neurons following ischemia, and deficiency of STAT1 was associated with reductions in cerebral infarct size and TUNEL positive staining and improved neurological scores (Takagi et al., 2002). Such findings suggest that STAT1 expression in neurons plays a functional role in cerebral injury following ischemia.

In contrast, and somewhat analogous to findings in the heart, STAT3 appears to confer protection against cerebral injury following ischemic stroke in several different animal models (Hu et al., 2017; Jung et al., 2009; Liang et al., 2016; Suzuki et al., 2001; Tang et al., 2018; Wang et al., 2017). For example, middle cerebral artery occlusion is associated with activation of JAK2 and STAT3, which was associated with neuroprotection and angiogenesis (Tang et al., 2018; Wang et al., 2017). The neuroprotective effects of JAK2 and STAT3 were associated with increases in epidermal growth factor receptor signaling, which could be inhibited with the JAK inhibitor, AG-490 (Tang et al., 2018). Neuronal cell death as a result of ischemia is associated with loss of STAT3 phosphorylation and neuroprotection due in part to loss of manganese superoxide dismutase, a STAT3 target gene (Jung et al., 2009).

Ischemic stroke produced by middle cerebral artery occlusion is associated with a significant increase in STAT3 phosphorylation within 6 hours of ischemia in cortical neurons, but not in astrocytes or microglia (Wen et al., 2001). More recently, it has been suggested that the protective and detrimental effects of JAK-STAT activation in stroke are most likely related of the temporal and spatial activation of the various STAT isoforms. For example, the degree of cerebral injury produced by middle cerebral artery occlusion may actually by reflective of a balance of increased STAT3 phosphorylation and reduced phosphorylation of STAT6 (Jang et al., 2014). Thus, it may be possible to pharmacologically manipulate the balance and sequence of STAT activation to limit the clinical impact associated with stroke. Future studies directed toward understanding the role of JAK-STAT in stroke are clearly warranted.

7.9 Conclusion

Based on the discussion provided here, it is readily apparent that JAK-STAT plays a critical role in a number of cardiovascular diseases. While cytokines and cytokine receptors play key roles in activation of JAK-STAT signaling, growth factors as well as G protein-coupled receptors, such as angiotensin II and AT1R are also capable of activating JAK-STAT, albeit through slightly different mechanisms of JAK-STAT recruitment and activation. JAK-STAT is associated with cardiac and vascular fibrosis and dysfunction. Similarly, JAK-STAT appears to contribute to increases in arterial blood pressure, which also can have marked effects on cardiovascular structure and function. While pharmacological inhibitors provided much insight into the functional effects of JAK-STAT signaling, more selective inhibitors of JAK-STAT are being developed and tested clinically. Similarly, genetically modified, cell-specific animals have begun to shed some insight into JAK-STAT in a number of cardiovascular disease models. Finally, while JAK-STAT gain- and loss-of-function polymorphisms have provided some insight into the role of JAK-STAT in immunity and cancer, to date, there is little information regarding the effects of JAK-STAT mutations on cardiovascular disease outcomes and is thus an important and attractive area for additional exploration.

REFERENCES

Ai, D., Jiang, H., Westerterp, M. et al. (2014). Disruption of mammalian target of rapamycin complex 1 in macrophages decreases chemokine gene expression and atherosclerosis. *Circ Res* 114:1576–1584.

Ali, M.S., Sayeski, P.P., Bernstein, K.E. (2000). JAK2 acts as both a STAT1 kinase and as a molecular bridge linking STAT1 to the angiotensin II AT1 receptor. *J Biol Chem* 275:15586–15593.

Ali, M.S., Sayeski, PP, Dirksen, L.B., Hayzer, D.J., Marrero, M.B., Bernstein, K.E. (1997). Dependence on the motif YIPP for the physical association of JAK2 kinase with the intracellular carboxy tail of the angiotensin II AT1 receptor. *J Biol Chem* 272:23382–23388.

Ali, M.S., Sayeski, P.P., Safavi, A., Lyles, M., Bernstein, K.E. (1998). Janus kinase 2 (JAK2) must be catalytically active to associate with the AT1 receptor in response to angiotensin II. *Biochem Biophys Res Comm* 249:672–677.

Alturkmani, H.J., Zgheib, C., Zouein, F.A., Alshaaer, N.E., Kurdi, M., Booz, G.W. (2012). Selenate enhances STAT3 transcriptional activity in endothelial cells: differential actions of selenate and selenite on LIF cytokine signaling and cell viability. *J Inorg Biochem* 109:9–15.

Baird, A.M., Thomis, D.C., Berg, L.J. (1998). T cell development and activation in JAK3-deficient mice. *J Leukoc Biol* 63:669–677.

Bhat, G.J., Baker, K.M. (1997). Angiotensin II stimulates rapid serine phosphorylation of transcription factor Stat3. *Mol Cell Biochem* 170:171–176.

Boehme, A.K., Esenwa, C., Elkind, M.S. (2017). Stroke risk factors, genetics, and prevention. *Circ Res* 120:472–495.

Bolli, R., Stein, A.B., Guo, Y. et al. (2011). A murine model of inducible, cardiac-specific deletion of STAT3: its use to determine the role of STAT3 in the upregulation of cardioprotective proteins by ischemic preconditioning. *J Mol Cell Cardiol* 50:589–597.

Booz, G.W. (2005). Putting the brakes on cardiac hypertrophy: exploiting the NO-cGMP counter-regulatory system. *Hypertension* 45:341–346.

Booz, G.W., Day, J.N., Baker, K.M. (2002). Interplay between the cardiac renin angiotensin system and JAK–STAT signaling: role in cardiac hypertrophy, ischemia/reperfusion dysfunction, and heart failure. *J Mol Cell Cardiol* 34:1443–1453.

Buford, T.W. (2016). Hypertension and aging. *Ageing Res Rev* 26:96–111.

Chae, C.U., Lee, R.T., Rifai, N., Ridker, P.M. (2001). Blood pressure and inflammation in apparently healthy men. *Hypertension* 38:399–403.

Chang, W.T., Cheng, J.T., Chen, Z.C. (2016). Telmisartan improves cardiac fibrosis in diabetes through peroxisome proliferator activated receptor delta (PPAR delta): from bedside to bench. *Cardiovasc Diabetol* 15:113.

Chen, D., Liu, J., Rui, B. et al. (2014). GSTpi protects against angiotensin II-induced proliferation and migration of vascular smooth muscle cells by preventing signal transducer and activator of transcription 3 activation. *Biochim Biophys Acta* 1843:454–463.

Chen, F., Chen, D., Zhang, Y. et al. (2017). Interleukin-6 deficiency attenuates angiotensin II-induced cardiac pathogenesis with increased myocyte hypertrophy. *Biochem Biophys Res Comm* 494:534–541.

Chung, C.D., Liao, J., Liu, B. et al. (1997). Specific inhibition of Stat3 signal transduction by PIAS3. *Science* 278:1803–1805.

Darnell, J.E. Jr., Kerr, I.M., Stark, G.R. (1994). JAK–STAT pathways and transcriptional activation in response to IFNs and other extracellular signaling proteins. *Science* 264:1415–1421.

Das, A., Salloum, F.N., Filippone, S.M. et al. (2015). Inhibition of mammalian target of rapamycin protects against reperfusion injury in diabetic heart through STAT3 signaling. *Basic Res Cardiol* 110:31.

Dasgupta, M., Unal, H., Willard, B., Yang, J., Karnik, S.S., Stark, G.R. (2014). Critical role for lysine 685 in gene expression mediated by transcription factor unphosphorylated STAT3. *J Biol Chem* 289:30763–30771.

Didion, S.P. (2017). Cellular and oxidative mechanisms associated with interleukin-6 signaling in the vasculature. *Int J Mol Sci* 18:e2563.

Didion, SP, Faraci, F.M. (2003). Angiotensin II produces superoxide-mediated impairment of endothelial function in cerebral arterioles. *Stroke* 34:2038–2042.

Didion, S.P., Kinzenbaw, D.A., Faraci, F.M. (2005). Critical role for CuZn-superoxide dismutase in preventing angiotensin II-induced endothelial dysfunction. *Hypertension* 46:1147–1153.

Didion, S.P., Kinzenbaw, D.A., Schrader, L.I., Chu, Y., Faraci, F.M. (2009). Endogenous interleukin-10 inhibits angiotensin II-induced vascular dysfunction. *Hypertension* 54:619–624.

Didion, S.P., Sigmund, C.D., Faraci, F.M. (2000). Impaired endothelial function in transgenic mice expressing both human renin and human angiotensinogen. *Stroke* 31:760–764.

Drazner, M.H. (2011). The progression of hypertensive heart disease. *Circulation* 123:327–334.

Drenger, B., Ostrovsky, I.A., Barak, M., Nechemia-Arbely, Y., Ziv, E., Axelrod, J.H. (2011). Diabetes blockade of sevoflurane postconditioning is not restored by insulin in the rat heart: phosphorylated signal transducer and activator of transcription 3- and phosphatidylinositol 3-kinase-mediated inhibition. *Anesthesiology* 114:1364–1372.

Durbin, J.E., Hackenmiller, R., Simon, M.C., Levy, D.E. (1996). Targeted disruption of the mouse Stat1 gene results in compromised innate immunity to viral disease. *Cell* 84:443–450.

Dutzmann, J., Daniel, J.M., Bauersachs, J., Hilfiker-Kleiner, D., Seding, D.G. (2015). Emerging translational approaches to target STAT3 signaling and its impact on vascular disease. *Cardiovasc Res* 106:365–374.

Enomoto, D., Obana, M., Miyawaki, A., Maeda, M., Nakayama, H., Fujio, Y. (2015). Cardiac-specific ablation of the STAT3 gene in the subacute phase of myocardial infarction exacerbated cardiac remodeling. *Am J Physiol Heart Circ Physiol* 309:H471–480.

Feldman, G.M., Rosenthal, L.A., Liu, X. et al. (1997). STAT5A-deficient mice demonstrate a defect in granulocyte-macrophage colony-stimulating factor-induced proliferation and gene expression. *Blood* 90:1768–1776.

Feuerborn, R., Becker, S., Poti, F. et al. (2017). High density lipoprotein (HDL)-associated sphingosine 1-phosphate (S1P) inhibits macrophage apoptosis by stimulating STAT3 activity and survivin expression. *Atherosclerosis* 257:29–37.

Fiaschi, T., Magherini, F., Gamberi, T. et al. (2014). Hyperglycemia and angiotensin II cooperate to enhance collagen I deposition by cardiac fibroblasts through a ROS-STAT3-dependent mechanism. *Biochim Biophys Acta* 1843:2603–2610.

Frank, G.D., Saito, S., Motley, E.D. et al. (2002). Requirement of Ca(2+) and PKCdelta for Janus kinase 2 activation by angiotensin II: involvement of PYK2. *Mol Endocrinol* 16:367–377.

Frenzel, K., Wallace, T.A., McDoom, I. et al. (2006). A functional JAK2 tyrosine kinase domain is essential for mouse development. *Exp Cell Res* 312:2735–2744.

Fulop, T., Larbi, A., Douziech, N., Levesque, I., Varin, A., Herbein, G. (2006). Cytokine receptor signalling and aging. *Mech Ageing Dev* 127:526–537.

Gan, X.T., Rajapurohitam, V., Xue, J. et al. (2015). Myocardial hypertrophic remodeling and impaired left ventricular function in mice with a cardiac-specific deletion of Janus kinase 2. *Am J Pathol* 185:3202–3210.

Gomolak, J.R., Didion, S.P. (2014). Angiotensin II-induced endothelial dysfunction is temporally linked with increases in interleukin-6 and vascular macrophage accumulation. *Front Physiol* 5:396.

Guilluy, C., Bregeon, J., Toumaniantz, G. et al. (2010). The rho exchange factor Arhgef1 mediates the effects of angiotensin on vascular tone and blood pressure. *Nature Med* 16:183–190.

Gupta, A., Baradaran, H., Schweitzer, A.D. et al. (2013). Carotid plaque MRI and stroke risk: a systematic review and meta-analysis. *Stroke* 44:3071–3077.

Guschin, D, Rogers, N, Briscoe, J et al. (1995). A major role for the protein tyrosine kinase JAK1 in the JAK/STAT signal transduction pathway in response to interleukin-6. *EMBO J* 14:1421–1429.

Harvey, A., Montezano, A.C., Lopes, R.A., Rios, F., Touyz, R.M. (2016). Vascular fibrosis in aging and hypertension: molecular mechanisms and clinical implications. *Can J Cardiol* 32:659–668.

Hilfiker-Kleiner, D., Hilfiker, A., Fuchs, M. et al. (2004). Signal transducer and activator of transcription 3 is required for myocardial capillary growth, control of interstitial matrix deposition, and heart protection from ischemic injury. *Circ Res* 95:187–195.

Hilfiker-Kleiner, D., Shukla, P., Klein, G. et al. (2010). Continuous glycoprotein-130-mediated signal transducer and activator of transcription-3 activation promotes inflammation, left ventricular rupture, and adverse outcome in subacute myocardial infarction. *Circulation* 122:145–155.

Hu, G.Q., Du, X., Li, Y.J., Gao, X.Q., Chen, B.Q., Yu, L. (2017). Inhibition of cerebral ischemia/reperfusion injury-induced apoptosis: nicotiflorin and JAK2/STAT3 pathway. *Neural Regener Res* 12:96–102.

Hubbard, S.R. (2017). Mechanistic insights into regulation of JAK2 tyrosine kinase. *Front Endocrinol (Lausanne)* 8:361.

Hussain, M., Awan, F.R. (2018). Hypertension regulating angiotensin peptides in the pathobiology of cardiovascular disease. *Clin Exp Hypertens* 40:344–352.

Jacoby, J.J., Kalinowski, A., Liu, M.G. et al. (2003). Cardiomyocyte-restricted knockout of STAT3 results in higher sensitivity to inflammation, cardiac fibrosis, and heart failure with advanced age. Pr*oc Natl Acad Sci USA* 100:12929–12934.

Jang, S.S., Choi, J.H., Im, D.S. et al. (2014). The phosphorylation of STAT6 during ischemic reperfusion in rat cerebral cortex. *Neuroreport* 25:18–22.

Johnson, A.W., Kinzenbaw, D.A., Modrick, M.L., Faraci, F.M. (2013). Small molecule inhibitors of STAT3 protect against angiotensin II-induced vascular dysfunction and hypertension. *Hypertension* 61:437–442.

Jung, J.E., Kim, G.S., Narasimhan, P., Song, Y.S., Chan, P.H. (2009). Regulation of Mn-superoxide dismutase activity and neuroprotection by STAT3 in mice after cerebral ischemia. *J Neurosci* 29:7003–7014.

Jusufovic, M., Sandset, E.C., Skagen, K., Skjelland, M. (2016). Blood pressure lowering treatment in patients with carotid artery stenosis. *Curr Hypertens Rev* 12:148–155.

Kakutani, Y., Shioi, A., Shoji, T. et al. (2015). Oncostatin M promotes osteoblastic differentiation of human vascular smooth muscle cells through JAK3-STAT3 pathway. *J Cell Biochem* 116:1325–1333.

Kano, A., Wolfgang, M.J., Gao, Q. et al. (2003). Endothelial cells require STAT3 for protection against endotoxin-induced inflammation. *J Exp Med* 198:1517–1525.

Kaplan, M.H., Schindler, U., Smiley, S.T., Grusby, M.J. (1996a). Stat6 is required for mediating responses to IL-4 and for development of Th2 cells. *Immunity* 4:313–319.

Kaplan, M.H., Sun, Y.L., Hoey, T., Grusby, M.J. (1996b). Impaired IL-12 responses and enhanced development of Th2 cells in Stat4-deficient mice. *Nature* 382:174–177.

Kaptein, A., Paillard, V., Saunders, M. (1996). Dominant negative STAT3 mutant inhibits interleukin-6-induced JAK–STAT signal transduction. *J Biol Chem* 271:5961–5964.

Kawai, T., Forrester, S.J., O'Brien, S., Baggett, A., Rizzo, V., Eguchi, S. (2017). AT1 receptor signaling pathways in the cardiovascular system. *Pharmacol Res* 125:4–13.

Kile, B.T., Alexander, W.S. (2001). The suppressors of cytokine signalling (SOCS). *Cell Mol Life Sci* 58:1627–1635.

Kirabo, A., Kearns, P.N., Jarajapu, Y.P. et al. (2011a). Vascular smooth muscle JAK2 mediates angiotensin II-induced hypertension via increased levels of reactive oxygen species. *Cardiovasc Res* 91:171–179.

Kirabo, A., Oh, S.P., Kasahara, H., Wagner, K.U., Sayeski, P.P. (2011b). Vascular smooth muscle JAK2 deletion prevents angiotensin II-mediated neointima formation following injury in mice. *J Mol Cell Cardiol* 50:1026–1034.

Kirchmer, M.N., Franco, A., Albasanz-Puig, A. et al. (2014). Modulation of vascular smooth muscle cell phenotype by STAT-1 and STAT-3. *Atherosclerosis* 234:169–175.

Krebs, D.L., Hilton, D.J. (2001). SOCS proteins: negative regulators of cytokine signaling. *Stem Cells* 19:378–387.

Krempler, A., Qi, Y., Triplett, A.A., Zhu, J., Rui, H., Wagner, K.U. (2004). Generation of a conditional knockout allele for the Janus kinase 2 (JAK2) gene in mice. *Genesis* 40:52–57.

Kunisada, K., Negoro, S., Tone, E. et al. (2000). Signal transducer and activator of transcription 3 in the heart transduces not only a hypertrophic signal but a protective signal against doxorubicin-induced cardiomyopathy. *Proc Natl Acad Sci USA* 97:315–319.

Kurdi, M., Booz, G.W. (2007). Can the protective actions of JAK–STAT in the heart be exploited therapeutically? Parsing the regulation of interleukin-6-type cytokine signaling. *J Cardiovasc Pharmacol* 50:126–141.

Lai, N.C., Gao, M.H., Tang, E. et al. (2012). Pressure overload-induced cardiac remodeling and dysfunction in the absence of interleukin 6 in mice. *Lab Invest* 92:1518–1526.

Li, H., Liu, W., Xie, J. (2017). Circulating interleukin-6 levels and cardiovascular and all-cause mortality in the elderly population: a meta-analysis. *Arch Gerontol Geriatr* 73:257–262.

Li, H., Yao, W., Liu, Z. et al. (2016a). Hyperglycemia abrogates ischemic postconditioning cardioprotection by impairing AdipoR1/Caveolin-3/STAT3 signaling in diabetic rats. *Diabetes* 65:942–955.

Li, S., Geng, Q., Chen, H. et al. (2018). The potential inhibitory effects of miR19b on vulnerable plaque formation via the suppression of STAT3 transcriptional activity. *Int J Mol Med* 41:859–867.

Li, Y., Kinzenbaw, D.A., Modrick, M.L., Pewe, L.L., Faraci, F.M. (2016b). Context-dependent effects of SOCS3 in angiotensin II-induced vascular dysfunction and hypertension in mice: mechanisms and role of bone marrow-derived cells. *Am J Physiol Heart Circ Physiol* 311:H146–56.

Liang, Z., Wu, G., Fan, C. et al. (2016). The emerging role of signal transducer and activator of transcription 3 in cerebral ischemic and hemorrhagic stroke. *Prog Neurobiol* 137:1–16.

Liu, S., Liu, L.X., Zhang, Y.L. et al. (2018). Cardiac ablation of SOCS3 aggravates DOCA-salt-induced hypertrophic remodeling by activation of gp130-dependent signaling in mice. *Cell Physiol Biochem* 47:140–150.

Liu, X., Robinson, G.W., Wagner, K.U., Garrett, L., Wynshaw-Boris, A., Hennighausen, L. (1997). Stat5a is mandatory for adult mammary gland development and lactogenesis. *Genes Dev* 11:179–186.

Lo Coco, D., Lopez, G., Corrao, S. (2016). Cognitive impairment and stroke in elderly patients. *Vasc Health Risk Manag* 12:105–116.

Lo, S.H., Hsu, C.T., Niu, H.S., Niu, C.S., Cheng, J.T., Chen, Z.C. (2017). Cryptotanshinone inhibits STAT3 signaling to alleviate cardiac fibrosis in type 1-like diabetic rats. *Phytother Res* 31:638–646.

Longo, D.M., Louie, B., Putta, S. et al. (2012). Single-cell network profiling of peripheral blood mononuclear cells from healthy donors reveals age- and race-associated differences in immune signaling pathway activation. *J Immunol* 188:1717–1725.

Madamanchi, N.R., Li, S., Patterson, C., Runge, M.S. (2001). Thrombin regulates vascular smooth muscle cell growth and heat shock proteins via the JAK–STAT pathway. *J Biol Chem* 276:18915–18924.

Maggio, M., Guralnik, J.M., Longo, D.L., Ferrucci, L. (2006). Interleukin-6 in aging and chronic disease: a magnificent pathway. *J Gerontol A Biol Sci Med Sci* 61:575–584.

Marrero, M.B. (2005). Introduction to JAK/STAT signaling and the vasculature. *Vascul Pharmacol* 43:307–309.

Marrero, M.B., Paxton, W.G., Schieffer, B., Ling, B.N., Bernstein, K.E. (1996). Angiotensin II signalling events mediated by tyrosine phosphorylation. *Cell Signal* 8:21–26.

Marrero, M.B., Schieffer, B., Li, B., Sun, J., Harp, J.B., Ling, B.N. (1997). Role of Janus kinase/signal transducer and activator of transcription and mitogen-activated protein kinase cascades in angiotensin II- and platelet-derived growth factor-induced vascular smooth muscle cell proliferation. *J Biol Chem* 272:24684–24690.

Marrero, M.B., Schieffer, B., Paxton, W.G. et al. (1995). Direct stimulation of JAK/STAT pathway by the angiotensin II AT1 receptor. *Nature* 375:247–250.

McCormick, J., Suleman, N., Scarabelli, T.M., Knight, R.A., Latchman, D.S., Stephanou, A. (2012). STAT1 deficiency in the heart protects against myocardial infarction by enhancing autophagy. *J Cell Mol Med* 16:386–393.

McWhinney, C.D., Hunt, R.A., Conrad, K.M., Dostal, D.E., Baker, K.M. (1997). The type I angiotensin II receptor couples to Stat1 and Stat3 activation through JAK2 kinase in neonatal rat cardiac myocytes. *J Mol Cell Cardiol* 29:2513–2524.

Mehta, P.K., Griendling, K.K. (2007). Angiotensin II cell signaling: physiological and pathological effects in the cardiovascular system. *Am J Physiol Cell Physiol* 292:C82–897.

Meraz, M.A., White, J.M., Sheehan, K.C. et al. (1996). Targeted disruption of the Stat1 gene in mice reveals unexpected physiologic specificity in the JAK–STAT signaling pathway. *Cell* 84:431–442.

Nagai, M., Hoshide, S., Kario, K. (2010). Hypertension and dementia. *Am J Hypertens* 23:116–124.

Nagata, T., Yasukawa, H., Kyogoku, S. et al. (2015). Cardiac-specific SOCS3 deletion prevents in vivo myocardial ischemia reperfusion injury through sustained activation of cardioprotective signaling molecules. *PLoS One* 10:e0127942.

Nakagawa, K., and Nakashima, Y. (2018). Pathologic intimal thickening in human atherosclerosis is formed by extracellular accumulation of plasma-derived lipids and dispersion of intimal smooth muscle cells. *Atherosclerosis* 274:235–242.

Neubauer, H., Cumano, A., Müller, M., Wu, H., Huffstadt, U., Pfeffer, K. (1998). JAK2 deficiency defines an essential developmental checkpoint in definitive hematopoiesis. *Cell* 93:397–409.

Nosaka, T., van Deursen, J.M., Tripp, R.A. et al. (1995). Defective lymphoid development in mice lacking JAK3. *Science* 270:800–802.

O'Brown, Z.K., Van Nostrand, E.L., Higgins, J.P., Kim, S.K. (2015). The inflammatory transcription factors NFκB, STAT1 and STAT3 drive age-associated transcriptional changes in the human kidney. *PLoS Genet* 11:e1005734.

Oba, T., Yasukawa, H., Hoshijima, M. et al. (2012). Cardiac-specific deletion of SOCS-3 prevents development of left ventricular remodeling after acute myocardial infarction. *J Am Coll Cardiol* 59:838–852.

Okuda, K., Foster, R., Griffin, J.D. (1999). Signaling domains of the beta c chain of the GM-CSF/IL-3/IL-5 receptor. *Ann N Y Acad Sci* 872:305–312.

Oliverio, M.I., Best, C.F., Kim, H.S., Arendshorst, W.J., Smithies, O., Coffman, T.M. (1997). Angiotensin II responses in AT1A receptor-deficient mice: a role for AT1B receptors in blood pressure regulation. *Am J Physiol* 272:F515–20.

Owais, K., Huang, T., Mahmood, F. et al. (2015). Cardiopulmonary bypass decreases activation of the signal transducer and activator of transcription 3 (STAT3) pathway in diabetic human myocardium. *Ann Thorac Surg* 100:1636–1645.

Pan, J., Fukuda, K., Saito, M. et al. (1999). Mechanical stretch activates the JAK/STAT pathway in rat cardiomyocytes. *Circ Res* 84:1127–1136.

Parganas, E., Wang, D., Stravopodis, D. et al. (1998). JAK2 is essential for signaling through a variety of cytokine receptors. *Cell* 93:385–395.

Park, C., Li, S., Cha, E., Schindler, C. (2000). Immune response in Stat2 knockout mice. *Immunity* 13:795–804.

Park, S.O., Wamsley, H.L., Bae, K. et al. (2013). Conditional deletion of JAK2 reveals an essential role in hematopoiesis throughout mouse ontogeny: implications for JAK2 inhibition in humans. *PLoS One* 8:e59675.

Pasquin, S., Laplante, V., Kouadri, S. et al. (2018). Cardiotrophin-like cytokine increases macrophage-foam cell transition. *J Immunol* 201:2462–2471.

Patel, P., Ordunez, P., DiPette, D. et al. (2016). Improved blood pressure control to reduce cardiovascular disease morbidity and mortality: The standardized hypertension treatment and prevention project. *J Clin Hypertens* 18:1284–1294.

Paulus, W.J., Tschope, C. (2013). A novel paradigm for heart failure with preserved ejection fraction: comorbidities drive myocardial dysfunction and remodeling through coronary microvascular endothelial inflammation. *J Am Coll Cardiol* 62:263–271.

Pugach, E.K., Richmond, P.A., Azofeifa, J.G., Dowell, R.D., Leinwand, L.A. (2015). Prolonged Cre expression driven by the alpha-myosin heavy chain promoter can be cardiotoxic. *J Mol Cell Cardiol* 86:54–61.

Recinos, A. III, LeJeune, W.S., Sun, H. et al. (2007). Angiotensin II induces IL-6 expression and the JAK–STAT3 pathway in aortic adventitia of LDL receptor-deficient mice. *Atherosclerosis* 194:125–133.

Rodig, S.J., Meraz, M.A., White, J.M. et al. (1998). Disruption of the JAK1 gene demonstrates obligatory and nonredundant roles of the JAKs in cytokine-induced biologic responses. Cell 93:373–383.

Rosendorff, C., Lackland, D.T., Allison, M. et al. (2015). Treatment of hypertension in patients with coronary artery disease: a scientific statement from the American heart association, American college of cardiology, and American society of hypertension. *Circulation* 131:e435–70.

Ryan, M.J., Didion, S.P., Mathur, S., Faraci, F.M., Sigmund, C.D. (2004). Angiotensin II-induced vascular dysfunction is mediated by the AT1A receptor in mice. *Hypertension* 43:1074–1079.

Sakamoto, K., Wehde, B.L., Rädler, P.D., Triplett, A.A., Wagner, K.U. (2016). Generation of janus kinase 1 (JAK1) conditional knockout mice. *Genesis* 54:582–588.

Sayeski, P.P., Ali, M.S., Bernstein, K.E. (2000). The role of Ca2+ mobilization and heterotrimeric G protein activation in mediating tyrosine phosphorylation signaling patterns in vascular smooth muscle cells. *Mol Cell Biochem* 212:91–98.

Sayeski, P.P., Ali, M.S., Frank, S.J., Bernstein, K.E. (2001). The angiotensin II-dependent nuclear translocation of Stat1 is mediated by the JAK2 protein motif 231YRFRR. *J Biol Chem* 276:10556–10563.

Sayeski, P.P., Ali, M.S., Safavi, A. et al. (1999). A catalytically active JAK2 is required for the angiotensin II-dependent activation of fyn. *J Biol Chem* 274:33131–33142.

Schrader, L.I., Kinzenbaw, D.A., Johnson, A.W., Faraci, F.M., Didion, S.P. (2007). IL-6 deficiency protects against angiotensin II induced endothelial dysfunction and hypertrophy. *Arterioscler Thromb Vasc Biol* 27:2576–2581.

Schust, J., Sperl, B., Hollis, A., Mayer, T.U., Berg, T. (2006). Stattic: a small-molecule inhibitor of STAT3 activation and dimerization. *Chem Biol* 13:1235–1242.

Shen, Y., Schlessinger, K., Zhu, X. et al. (2004). Essential role of STAT3 in postnatal survival and growth revealed by mice lacking STAT3 serine 727 phosphorylation. *Mol Cell Biol* 24:407–419.

Shen-Orr, S.S., Furman, D., Kidd, B.A. et al. (2016). Defective signaling in the JAK–STAT pathway tracks with chronic inflammation and cardiovascular risk in aging humans. *Cell Syst* 3:374–84.e4.

Shimoda, K., van Deursen, J., Sangster, M.Y. et al. (1996). Lack of IL-4-induced Th2 response and IgE class switching in mice with disrupted Stat6 gene. *Nature* 380:630–633.

Shuai, K. (2006). Regulation of cytokine signaling pathways by PIAS proteins. *Cell Res* 16:196–202.

Siddiquee, K., Zhang, S., Guida, W.C. et al. (2007). Selective chemical probe inhibitor of Stat3, identified through structure-based virtual screening, induces antitumor activity. *Proc Natl Acad Sci USA* 104:7391–7396.

Smit, L.S., Vanderkuur, J.A., Stimage, A. et al. (1997). Growth hormone-induced tyrosyl phosphorylation and deoxyribonucleic acid binding activity of Stat5A and Stat5B. *Endocrinology* 138:3426–3434.

Smith, R.M., Suleman, N., Lacerda, L. et al. (2004). Genetic depletion of cardiac myocyte STAT-3 abolishes classical preconditioning. *Cardiovasc Res* 63:611–616.

Su, S.A., Yang, D., Wu, Y. et al. (2017). EphrinB2 regulates cardiac fibrosis through modulating the interaction of Stat3 and TGF-beta/Smad3 signaling. *Circ Res* 121:617–627.

Sun, J., Zhang, M., Chen, K. et al. (2018). Suppression of TLR4 activation by resveratrol is associated with STAT3 and Akt inhibition in oxidized low-density lipoprotein-activated platelets. *Eur J Pharmacol* 836:1–10.

Suzuki, S., Tanaka, K., Nogawa, S. et al. (2001). Phosphorylation of signal transducer and activator of transcription-3 (Stat3) after focal ischemia in rats. *Exp Neurol* 170:63–71.

Szczepanek, K., Chen, Q., Derecka, M. et al. (2011). Mitochondrial-targeted Signal transducer and activator of transcription 3 (STAT3) protects against ischemia-induced changes in the electron transport chain and the generation of reactive oxygen species. *J Biol Chem* 286:29610–29620.

Szczepanek, K., Xu, A., Hu, Y. et al. (2015). Cardioprotective function of mitochondrial-targeted and transcriptionally inactive STAT3 against ischemia and reperfusion injury. *Basic Res Cardiol* 110:53.

Sztriha, L.K., Nemeth, D., Sefcsik, T., Vecsei, L. (2009). Carotid stenosis and the cognitive function. *J Neurol Sci* 283:36–40.

Takagi, Y., Harada, J., Chiarugi, A., Moskowitz, M.A. (2002). STAT1 is activated in neurons after ischemia and contributes to ischemic brain injury. *J Cereb Blood Flow Metabol* 22:1311–1318.

Takeda, K., Noguchi, K., Shi, W. et al. (1997). Targeted disruption of the mouse Stat3 gene leads to early embryonic lethality. *Proc Natl Acad Sci USA* 94:3801–3804.

Tang, Y., Tong, X., Li, Y. et al. (2018). JAK2/STAT3 pathway is involved in the protective effects of epidermal growth factor receptor activation against cerebral ischemia/reperfusion injury in rats. *Neurosci Lett* 662:219–226.

Te Riet, L., van Esch, J.H., Roks, A.J., van Den Meiracker, A.H., Danser, A.H. (2015). Hypertension: renin-angiotensin-aldosterone system alterations. *Circ Res* 116:960–975.

Thomis, D.C., Gurniak, C.B., Tivol, E., Sharpe, A.H., Berg, L.J. (1995). Defects in B lymphocyte maturation and T lymphocyte activation in mice lacking JAK3. *Science* 270:794–797.

Tschope, C., Van Linthout, S. (2014). New insights in (inter)cellular mechanisms by heart failure with preserved ejection fraction. *Curr Heart Fail Rep* 11:436–444.

Udy, G.B., Towers, R.P., Snell, R.G. et al. (1997). Requirement of STAT5b for sexual dimorphism of body growth rates and liver gene expression. *Proc Natl Acad Sci USA* 94:7239–7244.

Ungureanu, D., Wu, J., Pekkala, T. et al. (2011). The pseudokinase domain of JAK2 is a dual-specificity protein kinase that negatively regulates cytokine signaling. *Nat Struct Mol Biol* 18:971–976.

van Keulen, D., Pouwer, M.G., Pasterkamp, G. et al. (2018). Inflammatory cytokine oncostatin M induces endothelial activation in macro- and microvascular endothelial cells and in APOE*3Leiden.CETP mice. *PLoS One* 13:e0204911.

Verma, S.K., Krishnamurthy, P., Barefield, D. et al. (2012). Interleukin-10 treatment attenuates pressure overload-induced hypertrophic remodeling and improves heart function via signal transducers and activators of transcription 3-dependent inhibition of nuclear factor-kappaB. *Circulation* 126:418–429.

Wang, L., Li, J., Li, D. (2015). Losartan reduces myocardial interstitial fibrosis in diabetic cardiomyopathy rats by inhibiting JAK/STAT signaling pathway. *Int J Clin Exp Pathol* 8:466–473.

Wang, R., Zhang, Y., Xu, L. et al. (2016). Protein inhibitor of activated STAT3 suppresses oxidized LDL-induced cell responses during atherosclerosis in apolipoprotein E-deficient mice. *Sci Rep* 6:36790.

Wang, T., Mao, X., Li, H. et al. (2013). N-Acetylcysteine and allopurinol up-regulated the JAK/STAT3 and PI3K/Akt pathways via adiponectin and attenuated myocardial postischemic injury in diabetes. *Free Radic Biol Med* 63:291–303.

Wang, X.L., Qiao, C.M., Liu, J.O., Li, C.Y. (2017). Inhibition of the SOCS1-JAK2-STAT3 signaling pathway confers neuroprotection in rats with ischemic stroke. *Cell Physiol Biochem* 44:85–98.

Wen, S.W., Wong, C.H. (2018). Aging- and vascular-related pathologies. *Microcirculation* 30:e12463.

Traditional, non-pharmacological methods to increase mitochondrial activity is through physical exercise, subsequently increasing SKM mass (Yan et al., 2011). However, as noted previously, maintenance of exercise routines is not an effective long-term strategy for weight management in humans (Theofilou and Saborit, 2013). Therefore inducing molecular pathways mediating mitochondrial activity could be an alternative route to reduce fat mass. Because of the necessity for energy production, mitochondrial processes are tightly regulated to maintain homeostasis. STAT3 is the most widely studied STAT in regards to its effects on mitochondrial function. Indeed, STAT3 has been confirmed to be imported to the mitochondria and associate with complex I and II of the electron transport chain, seemingly to increase activity of ATP generation (Tammineni et al., 2013; Wegrzyn et al., 2009). Altogether, the study of STAT-mediated effects on mitochondrial activity and function remains an open field.

8.10 Potential Pharmacological Therapies

There are many pharmacological inhibitors/activators of JAKs and STATs that have been used in various experimental models. One such inhibitor, ruxolitinib, inhibits JAK1/2 and has been studied in clinical trials for the treatment of certain cancers. Itacitinib, another JAK1 selective inhibitor, has been used in clinical trials for cancer treatment as well (Beatty et al., 2018). But it is not known if ruxolitinib and/or itacitinib has effects on mediating obesity and/or diabetes. It does not seem likely, since knockout of JAK2 in adipose tissue results in an increase in fat expansion and a reduction in energy expenditure. Extensive studies, aimed at discovering novel pharmacological JAK/STAT activators, have been carried out (Tai et al., 2012). Researchers screened over 1,400 naturally derived compounds and only four were identified (emodin, quercetin, apigenin and luteolin) as having the ability to activate JAK/STAT signaling. However, it does not appear that any of these compounds have been widely studied and certainly not in context related to metabolic processes.

Recent studies have developed methodologies to screen small molecule inhibitors of STAT5B (de Araujo et al., 2019). Identifying inhibitors of STAT5B is of importance as potential therapies for obesity, as STAT5B is a stimulator of early adipogenesis. Conversely, STAT5B reduces lipogenesis in mature adipocytes, which could be a potential downfall to utilizing inhibitors. Another major hurdle to developing pharmacological mediators of JAK/STAT signaling is their ubiquitous expression and often contradictory effects in various tissues, according to rodent studies and in vitro studies.

8.11 Summary

It is inarguable that obesity is a growing and serious disease that which results in the progression of many metabolic diseases, such as T2D. It is known that lifestyle alterations can counteract obesity and its associated disease processes, but, as previously mentioned, humans have difficulty maintaining lifestyle modifications for the long term. In this regard, many metabolic signaling pathways have been studied for their potential to act as therapies for preventable diseases, such as obesity. JAK/STAT signaling is a pathway that is under investigation for the treatment of obesity and diabetes. Here it is shown that proper JAK/STAT signaling is relevant in various tissues (adipose, SKM, liver, pancreas, hypothalamus) for mediating metabolic processes. It is also clear that targeting JAK/STAT pathways for the treatment of obesity is a difficult endeavor, owing to the complex and contradictory signaling in various tissues. Further, utilizing genetically manipulated mouse models to examine signaling pathways does not always provide direct translation to human models. In this regard, further studies examining the role of JAK/STATs in a human context are warranted.

REFERENCES

Baik, M., Lee, M. S., Kang, H. J., Park, S. J., Piao, M. Y., Nguyen, T. H., & Hennighausen, L. 2017. Muscle-specific deletion of signal transducer and activator of transcription 5 augments lipid accumulation in skeletal muscle and liver of mice in response to high-fat diet. *European Journal of Nutrition*, *56*(2), 569–579. doi:10.1007/s00394-015-1101-0.

Bates, S. H., Stearns, W. H., Dundon, T. A., Schubert, M., Tso, A. W., Wang, Y., … Myers, M. G. 2003. STAT3 signalling is required for leptin regulation of energy balance but not reproduction. *Nature*, *421*, 856–859.

Beatty, G. L., Shahda, S., Beck, T., Uppal, N., Cohen, S. J., Donehower, R., … Von Hoff, D. D. 2018. A phase Ib/II study of the JAK1 inhibitor, itacitinib, plus nab-paclitaxel and gemcitabine in advanced solid tumors. *The Oncologist*. doi:10.1634/theoncologist.2017-0665.

Biener, A., Cawley, J., & Meyerhoefer, C. 2017. The high and rising costs of obesity to the US health care system. *Journal of General Internal Medicine*, *32*(Suppl 1), 6–8. doi:10.1007/s11606-016-3968-8.

Bjørbæk, C., Uotani, S., Da Silva, B., & Flier, J. S. 1997. Divergent signaling capacities of the long and short isoforms of the leptin receptor. *Journal of Biological Chemistry*, *272*(51), 32686–32695. doi:10.1074/jbc.272.51.32686.

Bódis, K., & Roden, M. 2018. Energy metabolism of white adipose tissue and insulin resistance in humans. *European Journal of Clinical Investigation*, *0*(0), e13017. doi:10.1111/eci.13017.

Buettner, C., Pocai, A., Muse, E. D., Etgen, A. M., Myers, M. G., & Rossetti, L. 2006. Critical role of STAT3 in leptin's metabolic actions. *Cell Metab*, *4*(1), 49–60.

Cannon, B., & Nedergaard, J. A. N. 2004. Brown adipose tissue: Function and physiological significance. *Physiological Reviews*, *84*(1), 277–359. doi:10.1152/physrev.00015.2003.

Cernkovich, E. R., Deng, J., Bond, M. C., Combs, T. P., & Harp, J. B. 2008. Adipose-specific disruption of signal transducer and activator of transcription 3 increases body weight and adiposity. *Endocrinology*, *149*(4), 1581–1590.

Choi, D., Cai, E. P., & Woo, M. 2011. The redundant role of JAK2 in regulating pancreatic β-cell mass. *Islets*, *3*, 389–392.

Cui, Y., Hosui, A., Sun, R., Shen, K., Gavrilova, O., Chen, W., … Hennighausen, L. 2007. Loss of signal transducer and activator of transcription 5 leads to hepatosteatosis and impaired liver regeneration. *Hepatology*, *46*(2), 504–513. doi:10.1002/hep.21713.

Cui, Y., Huang, L., Elefteriou, F., Yang, G., Shelton, J. M., Giles, J. E., … Li, C. 2004. Essential role of STAT3 in body weight and glucose homeostasis. *Molecular and Cellular Biology*, *24*(1), 258–269. doi:10.1128/MCB.24.1.258-269.2004.

Darnell, J. E. 1997. STATs and gene regulation. *Science*, *277*(5332), 1630 LP–1635. Retrieved from http://science.sciencemag.org/content/277/5332/1630.abstract.

de Araujo, E. D., Manaswiyoungkul, P., Erdogan, F., Qadree, A. K., Sina, D., Tin, G., … Gunning, P. T. 2019. A functional in vitro assay for screening inhibitors of STAT5B phosphorylation. *Journal of Pharmaceutical and Biomedical Analysis*, *162*, 60–65. doi:10.1016/j.jpba.2018.08.036.

De Souza, C. T., Araujo, E. P., Bordin, S., Ashimine, R., Zollner, R. L., Boschero, A. C., … Velloso, L. A. 2005. Consumption of a fat-rich diet activates a proinflammatory response and induces insulin resistance in the hypothalamus. *Endocrinology*, *146*(10), 4192–4199. doi:10.1210/en.2004-1520.

Delahanty, L. M. 2002. Evidence-based trends for achieving weight loss and increased physical activity: Applications for diabetes prevention and treatment. *Diabetes Spectrum*, *15*(3), 183 LP–189. Retrieved from http://spectrum.diabetesjournals.org/content/15/3/183.abstract.

Deng, J., Hua, K., Caveney, E. J., Takahashi, N., & Harp, J. B. 2006. Protein inhibitor of activated STAT3 inhibits adipogenic gene expression. *Biochemical and Biophysical Research Communications*, *339*(3), 923–931. doi:10.1016/j.bbrc.2005.10.217.

Deng, J., Hua, K., Lesser, S. S., Greiner, A. H., Walter, A. W., Marrero, M. B., & Harp, J. B. 2000. Interleukin-4 mediates STAT6 activation in 3T3-L1 preadipocytes but not adipocytes. *Biochemical and Biophysical Research Communications*, *267*(2), 516–520. doi:10.1006/bbrc.1999.1993.

Derecka, M., Gornicka, A., Koralov, S. B., Szczepanek, K., Morgan, M., Raje, V., … Larner, A. C. 2012. Tyk2 and Stat3 regulate brown adipose tissue differentiation and obesity. *Cell Metabolism*, *16*(6), 814–824. doi:10.1016/j.cmet.2012.11.005.

9

JAK-STAT Signaling in Liver Fibrosis

Marwa K. Ibrahim and Noha G. Bader El Din
Department of Microbial Biotechnology, Genetic Engineering and Biotechnology Research Division
National Research Centre
Dokki, Egypt

9.1 Liver Fibrosis

Liver fibrosis is a multi-factorial pathological status resulting from excessive wound-healing process. The etiologies include excessive alcoholic intake, chronic infections with hepatitis viruses such as hepatitis-C virus (HCV) and-B virus (HBV), bile duct damage, and non-alcoholic fatty liver disease (NAFLD) or non-alcoholic steatohepatitis (NASH) (Younossi et al. 2018). The causative factors lead ultimately to liver fibrosis through induction of hepatocyte damage, uncontrolled proliferation of liver myofibroblasts, hyper-activation of inflammatory cells in the liver milieu, and over-secretion of extracellular matrix (ECM) components; which ends in the formation of fibrous scar. The presence of numerous fibrous scars within the liver alters the overall architecture, disrupts liver function, causes hepatocyte loss, and ultimately leads to liver failure (Kisseleva and Brenner 2008). Liver fibrosis can be reversed at its early stages as long as it did not progress to advanced stages of cirrhosis (Brenner 2013; Seki and Brenner 2015). This fact sheds light on the worth of studying the molecular bases of the signaling pathways that perform fibrotic activity. The latter will help discover new drugs with anti-fibrotic potentials.

Inflammation is a mutual driver of liver diseases, and the major trigger of liver tissue damage. The inflammatory pathways play a significant role in the fibrogenesis process in liver. They lead to the activation of hepatic stellate cells (HSCs), which then trans-differentiate into myofibroblasts that produce, in tremendous amounts, ECM components (Akçalı and Aydın 2018). Of the various inflammatory signaling pathways, the Janus kinase-signal transducer and activator of transcription (JAK-STAT) pathway, in one hand, orchestrates the innate immune response in this organ to different pathogens, and, on the other hand, plays a multitude of critical roles in the pathogenesis of hepatic fibrosis. Signaling through JAK-STAT pathways results in a robust inflammatory response; this notion makes this pathway as a great target for fibrosis hindrance.

9.2 JAK-STAT Pathway

The discovery of JAK-STAT pathway in mammals dates back to 1990s. The activation of JAK-STAT pathway in the liver is mainly triggered by growth hormone, wide range of cytokines (Gao 2005), growth factors (Ruff-Jamison et al. 1993), and viral proteins (Machida et al. 2010). Following hepatic injury, different cell subsets (Kupffer cells, HSCs, hepatocytes, natural killer (NK) cells, dendritic cells, and lymphocytes) localized in the liver produce array of cytokines either with inflammatory or hepatoprotective potentials. Of these cytokines, interferon-γ (IFN-γ), IFN-α/β, interleukin-6 (IL-6), and IL-22 can activate JAK-STAT signaling pathway (Gao 2005). In general, signaling cascade through JAK-STAT pathway is induced by the binding of the cytokines to their cognate receptors, which is followed by subsequent steps; receptor dimerization, dimerization of receptor-associated JAK, a series of phosphorylation of JAKs and receptor, and phosphorylation and activation of STAT proteins by the phosphorylated JAK–receptor complex. These activated STATs dimerize (form homo-or heterodimers) and translocate into the nucleus, where they induce the expression of genes involved in various cellular functions (Darnell et al. 1994; Schindler 1999). The family of JAKs encompasses of four non-receptor protein tyrosine kinases: JAK1, JAK2, JAK3, and Tyk2 (Ghoreschi et al. 2009), while STATs include a set of seven proteins (STAT—1, 2, 3, 4, 5a, 5b, and 6). According to the structure of receptor intracellular domains, each cytokine receptor activates specific group of individual JAKs and STATs (Gao et al. 2012). The activation of STATs by certain cytokines plays a diverse array of functions in liver pathophysiology (Mair et al. 2011; Gao et al. 2012). In this chapter, the updated information on the implication of the effector arms of JAK-STAT signaling pathway (STATs) in liver fibrogenesis is discussed at length in the below sections.

9.2.1 STAT1

STAT1 plays an important role in immune response and antiviral defense (Najjar and Fagard 2010; Gao et al. 2012). STAT1 signaling in the liver is usually activated by IFN-α/β and IFN- γ. Also, STAT1 is activated in the liver by IL-27 cytokine, which is secreted by activated dendritic cells and macrophages (Schroder et al. 2004). The activated STAT1 leads to the overexpression of many antiviral proteins. It was reported that STAT1-deficient mice had increased sensitivity to pathogens infection because they didn't respond to interferon (Meraz et al. 1996). During the interferon treatment, the expression and phosphorylation of STAT1 protein are increased in NK cells and induce killing of activated HSCs. Also, the STAT1 elevation in NK cells contributes to NK cell cytotoxicity and antiviral activity (Edlich et al. 2012). Moreover, IFN-λ (IL-28A and IL-28B, IL-29) causes prolonged STAT1 activation and expression of many interferon stimulated genes (ISG), which provides protective effects against viral infection and liver damage (Donnelly et al. 2011; Zhang et al. 2017, 2011).

In HSCs, STAT1 has anti-fibrotic effect by inhibiting the proliferation of HSCs, stimulating apoptosis and cell cycle arrest (Jeong et al. 2006). The STAT1 upon phosphorylation (induced by IFN-α) interacts with p300 and inhibits collagen gene transcription (Inagaki et al. 2003). Tanabe et al. reported that the activation of STAT1 by IFN-β treatment reduces hepatic fibrosis in concanavalin A (Con A)-induced hepatitis mice (Tanabe et al. 2007). IFN-γ has different anti-fibrotic effects via stimulation of STAT1 phosphorylation, upregulation of mothers against decapentaplegic homolog 7 (SMAD7) expression, and inhibition of transforming growth factor-beta (TGF-β) (Weng et al. 2009; Tang et al. 2017). Additionally, IL-27/STAT1 pathway plays anti-fibrotic role in HSCs cell line and inhibits liver fibrosis (Schroder et al. 2004). Jeong et al. reported that STAT1-deficient mice, which are treated with carbon tetrachloride (CCl4), are more susceptible to develop liver fibrosis and this finding demonstrated the protective effect of STAT1 against liver fibrosis (Jeong et al. 2006). Recently, it has been shown that STAT1 activation stimulates antitumor immunity and immune surveillance and STAT1 gene polymorphisms increase the risk of developing hepatocellular carcinoma (HCC) (Zhu et al. 2007; Meissl et al. 2017).

On the other hand, several studies proved that STAT1 in hepatocytes induces the transcription of several genes, which promote liver inflammation, injury, and fibrosis (Hong et al. 2002; Jaruga et al. 2003; Siebler et al. 2003; Wen et al. 2017). It was proved that the overexpression of STAT1 proteins by IFN-γ develops chronic hepatitis in transgenic mice (Toyonaga et al. 1994). Also, the IFN-γ induced cell death in hepatocellular carcinoma cell line Hep3B by overexpressing STAT1 protein. Recent study demonstrated that targeting of IFN-γ/STAT1 signaling by T helper 1 (Th1) cells accelerated the HSCs proliferation and secretion (Wen et al. 2017). Moreover, it was documented that The STAT1 activation in hepatocytes is considered a pro-apoptotic signal, which increases cell death and promotes liver damage (Hong et al. 2002; Gao et al. 2012). The disruption of STAT1 genes, IFN-γ, or IFN-γ receptor abolished liver damage in lipopolysaccharide (LPS)/D-galactosamine and Con A-induced hepatitis mice (Hong et al. 2002; Kim et al. 2003) (Figure 9.1).

It is well known that excessive and recurrent liver inflammation and injury can lead to liver damage, which then progress to fibrosis, cirrhosis, and HCC. It was shown that STAT1 activation

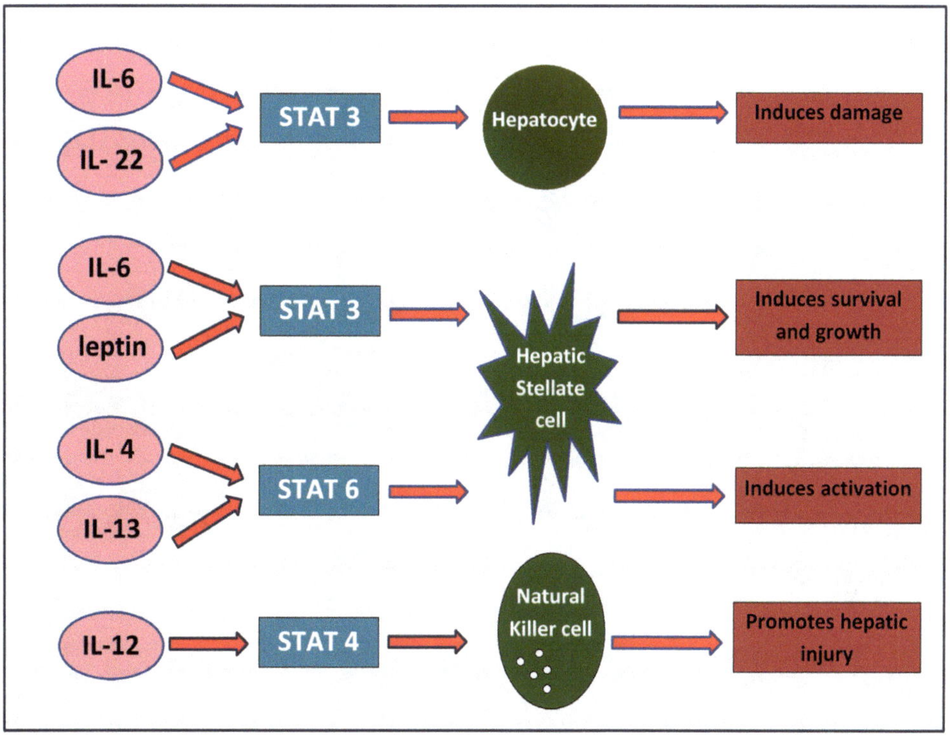

FIGURE 9.1 Role of STATs in promoting liver fibrosis. STAT3 enhances liver fibrosis by inducing both hepatocytes damage and hepatic stellate cells survival and growth. STAT6 increases the activation of hepatic stellate cells. STAT4 affects the IFN-γ production in natural killer cells, which acts as a pro-fibrotic signal to promote liver injury and fibrosis.

increased cyclin-dependent kinase inhibitor 1 (p21) and interferon regulatory factor (IRF) 1 expression, which inhibited liver regeneration (Sun and Gao 2004; Mair et al. 2011). As well as, disruption of STAT1 stopped the effect of polyinosinic-polycytidylic acid (poly I:C) treatment on the liver, which revealed that activation of STAT1-inhibited liver regeneration and promoted the liver fibrosis progression (Sun et al. 2006; Zhang et al. 2017). Moreover, Zhang et al. showed that STAT1 suppression plays critical role in the reversion of activated HSCs, which can lead to reversibility of liver fibrosis (Zhang et al. 2017).

In the hepatocytes, the STAT3 activation inhibits the pro-inflammatory STAT1 signaling and protects the liver from damage. Interestingly, activation of STAT1 and STAT3 has opposing roles in liver pathophysiology (inflammation, injury, regeneration, and fibrosis) where STAT1 activation in the liver is deleterious, whereas STAT3 activation is protective (Gao et al. 2012). Actually, STAT1 and STAT3 negatively regulate each other by the induction of suppressor of cytokine signaling-1 (SOCS1) and SOCS3 proteins. In Con A-induced hepatitis model, activated hepatic STAT1 stimulated Con A-induced hepatitis (Torisu et al. 2008), while activated hepatic STAT3 reduced it (Park et al. 2011). Therefore, the blockage of hepatic STAT1 activation by genetic modification of some genes prevents the liver injury, while obstruction of hepatic STAT3 increases it (Hong et al. 2002; Klein et al. 2005; Torisu et al. 2008; Lafdil et al. 2009) (Figure 9.2). Furthermore, it was reported recently that overexpression of Proline–serine–threonine–phosphatase-interacting protein2 (PSTPIP2) improves liver inflammation and fibrosis in mice by suppressing the activity of STAT1 activity, and enhancing the activity of STAT6 activity, which consequently regulates the macrophage polarization (Yang et al. 2018).

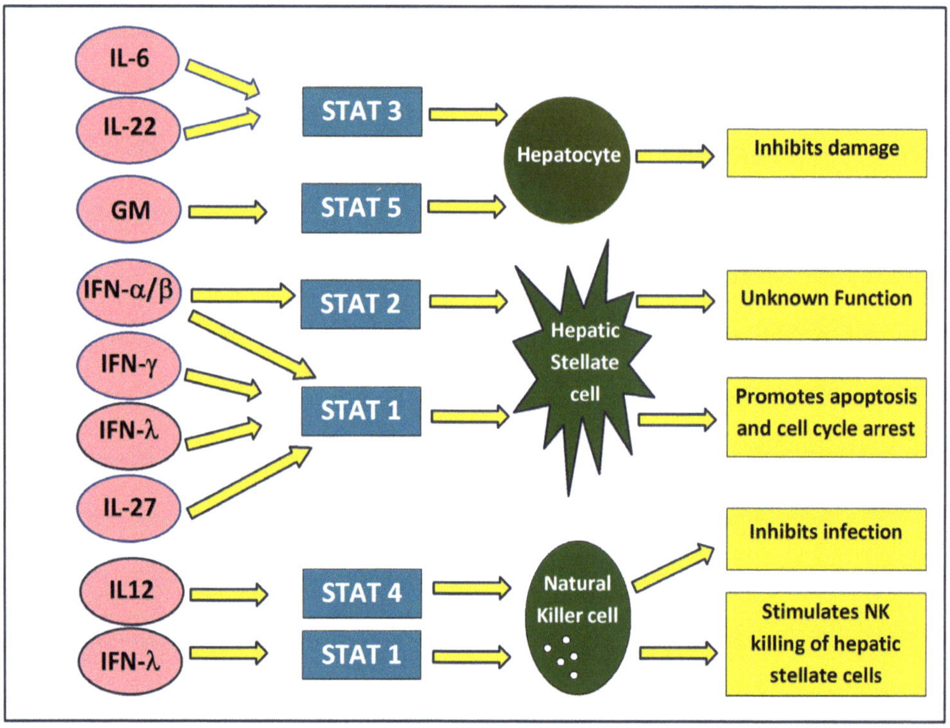

FIGURE 9.2 Role of STATs in preventing liver fibrosis. STAT3 and STAT5 have hepatoprotective effects by inhibiting hepatocytes injury and fibrosis. STAT1 has anti-fibrotic effects by inducing apoptosis, cell cycle arrest, and killing of activated hepatic stellate cells. The STAT1 and STAT4 elevation in natural killer cells contributes to NK cell cytotoxicity and provides protective effects against viral infection and liver damage.

9.2.2 STAT2

STAT2 is solely activated by the IFN family (IFN-α, IFN-β, and IFN-λ) (Orlent et al. 2011; Kong et al. 2012b). Albeit STAT2 is well-pronounced in antiviral immune response (Heim and Thimme 2014; Blaszczyk et al. 2015; Shrivastava et al. 2016), its role in liver pathophysiology (liver injury and fibrosis) is poorly studied. A study performed by Ibrahim et al. (Ibrahim et al. 2017) showed a significant upregulation of STAT2 and IRF7 in HCV patients with advanced stages of liver fibrosis when compared to HCV patients with early stages of liver fibrosis. In another study conducted to characterize the molecular changes associated with hepatic fibrosis at its early stages, HCV patients showed increased expression of STAT2/IRF9 in their diseased liver when compared to the healthy livers from controls (Bieche et al. 2005). In conclusion, both of the aforementioned studies suggest a positive role for STAT2 transcription factor in enhancing liver fibrosis progression.

9.2.3 STAT3

STAT3 plays key roles in the healthy liver; (1) it induces acute-phase response, (2) protects against hepatocellular damage, and (3) promotes liver regeneration (Wang et al. 2011a; Lee et al. 2016). However, while many studies suggest the anti-fibrogenic role of STAT3 in liver, some presented STAT3 as a fibrogenic factor and an excellent target for therapeutic intervention. STAT3 activation in the liver is provoked by a variety of cytokines in a cell type-dependent manner. The contradiction between different studies in defining the role of STAT3 in liver is mainly attributed to the nature of the cytokines driving the pathway and the responder cells. In hepatocyte, STAT3 activation is triggered by IL-6, IL-22, and IL-6 family cytokines where it induces an anti-inflammatory milieu to suppress liver inflammation under most conditions (Kroy et al. 2010; Wang et al. 2011b; Gao et al. 2012). In kupffer cells, STAT3 is a key downstream signaling protein of IL-10 (the major anti-inflammatory cytokine) and massive evidence confirms that STAT3 induces anti-inflammatory signals in myeloid cells, including kupffer cells to control liver inflammation (Murray 2006). Moreover, the conflicting data on the role of STAT3 in liver fibrosis is due to the kind of the liver injury models. For example, blockade of STAT3 in hepatocytes dramatically increased liver injury and inflammation upon chronic CCl_4 treatment (Wang et al. 2011b), while it decreased liver inflammation in a model of acute CCl_4 administration (Horiguchi et al. 2010). In T cells, STAT3 might promote or decelerate liver inflammation depending on the nature of liver injury model being investigated (Figure 9.1). While T cells conditional STAT3-knockout mice are resistant to liver inflammation induced by Con A injection and show reduced IL-17 production (Lafdil et al. 2009), inhibition of STAT3 in T cells through SOCS3 overexpression accelerated acetaminophen hepatotoxicity due to the over production of IFN-γ and tumor necrosis factor-alpha (TNF-α) (Numata et al. 2007).

Here are some examples for the literature that explored the pro-inflammatory role of STAT3 in liver. The co-treatment of CCl4-treated Wistar rats with anti-oxidants (Carvedilol and silymarin) succeeded to restore the normal architecture of the liver and the localization of collagen through reducing the elevated hepatic IL-6 and STAT-3 levels (Balaha et al. 2016). Primary HSCs isolated from mouse livers and treated with IL-6 demonstrated upregulation of the reticular fiber components α-Smooth muscle actin (αSMA) and alpha-1 type I collagen (Col1a) at mRNA and protein levels, which was associated with increased phosphorylation of STAT3. The fibrogenic phenotype of HSCs cells was reversed upon using STAT3 synthetic inhibitor (S3I-201) and, αSMA and Col1a expression levels returned to the normal levels similar to the untreated control samples (Kagan et al. 2017). Rats treated with diethylnitrosamine (DEN), used to induce a stepwise histopathological progression in liver (inflammation-fibrosis-HCC), showed a dramatic up-regulation of JAK2/STAT3 signaling in their liver in a time-dependent manner when compared to the control group (Ding et al. 2017). Liver tissue sections from cirrhotic and HCC patients showed high levels of both forms of STAT3 (phosphorylated and non-phosphorylated), with statistical significant differences between HCC and cirrhosis (Li et al. 2017). Liver injury either induced by adenovirus infection or CCl4 resulted in the increase of IL-6/IL-12 with the ensuing activation of STAT3 and STAT4 in naturally occurring regulatory T cells, thus reducing the suppressive function of regulatory T cells in the liver. The latter is one of the scenarios for the progression of liver inflammation into fibrosis and autoimmune hepatitis (Chi et al. 2018). Multiple lines of investigations evidenced that JAK1/STAT3 interact with SMAD pathway to enhance hepatic fibrosis through the upregulation of TGF-β expression (Ogata et al. 2006; Tang et al. 2017). S-allyl-cysteine (one of the major antioxidant compounds in aged garlic extract) was shown to attenuate liver fibrosis in Sprague-Dawley rats treated with CCl4, and that effect was associated with reduced phosphorylation of SMAD3 and STAT3 and diminished mRNA expression of TGF-β and IL-6 (Gong et al. 2018). In agreement with this study, cucurbitacin-B treatment in CCl4 treated mice ameliorated the fibrotic state by diminishing the expression of collagen-1α, αSMA, TGF-β, and matrix metalloproteinase (MMP) 2. All these effects were associated with the blockade of STAT3 phosphorylation (Sallam et al. 2018). On the other hand, some studies suggest a protective role of STAT3 in liver fibrosis. G-protein coupled receptor deficient mice (Gpr110-/- mice) during CCl4 challenge to induce liver fibrosis displayed lower liver fibrosis; which was attributed to augmented activation of IL-6/STAT3 pathway as the underlying mechanism (Ma et al. 2017). IL-22 downstream signaling pathway protects liver damage as evident by the finding on the positive relationship between the level of IL-22BP (an inhibitor of IL-22) and severity of liver fibrosis. The anti-fibrotic effect of IL-22 is suggested to be through the activation of STAT3 (Khawar et al. 2016). Kong X et al. reported that IL-22 inhibits liver fibrosis by the induction of HSC senescence, which was mediated by the activation of STAT3, SOCS3, and p53 pathways (Kong et al. 2012a). Tissue inhibitor of metalloproteinases-1 (TIMP-1) the well-known player in ECM maintenance that inhibits the activity of MMPs is one of the downstream gene targets of STAT3 (Sun 2010). TIMP-1 plays a protective role against acute and chronic liver injury. Following chronic CCl_4 treatment, the levels of the circulating and liver specific TIMP-1 were markedly diminished in conditional hepatocyte-specific STAT3 knockout mice (Wang et al. 2011c; Kasembeli et al. 2018).

9.2.4 STAT4

STAT4 is a key transcription factor in the process of Th1 cell differentiation and in promoting cellular-mediated immunity (Wang et al. 2015). STAT4 is expressed mainly in myeloid cells (Frucht et al. 2000), thymus, and testes and to a very low level in resting human T cells. Signaling through STAT4 is induced by arrays of cytokines, including IL-12 (Bacon et al. 1995; Wang et al. 2015), IL-2 (Wang et al. 1999), IFN-α/β (Cho et al. 1996; Miyagi et al. 2007), and IL-17 (Subramaniam et al. 1999). The activated STAT4 regulates the transcription of myriad genes. The most prominent STAT4-target gene is IFN-γ. IL-12-dependent STAT4 activation is well studied in liver diseases. IL-12/STAT4 signaling is a key regulator of tissue inflammation, fibrogenesis, and antiviral defense (Edlich et al. 2012; Kong et al. 2012b; Wang et al. 2015). In a mouse model of Schistosoma infection, the animals treated with anti-IL-12 had granuloma with larger size and intensified fibrosis (Cheng et al. 2012), which suggests a protective role of IL-12/STAT4 axis against liver fibrosis in these animals. In a mouse model of NASH, the knockdown of protein inhibitor of activated STAT4 (PIAS4) using short hairpin RNA (shRNA) dramatically attenuated liver fibrosis induced by high-fat high-carbohydrate (HFHC) diet through the downregulation of a panel of pro-fibrogenic ECM-related genes, including collagens (type I and type III), smooth muscle actin, and tissue inhibitors of metalloproteinase, the anti-fibrotic effect of PIAS4 silencing is mediated by the blockade of SMAD3-(a potent pro-fibrogenic transcription factor) related signal pathway (Xu et al. 2016).

The genetic variation in STAT4 may expose the host to the risk of fibrosis; confirming the indispensable role of this transcription factor in the liver health. Examples for the latter, genotyping of STAT4 single nucleotide polymorphism (SNP) (rs7574865) in 160 liver-transplanted patients with HCV recurrence revealed that STAT4-T-allele is highly linked to the development of advanced fibrosis (Eurich et al. 2013). For the same STAT4 SNP (rs7574865), Jiang et al. correlated G-allele with the risk of HBV-related liver cirrhosis (Jiang et al. 2015). In another study, the author elaborated the impact of STAT4 genetic variation on the outcome of STAT4 signaling with the ensuing consequence on the liver pathology. In this study, the Caucasian HBV-patients who are carriers of STAT4 rs7574865 GG genotype expressed at very low levels STAT4 in liver, peripheral blood mononuclear cells (PBMCs), and in NK cells. Upon stimulating NK cells (isolated from the patients with the risk allele) with IL-12/IL-18, the cells showed impaired STAT4 phosphorylation associated with a reduction in IFN-γ secretion. NK cells malfunction due to the impaired STAT4 phosphorylation-dependent IFN-γ production likely contributes to hepatic inflammation and fibrosis in HBV patients (El Sharkawy et al. 2018). On the other hand, some studies suggest a fibrotic role of STAT4 in the liver. Wang et al. used IL-35 gene-modified mesenchymal stem cells (IL-35-MSCs) to deliver IL-35 gene and investigated their protective effects in Con A-induced autoimmune hepatitis. The study showed the ability of these genetically modified cells to migrate to the diseased liver where it narrowed the necrosis area. The curative effect of these cells was attributed to their ability to interfere with JAK1-STAT1/STAT4 signaling cascade with the consequent decrease in IFN-γ production secreted by liver mononuclear cells (Wang et al. 2018). IL-12 deficiency rendered the mice to resist Con A-induced T cell hepatitis (Zhu et al. 2007), whereas IL-12 overexpression in the liver induces liver injury (Kong et al. 2012b). Similarly, IL-12 treatment has been shown to induce liver inflammation in multiple animal models (Harada et al. 2004; Chang et al. 2007), which is likely mediated by the subsequent production of IFN-γ in NK cells and natural killer T-cells (NKT) (Subleski et al. 2006). This suggests that IL-12/STAT4 acts as a pro-inflammatory signal pathway to promote liver injury and fibrosis (Figure 9.1). Collectively, STAT4-IL-12 dependent activation in immune cells has double sword edges, it may induce inflammation in the liver, which is likely to promote to liver injury and fibrosis, but may also protect against infection and hence, ameliorating liver injury and fibrosis.

9.2.5 STAT5

The STAT5a and STAT5b proteins are encoded by two different genes located on chromosome 17 of human and chromosome 11 of mice. The STAT5 proteins are activated mainly by growth hormone (GH) and interleukins (IL-2, IL-3, IL-5), as well as some other cytokines (Mair et al. 2011). In hepatocytes, the growth hormone activates STAT5, which controls the expression of many hepatic genes essential for growth, metabolism, and differentiation in the liver (Gao 2005). Additionally, STAT5 regulates many hepatoprotective factors that inhibit hepatocyte cell death. The GH/STAT5 signaling pathway regulates the expression of insulin-like growth factor 1 (IGF-1) and epidermal growth factor receptor (EGFR) in hepatocytes and muscle cells (Woelfle et al. 2003; Mair et al. 2011). It was reported that high GH and low IGF-1 serum levels are associated with human liver cirrhosis and the level of IGF-1 correlates with liver disease progression. IGF-1 low serum levels lead to hepatocellular dysfunction (Donaghy et al. 1995; Caregaro et al. 1998). It was demonstrated that loss of STAT5 or GH diminished the liver regeneration after partial hepatectomy (PHx). Moreover, STAT5 loss caused hepatosteatosis defective GH/STAT5/IGF-1 signaling, insulin resistance, and increased the activity of STAT1 (Cui et al. 2007).

In a cholestasis mouse model, IGF-1 is an important target of GH/STAT5 signals and the GH/STAT5/IGF-1 axis protects liver from injury (Sobrevals et al. 2010). It has been demonstrated that GH/STAT5 regulates the expression of hepatocyte nuclear factor 6 (HNF6) which stimulates liver regeneration (Wang et al. 2008). Also, STAT5 upregulates the expression of prolactin receptor (PRLP) and leukemia inhibitory factor receptor (LIFR), which play crucial role in liver regeneration and protection (Simon-Holtorf et al. 2006; Blaas et al. 2010). Deletion of STAT5 in mice caused GH resistance, and disruption of IGF-1 transcription induced by GH, which leads to cell damage and apoptosis (Blaas et al. 2010). Also, it has been shown that administration of IGF-1 improved the liver fibrosis by increasing the mitochondrial function, reducing the oxidative liver damage, and increasing the MMPs expression (Sobrevals et al. 2010). Consistently, IGF-1-stimulates HSCs to produce hepatocyte growth factor (HGF), which promotes HSCs apoptosis (Kim et al. 2005).

During liver inflammation or injury, hepatocytes secrete excessive amount of TGF-β, which induces extracellular matrix accumulation and ultimately leads to the development of liver fibrosis (Canbay et al. 2004; Jeong et al. 2006). STAT5 provides protection against fatty liver disease and liver injury (Mueller et al. 2011; Friedbichler et al. 2012; Baik et al. 2017) In cholestasis mouse model, it was confirmed that STAT5 has an anti-fibrotic effect as it protects the liver from injury and fibrosis (Kong et al. 2012b). STAT5 prevents the liver fibrosis induced by the deletion of multidrug resistance gene 2. Actually, the deletion of both STAT5 (hepatocyte and cholangiocyte) and multidrug resistance gene 2 leads to severe liver fibrosis and increases the number of apoptotic hepatocytes. Moreover, these deletions are associated with reduced expression of essential hepatoprotective genes such as HNF6, EGFR, PRLP, and LIFR (Blaas et al. 2010).

Hosui et al. discovered a relationship between STAT3, STAT5, and TGF-β in the pathogenesis of liver fibrosis. In STAT5-null mouse embryonic fibroblasts (MEF) cells, it was reported that TGF-β levels were highly elevated and treatment with STAT5 caused the suppression of TGF-β. Also, the STAT5 loss in hepatocytes after CCl4 treatment enhanced the TGF-β levels and GH-induced STAT3 activity. Therefore, it was proposed that STAT5 prevents liver fibrogenes by downregulating TGF-β and STAT3 (Figure 9.2). Nevertheless, it was reported that the deletion of STAT5 increases the sensitivity of HSC or Kupffer cells to TGF-β and promoted liver fibrosis (Hosui et al. 2009). Furthermore, hepatic STAT5 deletion promoted CCl4-induced HCC in wild-type mice (Hosui et al. 2009) and induced the development of HCC in hepatic glucocorticoid receptor knockout mice (Mueller et al. 2011) or in growth hormone transgenic mice (Friedbichler et al. 2012).

Several studies documented the role of STAT5 in tumorigenesis (Moriggl et al. 1999; Haddad et al. 2013). Interestingly, the role of STAT 5 in liver carcinogenesis is controversial. It was reported that STAT5 activation causes cellular transformation by stimulating progression of cell cycle and suppression of cell apoptosis. Activated STAT5 upregulates the expression of cell proliferation, invasion, metastasis, and anti-apoptotic-related genes, which promotes tumorigenesis (Ferbeyre and

Moriggl 2011). However, other studies proved that STAT5 has antioncogenic function in liver cancer as hepatic deletion of STAT5 led to HCC development after CCl4 treatment. Additionally, the loss of hepatic STAT5 is compensated by STAT3 upregulation, which stimulates hepatic tumorigenesis (Hosui et al. 2009). The activation of STAT5 induces the expression of cell cycle inhibitor (p15INK4B) in hepatocytes, which has a protective effect against liver carcinogenesis (Yu et al. 2011). Also, the hepatic STAT5 deletion enhanced liver cancer caused by hyper-activated growth hormone signaling (Friedbichler et al. 2012). Therefore, it was suggested that STAT5 has hepato-protective and anti-steatogenic effects and can be a tumor suppressor in liver carcinogensis (Mueller et al. 2011).

9.2.6 STAT6

STAT6 has both a pro and anti-inflammatory effect. Activation on of STAT6 in the liver is induced by Thl cytokine (IL-12) and Th2 cytokines mainly (IL-4, IL-13) (Wurster et al. 2000; Jaruga et al. 2003). STAT6 has important roles in the differentiation of Th2 lymphocyte (Kaplan et al. 1996; Murphy and Reiner 2002). Moreover, STAT6 plays key roles in controlling liver injury, inflammation, and fibrosis. The functions of STAT6 have been studied in different liver injury models including Con A-injection, ischemia/reperfusion, and LPS-treatment (Kato et al. 2000; Lentsch et al. 2001; Iff et al. 2009).

In HSCs, the STAT6 is predominately induced by both IL-4 and IL-13 and the blockage of STAT6 using small interfering RNA (siRNA) attenuates the activation of HSC *in vitro* (Aoudjehane et al. 2008). The IL-4 and IL-13 roles in liver fibrosis have been extensively examined and documented (Barron and Wynn 2011; Du et al. 2016; Gieseck et al. 2016), while the STAT6 roles in liver inflammation and fibrosis is controversial and needs comprehensive studies to be largely clear. Studies on different liver injury models suggested that STAT6 have either a harmful or protective effect on liver fibrogenesis.

Previously, it was shown that STAT6-deficient mice with *Schistosoma mansoni* infection have reduced collagen deposition in the liver and smaller granulomas than wild type mice (Kaplan et al. 1996). However, Liu et al. (2011) has confirmed that ERK1/2 pathway rather than STAT6 in HSCs contributes to IL-13 induction of connective tissue growth factor (CTGF), which stimulates deposition of extracellular matrix in the liver. IL-13 has been proved to promote liver fibrogenesis in *Schistosoma mansoni* infection model and the blockage of IL-13 reduces the liver fibrosis. The IL-13 upregulates several fibrotic proteins in HSCs and induces the HSCs activation (Fallon et al. 2000; Chiaramonte et al. 2001; Du et al. 2016).

Recently, it has been shown that IL-4 induces the activation of STAT6 in a different liver injury models (Njoku et al. 2009). IL-4 has pro-inflammatory and pro-fibrotic effects (Weng et al. 2018). IL-4 stimulates the activation of HSCs and collagens production (Aoudjehane et al. 2008; Jin et al. 2011). Upregulation of IL-4 expression has been detected is in the fibrotic liver of baboons infected with *Schistosoma mansoni*, and the liver fibrosis in *Schistosoma mansoni*-infected mice was significantly reduced by the blockage of IL-4 (Farah et al. 2000). The role of IL-4/STAT6 in hepatitis was confirmed in SOCS-1 deficiency-induced hepatitis (Naka et al. 2001). Moreover, the IL-4/STAT6 signaling has important functions in T-cell mediated hepatitis induced by Con A–treatment. It was suggested that IL-4/STAT6 enhances the eotaxins expression in sinusoidal endothelial cells and hepatocytes, stimulates IL-5 expression, and induces eosinophils and neutrophils infiltration into the liver and which leads to hepatitis. Also, it was documented that the activation of STAT6 in IL-4-deficient mice was significantly suppressed after Con A injection (Kaneko et al. 2000; Jaruga et al. 2003).

On the contrary, Ryan et al. recently showed that IL-4 deficient mice have increased liver injury and inflammation after acetaminophen injection (Ryan et al. 2012). Moreover, the severity of liver injury was reduced by the administrating of recombinant IL-4. It was proved in acetaminophen-induced injury model that IL-4 has hepatoprotective function by upregulating the synthesis of hepatic glutathione (GSH). In addition, several studies showed activation of STAT6 by both IL-4 and IL-13 protected the liver from ischemia/reperfusion injury. Moreover, STAT6 deficient mice were highly susceptible to endotoxin-induced liver injury (Lentsch et al. 2001). STAT6 has potential anti-inflammatory effect as it inhibited liver inflammation and protected endothelial cell and hepatocyte against damage (Kato et al. 2000) Concisely, the STAT6 activation promoted liver injury and fibrogenesis in T cell hepatitis model while it protected against ischemia/reperfusion and drug-induced liver injury (Gao et al. 2012).

9.3 Conclusion

Liver fibrosis is severe hepatocytes damage, which leads to end-stage liver disease. The chronic inflammation, continuous liver injury, and inadequate recovery from hepatocyte damage trigger numerous cytokines and cellular events that lead to the liver injury, the collagen deposition, and the interruption of the liver normal architecture. Several studies have reported the crucial role of JAK-STAT signaling pathway in hepatic fibrogenesis.

The current chapter summarizes the biological significance of different STATs signaling proteins in controlling liver fibrosis as shown in Figures 9.1 and 9.2. STAT1 has anti-fibrotic effect via inducing HSC apoptosis and cell cycle arrest or has pro-inflammatory signaling, which stimulates liver fibrosis. Interestingly, STAT1 and STAT3 negatively regulate each other as the activation of STAT3 prevents the deleterious effect of STAT1. STAT2 has proven antiviral effect but its role in liver injury and fibrosis is not well determined and needs to be further studied. STAT3 may have a pro-inflammatory or anti-fibrogenic role in liver. STAT3 induces hepatic stellate cells survival and growth, which promotes liver fibrosis; whereas STAT3, in hepatocytes, have anti-fibrotic and hepatoprotective effects via preventing the liver injury and fibrosis (Figure 9.1). STAT4 is very important for cellular-mediated immunity. Activation of STAT4 may induce inflammation and liver injury or may also protect against infection and hence, preventing liver damage and fibrosis. STAT5 controls many hepatoprotective factors that prevent hepatocyte cell death and increased liver regeneration. Moreover, STAT5 provides protection against fatty liver disease, liver injury, fibrosis, and hepatocellular carcinoma. STAT6 has both a pro and anti-inflammatory effect. However, STAT6 roles in liver inflammation and fibrosis are controversial and need comprehensive studies to be largely clear.

REFERENCES

Akçalı, K., and M. M. Aydın. 2018. Liver Fibrosis. *Turk J Gastroenterol* 29:8.

Aoudjehane, L., A. Pissaia, O. Scatton, P. Podevin, P.P. Massault, S. Chouzenoux, O. Soubrane, Y. Calmus, and F. Conti. 2008. Interleukin-4 Induces the Activation and Collagen Production of Cultured Human Intrahepatic Fibroblasts via the STAT-6 Pathway. *Lab Investig* 9:973–85.

Bacon, C. M., E. F. Petricoin, 3rd, J. R. Ortaldo, R. C. Rees, A. C. Larner, J. A. Johnston, and J. J. O'Shea. 1995. Interleukin 12 Induces Tyrosine Phosphorylation and Activation of STAT4 in Human Lymphocytes. *Proc Natl Acad Sci USA* 92(16):7307–11.

Baik, M., J. Kim, M. Y. Piao, H. J. Kang, S. J. Park, S.W. Na, S. H. Ahn, and J. H. Lee. 2017. Deletion of Liver-Specific STAT5 Gene Alters the Expression of Bile Acid Metabolism Genes and Reduces Liver Damage in Lithogenic Diet-Fed Mice. *J Nutr Biochem* 39:59–67.

Balaha, M., S. Kandeel, and W. Barakat. 2016. Carvedilol Suppresses Circulating and Hepatic IL-6 Responsible for Hepatocarcinogenesis of Chronically Damaged Liver in Rats. *Toxicol Appl Pharmacol* 311:1–11.

Barron, L., and T. A. Wynn. 2011. Fibrosis Is Regulated by Th2 and Th17 Responses and by Dynamic Interactions between Fibroblasts and Macrophages. *Am J Physiol Gastrointest Liver Physiol* 300(5): G723–8.

Bieche, I., T. Asselah, I. Laurendeau, D. Vidaud, C. Degot, V. Paradis, P. Bedossa, D. C. Valla, P. Marcellin, and M. Vidaud. 2005. Molecular Profiling of Early Stage Liver Fibrosis in Patients with Chronic Hepatitis C Virus Infection. *Virology* 332(1):130–44.

Blaas, L., J. Kornfeld, D. Schramek, M. Musteanu, J. Gumhold, F. Van Zijl, D. Schneller, and H. Esterbauer. 2010. Europe PMC Funders Group Disruption of the Growth Hormone — Signal Transducer and Activator of Transcription 5 — Insulin like Growth Factor 1 Axis Severely Aggravates Liver Fibrosis in a Mouse Model of Cholestasis. *Hepatology* 51(4):1319–26.

Blaszczyk, K., A. Olejnik, H. Nowicka, L. Ozgyin, Y. L. Chen, S. Chmielewski, K. Kostyrko, J. Wesoly, B. L. Balint, C. K. Lee, and H. A. Bluyssen. 2015. STAT2/IRF9 Directs a Prolonged ISGF3-like Transcriptional Response and Antiviral Activity in the Absence of STAT1. *Biochem J* 466(3):511–24.

Brenner, D. A. 2013. Reversibility of Liver Fibrosis. *Gastroenterol Hepatol (NY)* 9(11):737–39.

Canbay, A., S. Friedman, and G. J. Gores. 2004. Apoptosis: The Nexus of Liver Injury and Fibrosis. *Hepatology* 39(2):273–78.

Caregaro, L., F. Alberino, P. Angeli, and A. Gatta. 1998. Insulin-like Growth Factor 1 (IGF-1) in Liver Cirrhosis: A Marker of Hepatocellular Dysfunction? *J Hepatol* 29(2):342.

Chang, C. J., Y. H. Chen, K. W. Huang, H. W. Cheng, S. F. Chan, K. F. Tai, and L. H. Hwang. 2007. Combined GM-CSF and IL-12 gene Therapy Synergistically Suppresses the Growth of Orthotopic Liver Tumors. *Hepatology* 45(3):746–54.

Cheng, Y. L., W. J. Song, W. Q. Liu, J. H. Lei, Z. Kong, and Y. L. Li. 2012. The Effects of Interleukin (IL)-12 and IL-4 Deficiency on Worm Development and Granuloma Formation in Schistosoma Japonicum-Infected Mice. *Parasitol Res* 110(1):287–93.

Chi, G., X. X. Feng, Y. X. Ru, T. Xiong, Y. Gao, H. Wang, Z. L. Luo, R. Mo, F. Guo, Y. P. He, G. M. Zhang, D. A. Tian, and Z. H. Feng. 2018. TLR2/4 Ligand-Amplified Liver Inflammation Promotes Initiation of Autoimmune Hepatitis Due to Sustained IL-6/IL-12/IL-4/IL-25 Expression. *Mol Immunol* 99:171–81.

Chiaramonte, M. G., A.W. Cheever, J. D. Malley, D. D. Donaldson, and T. A. Wynn. 2001. Studies of Murine Schistosomiasis Reveal Interleukin-13 Blockade as a Treatment for Established and Progressive Liver Fibrosis. *Hepatology* 34(2):273–82.

Cho, S. S., C. M. Bacon, C. Sudarshan, R. C. Rees, D. Finbloom, R. Pine, and J. J. O'Shea. 1996. Activation of STAT4 by IL-12 and IFN-Alpha: Evidence for the Involvement of Ligand-induced Tyrosine and Serine Phosphorylation. *J Immunol* 157(11):4781–89.

Cui, Y., A. Hosui, R. Sun, K. Shen, O. Gavrilova, W. Chen, M.C. Cam, B. Gao, G.W. Robinson, and L. Hennighausen. 2007. Loss of Signal Transducer and Activator of Transcription 5 Leads to Hepatosteatosis and Impaired Liver Regeneration. *Hepatology* 46(2):504–13.

Darnell, J.E., Jr., I.M. Kerr, and G. R. Stark. 1994. JAK–STAT Pathways and Transcriptional Activation in Response to IFNs and Other Extracellular Signaling Proteins. *Science* 264(5164):1415–21.

Ding, Y. F., Z. H. Wu, Y. J. Wei, L. Shu, and Y. R. Peng. 2017. Hepatic Inflammation-fibrosis-cancer Axis in the Rat Hepatocellular Carcinoma Induced by Diethylnitrosamine. *J Cancer Res Clin Oncol* 143 (5):821–34.

Donaghy, A., R. Ross, A. Gimson, S.C. Hughes, J. Holly, and R. Williams. 1995. Growth Hormone, Insulinlike Growth Factor-1, and Insulinlike Growth Factor Binding Proteins 1 and 3 in Chronic Liver Disease. *Hepatology* 21(3):680–88.

Donnelly, R., H. Dickensheets, and T. O'Brien. 2011. Interferon-Lambda and Therapy for Chronic Hepatitis C Virus Infection. *Trends Immunol* 32(9):443–50.

Du, P., Q. Ma, Z. De Zhu, G. Li, Y. Wang, Q. Q. Li, Y. F. Chen, Z. Z. Shang, J. Zhang, and L. Zhao. 2016. Mechanism of Corilagin Interference with IL-13/STAT6 Signaling Pathways in Hepatic Alternative Activation Macrophages in Schistosomiasis-Induced Liver Fibrosis in Mouse Model. *Eur J Pharmacol* 793:119–126.

Edlich, B., G. Ahlenstiel, A.Z. Azpiroz, J. Stoltzfus, M. Noureddin, E. Serti, J. J. Feld, T. J. Liang, Y. Rotman, and B. Rehermann. 2012. Early Changes in Interferon Signaling Define Natural Killer Cell Response and Refractoriness to Interferon-Based Therapy of Hepatitis C Patients. *Hepatology* 55(1):39–48.

El Sharkawy, R., K. Thabet, P. Lampertico, S. Petta, A. Mangia, T. Berg, M. Metwally, A. Bayoumi, A. Boonstra, W. P. Brouwer, A. Smedile, M. L. Abate, A. Loglio, M. W. Douglas, A. Khan, R. Santoro, J. Fischer, D. J. Leeming, C. Liddle, J. George, and M. Eslam. 2018. A STAT4 Variant Increases Liver Fibrosis Risk in Caucasian Patients with Chronic Hepatitis B. *Aliment Pharmacol Ther* 48(5):564–73.

Eurich, D., S. Boas-Knoop, B. Struecker, R. Neuhaus, P. Neuhaus, and M. Bahra. 2013. Genetic Variants of STAT-4 Affect the Development of Graft Fibrosis after Liver Transplantation for HCV-induced Liver Disease. *Transplantation* 95(1):203–08.

Fallon, P. G., E. J. Richardson, G. J. McKenzie, and A. N. J. McKenzie. 2000. Schistosome Infection of Transgenic Mice Defines Distinct and Contrasting Pathogenic Roles for IL-4 and IL-13: IL-13 Is a Profibrotic Agent. *J Immunol* 164(5):2585–91.

Farah, I. O., P. W. Mola, T. M. Kariuki, M. Nyindo, R. E. Blanton, and C. L. King. 2000. Repeated Exposure Induces Periportal Fibrosis in Schistosoma Mansoni-Infected Baboons: Role of TGF- and IL-4. *J Immunol* 164(10):5337–43.

Ferbeyre, G., and R. Moriggl. 2011. The Role of Stat5 Transcription Factors as Tumor Suppressors or Oncogenes. *Biochim Biophys Acta Rev Cancer* 1815(1):104–14.

Friedbichler, K., M. Themanns, K. M. Mueller, M. Schlederer, J. Kornfeld, L. M. Terracciano, A. V. Kozlov, S. Haindl, L. Kenner, T. Kolbe, M. Mueller, K. J. Snibson, M. H. Heim, and R. Moriggl. 2012. Growth-Hormone–Induced Signal Transducer and Activator of Transcription 5 Signaling Causes Gigantism, Inflammation, and Premature Death but Protects Mice From Aggressive Liver Cancer. *Hepatology* 55(3):941–52.

Frucht, D. M., M. Aringer, J. Galon, C. Danning, M. Brown, S. Fan, M. Centola, C. Y. Wu, N. Yamada, H. El Gabalawy, and J. J. O'Shea. 2000. Stat4 is Expressed in Activated Peripheral Blood Monocytes, Dendritic Cells, and Macrophages at Sites of Th1-mediated Inflammation. *J Immunol* 164(9):4659–64.

Gao, B. 2005. Cytokines, STATs and Liver Disease. *Cell Mol Immunol* 2(2):92–100.

Gao, B., H. Wang, F. Lafdil, and D. Feng. 2012. STAT Proteins – Key Regulators of Anti-Viral Responses, Inflammation, and Tumorigenesis in the Liver. *J Hepatol* 57(2):430–41.

Ghoreschi, K., A. Laurence, and J. J. O'Shea. 2009. Janus Kinases in Immune Cell Signaling. *Immunol Rev* 228(1):273–87.

Gieseck, R. L., T. R. Ramalingam, K. M. Hart, K. M. Vannella, D. A. Cantu, W. Lu, S. Ferreira-González, S. J. Forbes, and L. Vallier. 2016. Interleukin-13 Activates Distinct Cellular Pathways Leading to Ductular Reaction, Steatosis, and Fibrosis. *Immunity* 45(3):145–58.

Gong, Z., H. Ye, Y. Huo, L. Wang, Y. Huang, M. Huang, and X. Yuan. 2018. S-Allyl-Cysteine Attenuates Carbon Tetrachloride-Induced Liver Fibrosis in Rats by Targeting STAT3/SMAD3 Pathway. *Am J Transl Res* 10(5):1337–46.

Haddad, B. R., L. Gu, T. Mirtti, A. Dagvadorj, P. Vogiatzi, D.T. Hoang, R. Bajaj, B. Leiby, E. Ellsworth, S. Blackmon, C. Ruiz, M. Curtis, P. Fortina, A. Ertel, C. Liu, H. Rui, T. Visakorpi, L. Bubendorf, C. D. Lallas, E. J. Trabulsi, P. McCue, L. Gomella, and M. T. Nevalainen. 2013. STAT5A/B Gene Locus Undergoes Amplification during Human Prostate Cancer Progression. *Am J Pathol* 182 (6):2264–75.

Harada, N., M. Shimada, S. Okano, T. Suehiro, Y. Soejima, Y. Tomita, and Y. Maehara. 2004. IL-12 Gene Therapy is An Effective Therapeutic Strategy for Hepatocellular Carcinoma in Immunosuppressed Mice. *J Immunol* 173(11):6635–44.

Heim, M. H., and R. Thimme. 2014. Innate and Adaptive Immune Responses in HCV Infections. *J Hepatol* 61(1 Suppl):S14–25.

Hong, F., B. Jaruga, W. H. Kim, S. Radaeva, O. N. El-Assal, Z. Tian, V. A. Nguyen, and B. Gao. 2002. Opposing Roles of STAT1 and STAT3 in T Cell-Mediated Hepatitis: Regulation by SOCS. *J Clin Invest* 110(10):1503–13.

Horiguchi, N., F. Lafdil, A. M. Miller, O. Park, H. Wang, M. Rajesh, P. Mukhopadhyay, X. Y. Fu, P. Pacher, and B. Gao. 2010. Dissociation between Liver Inflammation and Hepatocellular Damage Induced by Carbon Tetrachloride in Myeloid Cell-Specific Signal Transducer and Activator of Transcription 3 Gene Knockout Mice. *Hepatology* 51(5):1724–34.

Hosui, A., A. Kimura, D. Yamaji, B. M. Zhu, R. Na, and L. Hennighausen. 2009. "Loss of STAT5 Causes Liver Fibrosis and Cancer Development through Increased TGF-β and STAT3 Activation." *J Exp Med* 206(4):819–31. doi: 10.1084/jem.20080003.

Ibrahim, M. K., A. Khedr, N. G. Bader El Din, A. Khairy, and M. K. El Awady. 2017. Increased Incidence of Cytomegalovirus Coinfection in HCV-Infected Patients with Late Liver Fibrosis is Associated with Dysregulation of JAK–STAT Pathway. *Sci Rep* 7(1):10364.

Iff, J., W. Wang, T. Sajic, N. Oudry, E. Gueneau, G. Hopfgartner, E. Varesio, and I. Szanto. 2009. Differential Proteomic Analysis of STAT6 Knockout Mice Reveals New Regulatory Function in Liver Lipid Homeostasis. *J Proteome Res* 8(10):4511–24.

Inagaki, Y., T. Nemoto, Z. Miwa Kushida, Y. Sheng, K. Higa, K. Ikeda, F. Shirasaki, K. Takehara, K. Sugiyama, M. Fujii, H. Yamauchi, A. Nakao, B. de Crombrugghe, T. Watanabe, and I. Okazaki. 2003. Interferon Alfa Down-Regulates Collagen Gene Transcription and Suppresses Experimental Hepatic Fibrosis in Mice. *Hepatology* 38(4):890–99.

Jaruga, B., F. Hong, R. Sun, S. Radaeva, and B. Gao. 2003. Crucial Role of IL-4/STAT6 in T Cell-Mediated Hepatitis: Up-Regulating Eotaxins and IL-5 and Recruiting Leukocytes. *J Immunol* 171(6):3233–44.

Jeong, W. I., O. Park, S. Radaeva, and B. Gao. 2006. STAT1 Inhibits Liver Fibrosis in Mice by Inhibiting Stellate Cell Proliferation and Stimulating NK Cell Cytotoxicity. *Hepatology* 44(6):1441–51.

Jiang, D. K., X. P. Ma, X. Wu, L. Peng, J. Yin, Y. Dan, H. X. Huang, D. L. Ding, L. Y. Zhang, Z. Shi, P. Zhang, H. Yu, J. Sun, S. Lilly Zheng, G. Deng, J. Xu, Y. Liu, J. Guo, G. Cao, and L. Yu. 2015. Genetic Variations in STAT4,C2,HLA-DRB1 and HLA-DQ Associated with Risk of Hepatitis B Virus-related Liver Cirrhosis. *Sci Rep* 5:16278.

Jin, Z., R. Sun, H. Wei, X. Gao, Y. Chen, and Z. Tian. 2011. Accelerated Liver Fibrosis in Hepatitis B Virus Transgenic Mice: Involvement of Natural Killer T Cells. *Hepatology* 53(1):219–29.

Kagan, P., M. Sultan, I. Tachlytski, M. Safran, and Z. Ben-Ari. 2017. Both MAPK and STAT3 Signal Transduction Pathways are Necessary for IL-6-Dependent Hepatic Stellate Cells Activation. *PLoS One* 12(5):e0176173.

Kaneko, B.Y., M. Harada, T. Kawano, M. Yamashita, Y. Shibata, F. Gejyo, T. Nakayama, and M. Taniguchi. 2000. Augmentation of Valpha14 NKT Cell – Mediated Cytotoxicity by Interleukin 4 in an Autocrine Mechanism Resulting in the Development of Concanavalin A – Induced Hepatitis. *J Exp Med* 191(1):105–14.

Kaplan, M. H., U. Schindler, S. T. Smiley, and M. J. Grusby. 1996. Stat6 Is Required for Mediating Responses to IL-4 and for the Development of Th2 Cells. *Immunity* 4(3):313–19.

Kasembeli, M. M., U. Bharadwaj, P. Robinson, and D. J. Tweardy. 2018. Contribution of STAT3 to Inflammatory and Fibrotic Diseases and Prospects for its Targeting for Treatment. *Int J Mol Sci* 19 (8):pii: E2299.

Kato, A., H. Yoshidome, M. J. Edwards, and A. B. Lentsch. 2000. Reduced Hepatic Ischemia/Reperfusion Injury by IL-4: Potential Anti- Inflammatory Role of STAT6. *Inflamm Res* 49(6):275–79.

Khawar, M. B., F. Azam, N. Sheikh, and K. Abdul Mujeeb. 2016. How Does Interleukin-22 Mediate Liver Regeneration and Prevent Injury and Fibrosis? *J Immunol Res* 2016:2148129.

Kim, W., F. Hong, S. Radaeva, B. Jaruga, S. Fan, and B. Gao. 2003. STAT1 Plays an Essential Role in LPS/D-Galactosamine-Induced Liver Apoptosis and Injury. *Am J Physiol Gastrointest Liver Physiol* 20892:761–68.

Kim, W. H., K. Matsumoto, K. Bessho, and T. Nakamura. 2005. Growth Inhibition and Apoptosis in Liver Myofibroblasts Promoted by Hepatocyte Growth Factor Leads to Resolution from Liver Cirrhosis. *Am J Pathol* 166(4):1017–28.

Kisseleva, T., and D. A. Brenner. 2008. Mechanisms of Fibrogenesis. *Exp Biol Med (Maywood)* 233 (2):109–22.

Klein, C., T. Wüstefeld, U. Assmus, T. Roskams, S. Rose-John, M. Müller, M. P. Manns, M. Ernst, and C. Trautwein. 2005. Triggers Liver Protection in T Cell – Mediated Liver Injury Find the Latest Version: The IL-6–Gp130–STAT3 Pathway in Hepatocytes Triggers Liver Protection in T Cell – Mediated Liver Injury. *J Clin Invest* 115(4):860–69.

Kong, X., D. Feng, H. Wang, F. Hong, A. Bertola, F. S. Wang, and B. Gao. 2012a. Interleukin-22 Induces Hepatic Stellate Cell Senescence and Restricts Liver Fibrosis in Mice. *Hepatology* 56 (3):1150–59.

Kong, X., N. Horiguchi, M. Mori, and B. Gao. 2012b. Cytokines and STATs in Liver Fibrosis. *Front Physiol* 3:1–7.

Kroy, D. C., N. Beraza, D. F. Tschaharganeh, L. E. Sander, S. Erschfeld, A. Giebeler, C. Liedtke, H. E. Wasmuth, C. Trautwein, and K. L. Streetz. 2010. Lack of Interleukin-6/glycoprotein 130/ signal Transducers and Activators of Transcription-3 Signaling in Hepatocytes Predisposes to Liver Steatosis and Injury in Mice. *Hepatology* 51(2):463–73.

Lafdil, F., H. Wang, O. Park, W. Zhang, Y. Moritoki, S. Yin, X. Y. Fu, M. E. Gershwin, Z. X. Lian, and B. Gao. 2009. Myeloid STAT3 Inhibits T cell-mediated Hepatitis by Regulating T Helper 1 Cytokine and Interleukin-17 Production. *Gastroenterology* 137(6):2125–35 e1–2.

Lee, S. C., H. J. Jeong, S. K. Lee, and S. J. Kim. 2016. Hypoxic Conditioned Medium From Human Adipose-Derived Stem Cells Promotes Mouse Liver Regeneration through JAK/STAT3 Signaling. *Stem Cells Transl Med* 5(6):816–25.

Lentsch, A. B., A. Kato, B. Davis, W. Wang, C. Chao, and M. J. Edwards. 2001. STAT4 and STAT6 Regulate Systemic Inflammation and Protect against Lethal Endotoxemia. *J Clin Invest* 108 (10):1475–82.

Li, Y. P., W. Z. Wang, X. Q. Chen, L. B. Li, Z. Y. Liang, K. Ru, and J. N. Li. 2017. Signal Transducer and Activator of Transcription 3 for the Differentiation of Hepatocellular Carcinoma from Cirrhosis. *Chin Med J (Engl)* 130(22):2686–90.

Ma, B., J. Zhu, J. Tan, Y. Mao, L. Tang, C. Shen, H. Zhang, Y. Kuang, J. Fei, X. Yang, and Z. Wang. 2017. Gpr110 Deficiency Decelerates Carcinogen-induced Hepatocarcinogenesis via Activation of the IL-6/STAT3 Pathway. *Am J Cancer Res* 7(3):433–47.

Machida, K., H. Tsukamoto, J. C. Liu, Y. P. Han, S. Govindarajan, M. M. Lai, S. Akira, and J. H. Ou. 2010. c-Jun Mediates Hepatitis C Virus Hepatocarcinogenesis through Signal Transducer and Activator of Transcription 3 and Nitric Oxide-dependent Impairment of Oxidative DNA Repair. *Hepatology* 52(2):480–92.

Mair, M., L. Blaas, C. H. Osterreicher, E. Casanova, and R. Eferl. 2011. JAK–STAT Signaling in Hepatic Fibrosis. *Front Biosci (Landmark Ed)* 16:2794–811.

Meissl, K., S. Macho-Maschler, M. Müller, and B. Strobl. 2017. The Good and the Bad Faces of STAT1 in Solid Tumours. *Cytokine* 89:12–20.

Meraz, M. a, J. M. White, K. C. Sheehan, E. Bach, S.J. Rodig, S. Dighe, D.H. Kaplan et al. 1996. Targeted Disruption of the Stat1 Gene in Mice Reveals Unexpected Physiologic Specificity in the JAK–STAT Signaling Pathway. *Cell* 84:422–31.

Miyagi, T., M. P. Gil, X. Wang, J. Louten, W. M. Chu, and C. A. Biron. 2007. High Basal STAT4 Balanced by STAT1 Induction to Control Type 1 Interferon Effects in Natural Killer Cells. *J Exp Med* 204(10):2383–96.

Moriggl, R., D.J. Topham, S. Teglund, V. Sexl, C. Mckay, D. Wang, A. Hoffmeyer et al. 1999. Stat5 Is Required for IL-2-Induced Cell Cycle Progression of Peripheral T Cells. *Immunity* 10:249–59.

Mueller, K.M., J. Kornfeld, K. Friedbichler, L. Blaas, G. Egger, H. Esterbauer, P. Hasselblatt et al. 2011. Impairment of Hepatic Growth Hormone and Glucocorticoid Receptor Signaling Causes Steatosis and Hepatocellular Carcinoma in Mice. *Hepatology* 54(4):1398–409.

Murphy, K.M., and S.L. Reiner. 2002. The Lineage Decisions of Helper T Cells. *Nat Rev Immunol* 2 (12):933–44.

Murray, P. J. 2006. Understanding and Exploiting the Endogenous Interleukin-10/STAT3-mediated Anti-inflammatory Response. *Curr Opin Pharmacol* 6(4):379–86.

Najjar, I., and R. Fagard. 2010. STAT1 and Pathogens, Not a Friendly Relationship. *Biochimie* 92 (5):425–44.

Naka, T., H. Tsutsui, M. Fujimoto, Y. Kawazoe, H. Kohzaki, Y. Morita, R. Nakagawa, M. Narazaki, K. Adachi, T. Yoshimoto, K. Nakanishi, and T. Kishimoto. 2001. SOCS-1/SSI-1-Deficient NKT Cells Participate in Severe Hepatitis through Dysregulated Cross-Talk Inhibition of IFN-γ and IL-4 Signaling in Vivo. *Immunity* 14(5):535–45.

Njoku, D. B., Z. Li, N. D. Washington, J. L. Mellerson, V. Monica, R. Sharma, and N. R. Rose. 2009. Suppressive and Pro-inflammatory Roles for IL-4 in the Pathogenesis of Experimental Drug-induced Liver Injury. *Eur J Immunol* 39(6):1652–63.

Numata, K., M. Kubo, H. Watanabe, K. Takagi, H. Mizuta, S. Okada, S. L. Kunkel, T. Ito, and A. Matsukawa. 2007. Overexpression of Suppressor of Cytokine Signaling-3 in T Cells Exacerbates Acetaminophen-induced Hepatotoxicity. *J Immunol* 178(6):3777–85.

Ogata, H., T. Chinen, T. Yoshida, I. Kinjyo, G. Takaesu, H. Shiraishi, M. Iida, T. Kobayashi, and A. Yoshimura. 2006. Loss of SOCS3 in the Liver Promotes Fibrosis by Enhancing STAT3-mediated TGF-Beta1 Production. *Oncogene* 25(17):2520–30.

Orlent, H., H. Reynaert, S. Bourgeois, V. Dideberg, M. Adler, I. Colle, S. De Maeght, W. Laleman, P. Michielsen, C. Moreno, J. P. Mulkay, P. Starkel, and J. Delwaide. 2011. IL28B Polymorphism and the Control of Hepatitis C Virus Infection: Ready for Clinical Use? *Acta Gastroenterol Belg* 74 (2):317–22.

Park, O., H. Wang, H. Weng, L. Feigenbaum, H. Li, S. Yin, S.H. Ki et al. 2011. In Vivo Consequences of Liver-Specific Interleukin-22 Expression in Mice: Implications for Human Liver Disease Progression. *Hepatology* 54(1):252–61.

Ruff-Jamison, S., K. Chen, and S. Cohen. 1993. Induction by EGF and Interferon-gamma of Tyrosine Phosphorylated DNA Binding Proteins in Mouse Liver Nuclei. *Science* 261(5129):1733–36.

Ryan, P. M., M. Bourdi, M. C. Korrapati, W. R. Proctor, R. A. Vasquez, S. B. Yee, T. D. Quinn, M. Chakraborty, and L. R. Pohl. 2012. "Endogenous Interleukin-4 Regulates Glutathione Synthesis Following Acetaminophen-induced Liver Injury in Mice." *Chem Res Toxicol* 25(1):83–93. doi: 10.1021/tx2003992.

Sallam, A. M., A. Esmat, and A. B. Abdel-Naim. 2018. Cucurbitacin-B Attenuates CCl4-induced Hepatic Fibrosis in Mice through Inhibition of STAT-3. *Chem Biol Drug Des* 91(4):933–41.

Schindler, C. 1999. Cytokines and JAK–STAT Signaling. *Exp Cell Res* 253(1):7–14.

Schroder, K., P. J. Hertzog, T. Ravasi, and D. A. Hume. 2004. Interferon- Y: An Overview of Signals, Mechanisms and Functions. *J Leukoc Biol* 75:163–89.

Seki, E., and D. A. Brenner. 2015. Recent Advancement of Molecular Mechanisms of Liver Fibrosis. *J Hepatobiliary Pancreat Sci* 22(7):512–18.

Shrivastava, S., E. G. Meissner, E. Funk, S. Poonia, V. Shokeen, A. Thakur, B. Poonia, S. K. Sarin, N. Trehanpati, and S. Kottilil. 2016. Elevated Hepatic Lipid And Interferon Stimulated Gene Expression in HCV GT3 Patients Relative to Non-alcoholic Steatohepatitis. *Hepatol Int* 10(6):937–46.

Siebler, J., S. Wirtz, S. Klein, M. Protschka, M. Blessing, P.R. Galle, and M.F. Neurath. 2003. A Key Pathogenic Role for the STAT1/T-Bet Signaling Pathway in T-Cell-Mediated Liver Inflammation. *Hepatology* 38(6):1573–80.

Simon-Holtorf, J., H. Mönig, H. J. Klomp, A. Reinecke-Lüthge, U.R. Fölsch, and S. Kloehn. 2006. Expression and Distribution of Prolactin Receptor in Normal, Fibrotic, and Cirrhotic Human Liver. *Exp Clin Endocrinol Diabetes* 114(10):584–89.

Sobrevals, L., C. Rodriguez, J. L. Romero-Trevejo, G. Gondi, Ĩ. Monreal, A. Pãneda, N. Juanarena, S. Arcelus, N. Razquin, L. Guembe, G. González-Aseguinolaza, and J. Prieto. 2010. Insulin-like Growth Factor I Gene Transfer to Cirrhotic Liver Induces Fibrolysis and Reduces Fibrogenesis Leading to Cirrhosis Reversion in Rats. *Hepatology* 51(3):912–21.

Subleski, J. J., V. L. Hall, T. C. Back, J. R. Ortaldo, and R. H. Wiltrout. 2006. Enhanced Antitumor Response by Divergent Modulation of Natural Killer and Natural Killer T Cells in the Liver. *Cancer Res no* 66(22):11005–122. doi:10.1158/0008-5472.CAN-06-0811.

Subramaniam, S. V., R. S. Cooper, and S. E. Adunyah. 1999. Evidence for the Involvement of JAK/STAT Pathway in the Signaling Mechanism of Interleukin-17. *Biochem Biophys Res Commun* 262(1):14–19.

Sun, J. 2010. Matrix Metalloproteinases and Tissue Inhibitor of Metalloproteinases Are Essential for the Inflammatory Response in Cancer Cells. *J Signal Transduct* 2010:985132.

Sun, R., and B. Gao. 2004. Negative Regulation of Liver Regeneration by Innate Immunity (Natural Killer Cells/Interferon-γ). *Gastroenterology* 127(5):1525–39.

Sun, R., O. Park, N. Horiguchi, S. Kulkarni, W. I. Jeong, H.Y. Sun, S. Radaeva, and B. Gao. 2006. STAT1 Contributes to DsRNA Inhibition of Liver Regeneration after Partial Hepatectomy in Mice. *Hepatology* 44(4):955–66.

Tanabe, J., A. Izawa, N. Takemi, Y. Miyauchi, Y. Torii, H. Tsuchiyama, T. Suzuki, S. Sone, and K. Ando. 2007. Interferon-β Reduces the Mouse Liver Fibrosis Induced by Repeated Administration of Concanavalin A via the Direct and Indirect Effects. *Immunology* 122(4):562–70.

Tang, L. Y., M. Heller, Z. Meng, L. R. Yu, Y. Tang, M. Zhou, and Y. E. Zhang. 2017. Transforming Growth Factor-beta (TGF-beta) Directly Activates the JAK1-STAT3 Axis to Induce Hepatic Fibrosis in Coordination with the SMAD Pathway. *J Biol Chem* 292(10):4302–12.

Torisu, T., M. Nakaya, S. Watanabe, M. Hashimoto, H. Yoshida, T. Chinen, R. Yoshida, F. Okamoto, T. Hanada, K. Torisu, G. Takaesu, T. Kobayashi, H. Yasukawa, and A. Yoshimura. 2008. Suppressor of Cytokine Signaling 1 Protects Mice against Concanavalin A-Induced Hepatitis by Inhibiting Apoptosis. *Hepatology* 47(5):1644–54.

Toyonaga, T., O. Hino, S. Sugai, S. Wakasugi, K. Abe, M. Shichiri, and K. Yamamura. 1994. Chronic Active Hepatitis in Transgenic Mice Expressing Interferon-Gamma in the Liver. *Proc Natl Acad Sci USA* 91(2):614–18.

Wang, G.L., E. Salisbury, X. Shi, L. Timchenko, E.E. Medrano, and N.A. Timchenko. 2008. HDAC1 Cooperates with C/EBPα In the Inhibition of Liver Proliferation in Old Mice. *J Biol Chem* 283 (38):26169–788.

Wang, H., F. Lafdil, X. Kong, and B. Gao. 2011a. Signal Transducer and Activator of Transcription 3 in Liver Diseases: A Novel Therapeutic Target. *Int J Biol Sci* 7(5):536–50.

Wang, H., F. Lafdil, L. Wang, O. Park, S. Yin, J. Niu, A. M. Miller, Z. Sun, and B. Gao. 2011b. Hepatoprotective versus Oncogenic Functions of STAT3 in Liver Tumorigenesis. *Am J Pathol* 179 (2):714–24.

Wang, H., F. Lafdil, L. Wang, S. Yin, D. Feng, and B. Gao. 2011c. Tissue Inhibitor of Metalloproteinase 1 (TIMP-1) Deficiency Exacerbates Carbon Tetrachloride-induced Liver Injury and Fibrosis in Mice: Involvement of Hepatocyte STAT3 in TIMP-1 Production. *Cell Biosci* 11:14.

Wang, K. S., J. Ritz, and D. A. Frank. 1999. IL-2 Induces STAT4 Activation in Primary NK Cells and NK Cell Lines, but Not in T Cells. *J Immunol* 162(1):299–304.

Wang, W., H. Guo, H. Li, Y. Yan, C. Wu, X. Wang, X. He, and N. Zhao. 2018. Interleukin-35 Gene-Modified Mesenchymal Stem Cells Protect Concanavalin A-Induced Fulminant Hepatitis by Decreasing the Interferon Gamma Level. *Hum Gene Ther* 29(2):234–41.

Wang, Y., A. Qu, and H. Wang. 2015. Signal Transducer and Activator of Transcription 4 in Liver Diseases. *Int J Biol Sci* 11(4):448–55.

Wen, J., Y. Zhou, J. Wang, J. Chen, W. Yan, J. Wu, J. Yan et al. 2017. Interactions between Th1 Cells and Tregs Affect Regulation of Hepatic Fibrosis in Biliary Atresia through the IFN- 3/STAT1 Pathway. *Cell Death Differ* 24(6):997–1006.

Weng, H. L., Y. Liu, J. L. Chen, T. Huang, L. J. Xu, P. Godoy, J. H. Hu, C. Zhou, F. Stickel, A. Marx, R. M. Bohle, V. Zimmer, F. Lammert, S. Mueller, M. Gigou, D. Samuel, P. R. Mertens, M. V. Singer, H. K. Seitz, and S. Dooley. 2009. The Etiology of Liver Damage Imparts Cytokines Transforming Growth Factor B1 or Interleukin-13 as Driving Forces in Fibrogenesis. *Hepatology* 50(1):230–43.

Weng, S. Y., X. Wang, S. Vijayan, Y. Tang, Y. O. Kim, K. Padberg, T. Regen, O. Molokanova, T. Chen, T. Bopp, H. Schild H. F. Brombacher, J. R. Crosby, M. L. McCaleb, A. Waisman, E. Bockamp, and D. Schuppan. 2018. IL-4 Receptor Alpha Signaling through Macrophages Differentially Regulates Liver Fibrosis Progression and Reversal. *EBioMedicine* 29:92–103.

Woelfle, J., J. Billiard, and P. Rotwein. 2003. Acute Control of Insulin-like Growth Factor-I Gene Transcription by Growth Hormone through Stat5b. *J Biol Chem* 278(25):22696–702.

Wurster, A. L., T. Tanaka, and M. J. Grusby. 2000. The Biology of Stat4 and Stat6. *Oncogene* 19 (21):2577–84.

Xu, H., Z. Fan, W. Tian, and Y. Xu. 2016. Protein Inhibitor of Activated STAT 4 (PIAS4) Regulates Liver Fibrosis through Modulating SMAD3 Activity. *J Biomed Res* 30(6):496–501.

Yang, Y., X. Q. Wu, W. X. Li, H. M. Huang, H. D. Li, X. Y. Pan, X. F. Li, C. Huang, X. M. Meng, L. Zhang, X. W. Lv, H. Wang, and J. Li. 2018. PSTPIP2 Connects DNA Methylation to Macrophage Polarization in CCL4-induced Mouse Model of Hepatic Fibrosis. *Oncogene* 37 (47):6119–35.

Younossi, Z. M., R. Loomba, Q. M. Anstee, M. E. Rinella, E. Bugianesi, G. Marchesini, B. A. Neuschwander-Tetri, L. Serfaty, F. Negro, S. H. Caldwell, V. Ratziu, K. E. Corey, S. L. Friedman, M. F. Abdelmalek, S. A. Harrison, A. J. Sanyal, J. E. Lavine, P. Mathurin, M. R. Charlton, Z. D. Goodman, N. P. Chalasani, K. V. Kowdley, J. George, and K. Lindor. 2018. Diagnostic Modalities for Nonalcoholic Fatty Liver Disease, Nonalcoholic Steatohepatitis, and Associated Fibrosis. *Hepatology* 68(1):349–60.

Yu, J. H., B. Zhu, M. Wickre, G. Riedlinger, W. Chen, A. Hosui, W. R. Gertraud, and L. Hennighausen. 2011. The Transcription Factors STAT5A and STAT5B Negatively Regulate Cell Proliferation through the Activation of Cdkn2b and Cdkn1a Expression. *Hepatology* 52 (5):1808–18.

Zhang, H., F. Chen, X. Fan, C. Lin, Y. Hao, H. Wei, W. Lin, Y. Jiang, and F. He. 2017. Quantitative Proteomic Analysis on Activated Hepatic Stellate Cells Reversion Reveal STAT1 as a Key Regulator between Liver Fibrosis and Recovery. *Sci Rep* 7:1–10.

Zhang, L., N. Jilg, R. Shao, W. Lin, and D. Fusco. 2011. IL28B Inhibits Hepatitis C Virus Replication through the JAK–STAT Pathway. *J Hepatol* 55(2):289–98.

Zhu, R., S. Diem, L. M. Araujo, A. Aumeunier, J. Denizeau, E. Philadelphe, D. Damotte, M. Samson, P. Gourdy, M. Dy, E. Schneider, and A. Herbelin. 2007. The Pro-Th1 Cytokine IL-12 Enhances IL-4 Production by Invariant NKT Cells: Relevance for T Cell-mediated Hepatitis. *J Immunol* 178 (9):5435–42.

10

Renal Disorders: Involvement of JAK-STAT Pathway

Yuji Nozaki
Department of Hematology and Rheumatology
Kindai University School of Medicine
Osaka-Sayama, Japan

10.1 Introduction

Many signaling pathways have been shown to be involved in the progression of renal disease in humans and in animal models, usually due to persistent activation of these pathways. Perhaps the best well known of such pathways include certain protein kinase pathways, such as protein kinase C and mitogen-activated kinase pathways, which lead to cellular growth and fibrosis, and those, such as angiotensin (ANG) II and transforming growth factor-β (TGF-β)/SMAD signaling pathways, which enhance fibrosis leading to kidney scarring (Chen, Chen, and Harris 2012; Lan and Chung 2012; Li and Gobe 2006; Siragy and Carey 2010). In addition, mesangial cells possess both contractile and mitogenic properties and contribute to the physiological regulation of glomerular dynamics (Johnson et al. 1993; Wenzel et al. 1995). Growth factors have a maladaptive role in the glomerular damage accompanying experimental and human glomerulonephritis.

One of the major pathways that responds to and transduces inflammatory signals is the Janus kinase–signal transducer and activator of transcription (JAK-STAT) pathway. The JAK-STAT pathway transmits signals from extracellular ligands, including many cytokines and chemokines as well as growth factors and hormones, directly to the nucleus to induce a variety of cellular responses (Lu et al. 2009). While many of these responses are present and best characterized in lymphoid cells, they have also been reported in many other cell types including kidney parenchymal cells such as podocytes, mesangial cells, and tubular cells (Berthier et al. 2009; Choudhury, Ghosh-Choudhury, and Abboud 1998). Given the role of JAK-STAT activation in response to cytokines and chemokines and the unequivocal role of inflammation in promoting progressive kidney injury, it is not surprising that JAK-STAT activation is involved in the pathogenesis of renal disease as well as of acute kidney injury. Since its discovery, JAK-STAT pathway has been intensively studied. Because of its strong immunomodulatory function, JAK-STAT pathway and its components are good candidates for immunological intervention for disease control. Indeed, several clinical trials administering JAK inhibitor have been performed for various diseases such as chronic kidney disease (CKD), rheumatoid arthritis (RA), psoriasis, inflammatory bowel disease, and atopic dermatitis (Boyle et al. 2015; Brosius, Tuttle, and Kretzler 2016; Levy et al. 2015; Olivera, Danese, and Peyrin-Biroulet 2017).

There is considerable interest in JAK-STAT as a therapeutic target, especially as a treatment for renal disease. CKD in renal disorders is one of the leading causes of morbidity and mortality in patients with diabetic nephropathy, lupus nephritis, and kidney transplantation, and ANG II and inflammatory cytokines have been implicated in the pathogenesis of maladaptive growth and inflammation in renal tissues in these patients (Ling, Seal, and Eaton 1993; Wang et al. 2010).

In this review, the JAK-STAT signaling mechanism and how the inhibition of JAK-STAT expression aids in the development of effective treatments for the renal diseases have been discussed.

10.2 JAK-STAT Signaling

The JAK-STAT pathway is the major signaling cascade downstream of type I and type II cytokine receptors. JAKs have been shown to be an effective therapeutic target for various autoimmune and inflammatory diseases driven by cytokines (Clark, Flanagan, and Telliez 2014; O'Shea et al. 2015). JAKs are cytoplasmic tyrosine kinases that participate in the signaling of a broad range of cell surface receptors, particularly members of the cytokine receptor common gamma (cg) chain family (Vincenti and Kirk 2008). There are four mammalian JAKs: JAK1, 2, 3, and tyrosine kinase 2 (Tyk2). The activation of JAK occurs by a ligand-receptor interaction, which results in signaling through the phosphorylation of cytokine receptors and the creation of docking sites for signaling proteins known as STAT (Podder and Kahan 2004). JAKs catalyze STAT phosphorylation, which facilitates STAT dimerization and nuclear transport. The result is the regulation of gene expression and transcription (Podder and Kahan 2004; Wiik et al. 1994). Although JAK proteins are structurally related, their differences in activation and downstream effects allow for a high degree of specificity. JAK1 is activated by the ligands binding to the class II receptors (interferon (IFN)-α/β, IFN-γ, and interleukin (IL)-10) and the γc receptors (IL-2, IL-4, IL-7, IL-9, IL-15, and IL-21) (Figure 10.1). JAK2 is activated mostly by thrombopoietin receptor, IL-3, granulocyte-macrophage-colony-stimulating factor (GM-CSF), and IFN-γ. While the majority of JAKs are ubiquitously expressed, JAK3 expression appears to be generally restricted to hematopoietic lineages and vascular muscle cells and JAK3 plays a critical role in lymphocyte development and function. Tyk2 mediates mostly the signaling induced by IL-12. JAK3, unlike the ubiquitous expression of other JAK subtypes, has a restricted tissue distribution and is primarily found on hematopoietic cells and uniquely associates with the cg chain (Witthuhn et al. 1994). The importance of this signaling pathway is evident when one considers that mice and humans, with genetic absence or mutation in either the cg subunit or JAK3, express defects in lymphoid development that give rise to a severe combined immunodeficiency syndrome phenotype (Buckley 2004).

Since the JAK-STAT pathways play major activating roles in a variety of disease processes, there has been a robust effort to develop specific inhibitors of this pathway. As it is relatively easy to identify inhibitors of protein kinases, development of JAK inhibitors has received most attention. At present, three JAK inhibitors have received FDA approval for clinical use. Ruxolitinib (Jakafi, Incyte) is a potent inhibitor of both JAK1 and JAK2, and it received FDA approval in November 2011 for the treatment of polycythemia vera and myelofibrosis (Verstovsek et al. 2010). Tofacitinib (Xeljanz, Pfizer) was initially designed to be a specific inhibitor of JAK1 and JAK3 kinase and it received FDA approval in November 2012, and therefore has been used as an immunosuppressant in transplantation and for the treatment of autoimmune diseases (Borie et al. 2003). Baricitinib as a JAK1 and JAK2 inhibitor (Olumiant, Eli Lilly) received FDA approval for the treatment of moderately to severely active RA in June 2018. Several other JAK inhibitors have been developed as immunosuppressive agents for RA and other autoimmune diseases. For example, upadacitinib (a JAK1 inhibitor), filgotininib (a JAK1 inhibitor), and peficinib (a JAK1 inhibitor) have been shown to be effective in the treatment of arthritis (Burmester et al. 2018; Westhovens et al. 2017). Therefore, it is likely that additional JAK inhibitors will become clinically available in the next few years.

STAT genes are also essential for proper immune function, and loss-of-function mutations in these proteins can be associated with immunodeficiency syndromes (Holland, DeLeo, and Elloumi 2007). Certain JAK-STAT polymorphisms are associated with an increased risk of developing autoimmune diseases (Schwartz et al. 2016). These observations have also led to the development of JAK inhibitors for the treatment of human disease (Clark, Flanagan, and Telliez 2014).

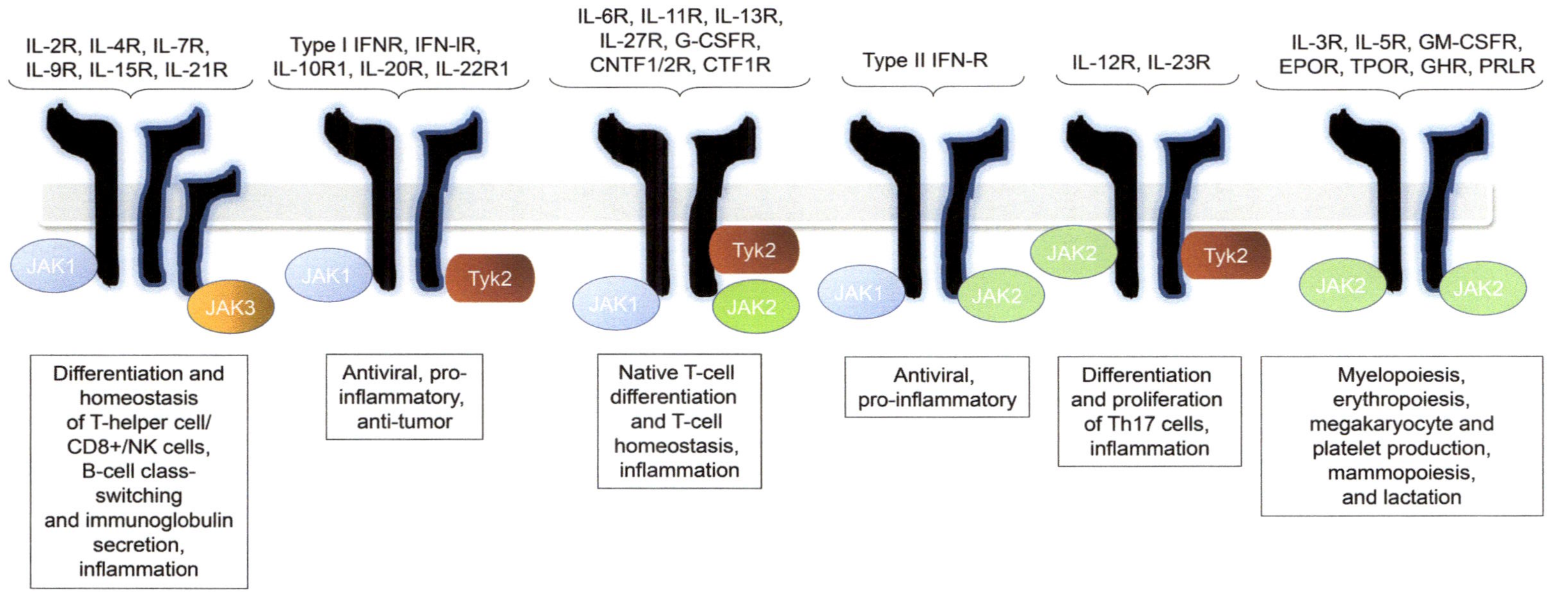

FIGURE 10.1 JAK inhibitions and immune regulation by JAK pathway. Inhibition of JAK3 would affect signaling mediated by only the common gamma chain-associated cytokine receptors (IL-2R, IL-4R, IL-9R, and IL-21R) and regulate T cell, NK cell, and B cell function. On the other hand, JAK2 or TYK2 inhibition would influence multiple cytokine receptor signaling pathways.

Abbreviation: EPO, erythropoietin: GH, growth hormone: TPO, thrombopoietin: GM-CSF, granulocyte macrophage colony-stimulating factor: PRL, prolactin receptor: CNTFR, ciliary neutrophic factor receptor: IL, interleukin: IFN, interferon: NK, natural killer.

10.3 JAK Inhibitor

10.3.1 Tofacitinib

Tofacitinib is a first-generation JAK inhibitor that inhibits JAK1 and JAK3 (with greater potency than JAK2) (Changelian et al. 2003; Furumoto and Gadina 2013). Since common γ chain-using cytokines use JAK1 and JAK3, tofacitinib efficiently blocks their signaling cascades. Tofacitinib also blocks gp130-using cytokines as well as signaling downstream of type I/II IFNs. Tofacitinib may, therefore, have diverse effects on cytokines and cells that contribute to SLE pathogenesis, including subsets of $CD4^+$ T cells (helper T cells including Th1 and pathogenic Th17 cells), $CD8^+$ T cells, B cells, and innate immune cells (Ghoreschi, Jesson, and Li 2011; Okiyama et al. 2014; Xing et al. 2014). Tofacitinib has been reported to increase HDL and a previous report indicated that this drug can restore lipoprotein homeostasis and improve cholesterol efflux capacity in RA patients (Boers et al. 2003; Charles-Schoeman et al. 2015). The effects of tofacitinib on cholesterol transport were proposed to be resulting from an increase in cholesterol ester production rate through augmented activity of the lecithin-cholesterol acyltransferase (Charles-Schoeman et al. 2015). Given the changes observed in lipoprotein profiles of tofacitinib-treated mice and, in particular, the reduction in the fraction of free cholesterol—which has recently been shown to be pro-atherogenic—the lipoprotein changes induced by tofacitinib could be an additional mechanism by which JAK inhibition could be vasculoprotective in systemic autoimmunity (Thacker et al. 2015).

10.3.2 Baricitinib

Baricitinib is a novel oral molecule that potently inhibits JAK1 and JAK2 (with greater potency than JAK3 and TyK2) (Richez et al. 2017). The drug was developed as an immunosuppressive agent for the treatment of RA, atopic dermatitis, and systemic lupus. Its development in psoriasis and diabetic chronic disease has been halted. In vitro, baricitinib inhibited generation of type I and type II IFNs (IFNα and IFNγ), the common γ-chain cytokines (IL-15 and IL-21), and IL-6 and IL-27. Baricitinib also showed some decrease in potency for IL-10, IL-12, IL-23, and erythropoietin (EPO). Baricitinib demonstrated dose- and time-dependent inhibition of cytokine-induced pSTAT3, with maximal inhibition of pSTAT3 occurring 1–2 h after administration for all doses (Shi et al. 2014). Efficacy in rat adjuvant-induced arthritis correlated with the level of pSTAT3 in a dose- and time-dependent manner. Efficacy was observed as soon as pSTAT3 levels were inhibited by 15% and even more than 24 h later, suggesting that baricitinib possesses significant activity even in the absence of complete and continuous pathway inhibition (Fridman et al. 2010). In cell-based assays using human T cells, the drug potently inhibited phosphorylation of STAT3 and subsequent production of monocyte chemoattractant protein-1, as well as IL-23, and induced STAT3 phosphorylation and the subsequent production of IL-17 and IL-22 (Markham 2017). Furthermore, JAK1 and JAK2 inhibition, and not JAK3 alone, has been shown to control STAT3-dependent signaling pathway (Migita et al. 2013).

10.3.3 Ruxolitinib

Ruxolitinib is unique among JAK kinase inhibitors in clinical development as it potently inhibits JAK1 in addition to JAK2 for the treatment of polycythemia vera and myelofibrosis (Verstovsek et al. 2010). JAK1 is hyperactivated in the peripheral blood of myelofibrosis patients (Quintás-Cardama et al. 2010). Furthermore, Jedidi et al. demonstrated that cytokines capable of signaling through JAK1 can convey resistance to JAK2-targeted therapies (Jedidi et al. 2009). JAK2 also mediates cytokine signaling from EPO, thrombopoietin (TPO), and GM-CSF receptors (Witthuhn et al. 1993). Germline deletion of JAK2 causes embryonic lethality by deficient definitive erythropoiesis, which was confirmed by a conditional JAK2 allele and germline Cre-recombinase (Krempler et al. 2004;

Neubauer et al. 1998; Parganas et al. 1998). Conditional JAK2 loss in hematopoietic stem cells/progenitor cells induces anemia and thrombocytopenia, whereas deletion of the TPO receptor MPL and MPL in platelets leads to thrombocytosis (Grisouard et al. 2014; Lannutti et al. 2009; Ng et al. 2014; Park et al. 2013; Pikman et al. 2006; Tiedt et al. 2009). Somatic mutations in the JAK2 pathway, including JAK2V617F, JAK2 exon 12 mutations, and in the MPL, are observed in myeloproliferative neoplasms (MPN) patients. JAK kinase inhibitors attenuate splenomegaly and constitutional symptoms in MPN but cannot induce molecular responses (Baxter et al. 2005; Harrison et al. 2012; James et al. 2005; Kralovics et al. 2005; Levine et al. 2005; Scott et al. 2007; Verstovsek et al. 2012). JAK inhibitor therapy results in the variable development of anemia and thrombocytopenia preventing dose intensification and maximal target inhibition (Mesa and Cortes 2013). Cytopenias have been considered on-target effects of JAK2 inhibitors inseparable from therapeutic efficacy. However, the risk of thrombocytopenia varies among JAK2 inhibitors, suggesting thrombocytopenia may not be a direct consequence of JAK2 inhibition (Tam and Verstovsek 2013). Thus, there may be both cell autonomous and non-autonomous benefits of inhibiting JAK1 as well as JAK2.

10.3.4 Next Generation of JAK Inhibitors

The various clinical trials in the new JAK inhibitors such as filgotinib (Galapagos/Gilead), upadacitinib (AbbVie), peficitinib (Smyraf, Astellas), and decernotinib (Vertex) are running in RA (Burmester et al. 2018; Genovese et al. 2016, 2017; Kavanaugh et al. 2017). These data are seemed to be very promising for treating RA. Filgotinib demonstrated a selective inhibition of JAK1 and JAK2 over JAK3 and Tyk2, and specifically in whole blood assays, a selectivity of 30-fold for JAK1 over JAK2 was revealed. Details of the in vitro assays used for determining JAK selectivity are discussed by Van Rompaey et al. in their preclinical work on filgotinib (Van et al. 2013). Upadacitinib demonstrated 74- and 58-fold greater selectivity for JAK1 over JAK2 and JAK3, respectively (Mohamed et al. 2016). Peficitinib has moderate selectivity for JAK3 and inhibited JAK1 and JAK3 with 50% inhibitory concentrations of 3.9 and 0.7 nM, respectively; it has also shown 7.1-fold selectivity for JAK3 relative to JAK2. Peficitinib phase III has performed in south-east Asia only. Decernotinib, in first evaluation, was demonstrated to be a selective and potent inhibitor of JAK3 in vitro and modulated pro-inflammatory responses in models of immune-mediated diseases, such as collagen-induced arthritis and delayed-type hypersensitivity (Mahajan et al. 2015). Both JAK1 and JAK3 inhibition apparently are effective and one needs to explore the mechanistic pathways leading to efficacy. Just labeling for certain specificity is not appropriate when one looks at the differences in ex vivo results and in vitro cellular pharmacology from baricitinib to upadacitinib, filgotinib, and tofacitinib (McInnes, Higgs, and Lee 2017). JAK inhibitors display different in vitro pharmacological profiles, which, coupled to their in vivo pharmacokinetics, suggests that they modulate distinct cytokine pathways to differing degrees and durations over 24 h. Baricitinib and filgotinib inhibited JAK1/3 signaling to a lesser extent than upadacitinib and tofacitinib. Differences between the drugs are certainly there and the differential clinical relevance needs specific study. In renal disorders, further studies also need to examine whether new JAK inhibitors could be effective on the progression of renal damage.

10.4 Renal Disease

10.4.1 Diabetic Nephropathy

Diabetic nephropathy is the most common cause of end-stage renal disease (ESRD) in many other countries today (Saran et al. 2015). Furthermore, its incidence continues to increase because of the epidemic of type 2 diabetes, despite the improvement in the rates of progressive diabetic nephropathy (Saran et al. 2015). Diabetic nephropathy is characterized by sustained inflammation that promotes and directs much of the chronic injury process (Wada and Makino

2013). One of the major pathways that transduce inflammatory signals in diabetic nephropathy is the JAK-STAT pathway. The JAK-STAT pathway transmits signals from extracellular ligands, including many cytokines and chemokines as well as growth factors and hormones, directly to the nucleus to induce a variety of cellular responses (O'Shea and Plenge 2012). While many of these responses are best characterized in lymphoid cells, they have also been reported in intrinsic kidney cells such as podocytes, mesangial cells, and tubular cells (Berthier et al. 2009; Choudhury, Ghosh-Choudhury, and Abboud 1998). Gene and protein expression studies of kidney biopsies from people with early- and late-stage diabetic nephropathy have shown increased activation and expression of the JAK-STAT signaling pathway across the spectrum of diabetic nephropathy (Berthier et al. 2009; Woroniecka et al. 2011). In particular, increased expression and activity of JAK1 and JAK2 appear to promote diabetic nephropathy development and progression (Brosius and He 2015; Navarro-González et al. 2011; Zhang et al. 2017). Moreover, studies have suggested interactions between JAK-STAT and angiotensin signaling, including evidence for activating JAK2 (Marrero et al. 2006). Marrero MB et al. also reported that activation of the JAK-STAT pathway is essential for ANG II-induced growth of glomerular mesangial cells (Amiri et al. 2002; Wang et al. 2002). In addition, they have also shown that high glucose augments ANG II-induced activation of the JAK-STAT pathway in rat kidney glomeruli and that this pathway contributes importantly to production of the cytokine TGF-β as well as the extracellular matrix proteins—collagen IV and fibronectin (Banes et al. 2004; Wang et al. 2002). Furthermore, Ha H et al. have also demonstrated that high glucose induces intracellular production of reactive oxygen species (ROS), which have been shown to augment ANG II-induced signaling cascades and activate JAK and STAT proteins (Ha and Lee 2000; Simon et al. 1998; Ushio-Fukai et al. 1999). Figure 10.2 show the schematic depiction of high glucose-induced augmentation of ANG II-induced growth via the JAK-STAT pathway in glomerular mesangial cells. In the context of diabetic complications, evaluating the molecular mechanisms responsible for modulating ANG II-induced activation of the JAK-STAT pathway in mesangial cells by high glucose is an important area. The mechanisms by which high glucose promotes JAK2 activation may be related to activation of JAK2 by ROS, and ROS are induced

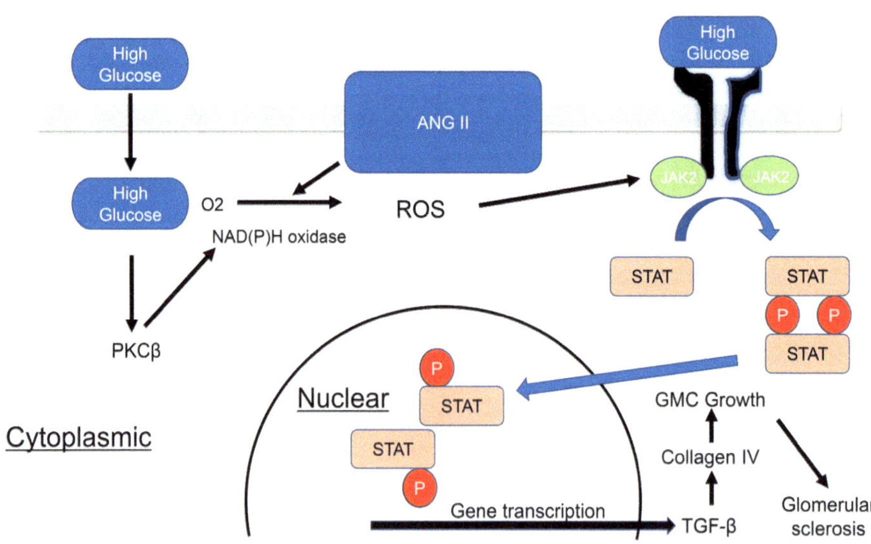

FIGURE 10.2 Schematic depiction of high glucose-induced augmentation of ANG II-induced growth via the JAK-STAT pathway in glomerular mesangial cells.

Abbreviations. ROS, reactive oxygen species: GMC, glomerular mesangial cell: ANG II, angiotensin II: STAT, signal transducer and activator of transcription: P, phosphorylation: JAK2, Janus kinase2: TGF-β, transforming growth factor-β.

by high glucose in mesangial cells (Ha and Lee 2000). A previous report has shown that high glucose, via the polyol pathway, induces a rapid increase in intracellular ROS, such as H2O2, which stimulates intracellular signaling events similar to those activated by ANG II, including phosphorylation of growth-promoting kinases such as JAK2 (Shaw et al. 2003). The polyol pathway generates ROS (Ha and Lee 2000; Shaw et al. 2003), which can then act as signaling mediators in the activation of downstream mitogenic pathways, such as the JAK-STAT cascade (Simon et al. 1998). It has also recently been reported that ANG II induces a rapid increase in intracellular H2O2, via NAD(P)H oxidase, which subsequently activates growth-related responses (Ushio-Fukai et al. 1999). In the addition, Bhat GJ et al. has shown that high glucose altered activation of the JAK-STAT pathway in vivo through ANG II in rat kidney glomeruli by inducing phosphorylation of JAK2 kinase and STAT proteins, namely, STAT1, STAT3, and STAT5A/B (Bhat et al. 1994, 1995; Chappey et al. 1997). The authors also reported that the streptozotocin-induced diabetic rats treated with the JAK2 inhibitor tyrphostin showed a significant reduction in urine output as well as a significant reduction in fluid intake. Furthermore, treatment with tyrphostin lowered systolic blood pressure and significantly reduced urinary protein excretion. These results provide further support for the hypothesis that the JAK2 contributes importantly to both acute and chronic glomerular dysfunction in diabetic nephropathy. Berthier CC et al. also found that expression of JAK2 protein in proximal tubule cells was high in patients with progressive kidney disease compared with that in patients without kidney disease and even compared with that in a few individuals with other glomerular diseases (Berthier et al. 2009). Li X et al. reported that Paeoniflorin attenuates renal lesions in diabetic mice and the protective effects may be associated with the prevention of macrophage infiltration and inhibition of the JAK2–STAT3 signaling pathway (Li et al. 2018). Other important regulators of JAK-STAT signaling are the suppressors of cytokine signaling (SOCS). SOCS1 and SOCS3 are downstream transcriptional targets of the JAK-STAT signaling pathway that bind to JAK proteins and interfere with their activity in a negative-feedback manner (Croker, Kiu, and Nicholson 2008). Rats injected with recombinant SOCS1 and SOCS3 adenovirus showed evidence of reduced JAK-STAT activation and amelioration in early diabetic changes (Õrtiz-Munoz et al. 2010). A previous study also reported that STAT3 acetylation is increased in both mouse and human diabetic kidneys (Liu et al. 2014). This has functional implications because acetylation of STAT proteins enhances STAT dimerization, which is critical for the translocation of STATs to the nucleus and their ability to modulate gene transcription (Zhuang 2013). Treatment with inhibitors that block acetylation-mediated association of STAT3 reduced both proteinuria and kidney injury in the db/db mouse model of diabetes.

Baricitinib, known as Olumiant, is a novel oral molecule that potently inhibits JAK1 and JAK2 (with greater potency than JAK3 and TyK2) (Richez et al. 2017). In diabetic nephropathy, baricitinib treatment resulted in a reduction in albuminuria at 3 and 6 months (Tuttle et al. 2018). After 4 weeks of study drug wash-out, the albuminuria reduction was sustained in the medium- and high-dose baricitinib groups. The reduction in albuminuria was approximately 40% versus placebo in the highest-dose baricitinib group. Importantly, there was evidence of target engagement and a reduction in renal inflammation with baricitinib treatment, as assessed by urinary levels of IFN-γ-induced protein 10 and plasma levels of TNF receptor 2 (TNFR2). Elevated plasma TNFR2 and TNFR1 levels have been reported as the biomarkers with high prognostic power for progression of diabetic nephropathy in both type 1 and type 2 diabetic populations (Gohda et al. 2012; Niewczas et al. 2012). However, there was a statistically significant decrease in hemoglobin compared to placebo in the high-dose baricitinib group at 6 months because EPO signaling is dependent on JAK2 activation (Kuhrt and Wojchowski 2015). Further studies are needed to examine the effects of baricitinib on progression of diabetic nephropathy for the treatment as determined by harder endpoints, such as estimated glomerular filtration (eGFR) decline, ESRD, and mortality events.

10.4.2 Lupus Nephritis

Systemic lupus erythematosus (SLE) is a chronic systemic autoimmune disorder characterized by loss of tolerance against nuclear autoantigens, anti-double-stranded DNA antibody production, immune complex (IC) deposition, and leukocyte infiltration in many target organs, as well as activated B and T cells (Bagavant and Fu 2009; Tsokos 2011). Lupus nephritis is an IC glomerulonephritis categorized as one of the most serious complications of SLE and is also one of the strongest predictors of a poor prognosis. Lupus nephritis can lead to severe proteinuria, hypertension, chronic renal failure, and, finally, ESRD. The lupus mouse model MRL/MpJ-Faslpr /lpr/J (MRL/lpr) recapitulates several of the clinical manifestations and immune dysregulation observed in human SLE. Indeed, these mice develop IC glomerulonephritis, inflammatory skin disease, aberrant T cell responses, dysregulated inflammatory cytokine synthesis, and an enhanced type I IFN signature (Cohen and Eisenberg 1991; Perry et al. 2011). Furthermore, MRL/lpr mice also represent a good model to study vascular dysfunction and neutrophil dysregulation characteristic of lupus (Andrews, Eisenberg, and Theofilopoulos 1978). Furumoto et al. reported the inhibition of the JAK-STAT pathway significantly ameliorated the lupus clinical phenotype and modulated features of dysregulated innate and adaptive immunity characteristic of this disease (Furumoto et al. 2017). Importantly, tofacitinib significantly improved vascular parameters suggesting a potential modulatory role in cardiovascular risk in this disease. While the treatments for SLE have improved over the last several decades, current therapies are still suboptimal, promote significant side effects and, to this date, have not shown to consistently modify both lupus disease activity, and the enhanced cardiovascular risk characteristic of this disease. Furumoto et al. also described tofacitinib has pleiotropic beneficial effects in murine lupus, which includes: (a) significant amelioration of clinical phenotype and organ damage; (b) modulation of innate and adaptive immune dysregulation; and (c) improvement in vascular functions and lipoprotein profiles. The observed effects of tofacitinib on innate and adaptive immune dysregulation likely play key roles in modifying clinical responses and vasculopathy. Indeed, pharmacologic JAK inhibition resulted in significant abrogation in IFN responses and other pro-inflammatory cytokines considered crucial in lupus pathogenesis, neutrophil extracellular traps (NETs), and T lymphocyte subsets associated with lupus pathogenicity. The main effects induced by tofacitinib on adaptive immune cells were a reduction of $CD8^+$ and dendritic (DN) T cells. This is consistent with other inflammatory models driven by cytotoxic $CD8^+$ lymphocytes, where JAK inhibition strongly suppressed these responses (Craiglow and King 2015; Okiyama et al. 2014; Xing et al. 2014). In SLE, higher numbers of cytotoxic, effector $CD8^+$ T cells are associated with disease activity and may lead to autoantigen generation through perforin/granzyme-related pathways (Blanco et al. 2005). DN T cells are expanded in the peripheral blood and tissues from SLE patients as well as murine models. These cells synthesize pro-inflammatory cytokines and induce antibody production through promotion of B-cell differentiation (Boggio et al. 2004; Crispin et al. 2008; Edgerton et al. 2009; Wong et al. 2008). Given the significant reductions in splenic DN T cells induced by tofacitinib and the decreases in circulating cytokine levels, it is likely that some of the beneficial effects of this drug in the lupus phenotype, including vasculopathy, are due to repression of such responses (Usui et al. 2012). Synthesis of type I IFNs driven by endogenous nucleic acids is considered an early event that primes the immune system for immune dysregulation and development of autoimmunity. As such, strategies that target the type I IFN pathway are actively being investigated (Kirou and Gkrouzman 2013). As type I IFNs signal through the JAK-STAT pathway, tofacitinib blocked the biological responses to these cytokines that was reflected in significant decreases in the type I IFN signature in these mice. It is likely that inhibition of this pathway by tofacitinib led to pleiotropic effects limiting dysregulation of innate and adaptive immune responses including roles in priming neutrophils to undergo NET formation, alterations in B cell ontogeny, as well as improvements in lupus vasculopathy. NETs have been proposed to represent an important source of immunostimulatory molecules and modified autoantigens that can trigger loss of tolerance in predisposed hosts. Furthermore, NETs are involved in the induction of type I IFN responses through activation of plasmacytoid DN cells. Cytokines, autoantibodies, and immune complexes can trigger enhanced NETs formation and may be involved in neutrophil dysregulation in SLE (Kaplan 2011). As tofacitinib treatment led to decreases in all of these lupus-associated features, it is possible that this is an important mechanism by which this drug improves the clinical phenotype as well as vascular dysfunction.

While the mechanisms leading to amelioration of the vasculopathy with tofacitinib are yet unclear, the significant modulation that this drug promoted in both type I IFN responses and NET formation, both considered important factors in premature endothelial damage, make it likely that these are the key downstream pathways by which this drug was effective (Denny et al. 2010; Knight et al. 2013; Thacker et al. 2010, 2012).

The specific blockade of JAK2 could also help treat several of the pathological manifestations of SLE, including arthritis and dermatitis, and may even help prevent or treat some comorbidities, such as atherosclerosis, as JAK2 has demonstrated a role in the development of atherosclerotic plaques and is necessary for EPO function in megakaryocytes (Ghoreschi, Laurence, and O'Shea 2009; López-Pedrera et al. 2010; Mazière, Conte, and Mazière 2001). Multiple cytokines have been implicated in playing key roles during the initiation, progression, and development of SLE including, but not limited to, IL-6, IL-12, and type I IFN (α/β) (Aringer and Smolen 2004; Chun et al. 2007; Niewold et al. 2007; Tucci et al. 2008). All three cytokines signal via receptors controlled by JAK kinases. Signaling of IL-6 via the IL-6R on activated B cells induces the dimerization with gp130 and activation of the receptor-associated JAK tyrosine kinases, JAK1 and JAK2. Most important is the role of IL-6 as this cytokine has been implicated in multiple autoimmune diseases and directly contributes to plasma cell survival in bone marrow niches (Winter et al. 2010). In addition, multiple studies in mouse models of SLE have repeatedly demonstrated the importance of IL-6 in driving SLE disease manifestations (Aringer and Smolen 2004; Cash et al. 2010; Chun et al. 2007; Jeon et al. 2010; Ripley et al. 2005). The activation of JAKs triggers the phosphorylation of IL-6R and gp130, followed by the activation of various secondary messengers and transcription factors including STAT3, MAPKs, and Akt providing growth and proliferation signals (Silver and Hunter 2010). Genetic ablation or polymorphisms in key suppressors of JAK-STAT signaling such as SOCS have been implicated in increased serum IL-6 levels and risk of developing SLE in humans (Fujimoto et al. 2004; Yu et al. 2003). In addition, JAKs play a critical role in transducing signals from the IL-6R, and IL-6 is involved in both SLE and the maintenance of the pool of potentially autoreactive plasma cells. Therefore, blockade of JAK signaling using a selective and potent JAK2 inhibitor could weaken the supportive effects of IL-6 on sustaining autoreactive plasma cells in SLE. Targeting cytokine/growth factor pathways important for plasma cell generation and SLE development is supported by literature, and targeting the IL-6 pathway and receptor for SLE treatment is currently being tested (Febbraio, Rose-John, and Pedersen 2010; Illei et al. 2010; Silver and Hunter 2010). Targeting of the BAFF pathway has been successful, and studies in the field are under way to look at the role of a proliferation-inducing ligand in plasma cell generation (Townsend, Monroe, and Chan 2010). A possible next step is to target the IL-6R pathway via the blockade of JAK2 to better target the development and survival of autoreactive, pathogenic plasma cells during early SLE.

However, it would be the possibility to progress the reduction of red blood cells, presumably due to JAK2-mediated EPO inhibition; in future studies, it will be important to assess the effect of JAK inhibitors on predisposition to infections and overall immune surveillance (Manshouri et al. 2008).

10.4.3 Kidney Transplantation

ESRD continues to be an expanding problem facing the healthcare system. The two primary treatment options for ESRD are dialysis and kidney transplantation. For many patients with ESRD, kidney transplantation improves the quality of life and patient survival compared to dialysis therapies (Wolfe et al. 1999). With the need for kidney transplant outpacing the supply of organs, it becomes paramount that we achieve the best outcome of the renal allografts, which includes maximizing renal allograft survival and minimizing the need for a second transplant. However, despite improved short-term success in allograft survival that has come with improved anti-rejection therapies, there has not been a commensurate improvement in long-term renal allograft survival (Lodhi, Lamb, and Meier-Kriesche 2011; Meier-Kriesche et al. 2004).

Chronic allograft nephropathy (CAN), a nonspecific term describing interstitial fibrosis and tubular atrophy, remains the most common cause of late allograft loss, while cardiovascular disease continues to be the leading cause of death with a functioning allograft post-transplantation (Ducloux et al. 2002; Jevnikar and Mannon 2008). The current challenge facing the transplant community is to develop immunosuppression that maintains short-term success but prolongs allograft survival. The advent of the calcineurin inhibitors (CNIs), cyclosporine (CsA) and tacrolimus, in the 1980s and 1990s, respectively, in conjunction with enhanced anti-proliferative agents such as mycophenolate mofetil (MMF) and the current widespread use of induction agents have contributed to lower acute rejection rates and improved 1-year kidney transplant allograft survival. The core of most current immunosuppressive protocols is CNI therapy. CNIs also contribute to worsening hypertension, diabetes, and dyslipidemia (Gaston et al. 2010; Mathis et al. 2004; Vincenti et al. 2007) and thus impart negative effects on the cardiometabolic risk profile. CNI minimization strategies to avoid nephrotoxicity may also have contributed to the development of chronic antibody-mediated rejection, which is an important cause of late graft failure (Karaman et al. 2008).

A previous study showed that tofacitinib has high affinity for JAK3, with little binding to unrelated kinases (Changelian et al. 2003). The in vivo effect of tofacitinib was first investigated in animal models of organ graft rejection. Efficacy could be shown for preventing heart or kidney rejection after transplantation without observing metabolic abnormalities or severe side effects due to immunosuppression (Kudlacz et al. 2004). Marked reduction in lymphocyte subsets was observed in a dose-dependent manner but was transient since after dosing cessation lymphocyte numbers began to normalize (Conklyn et al. 2004). In cynomolgus monkeys, oral dosing of JAK3 mainly reduced numbers of natural killer (NK) cells and effector memory $CD8^+$ T cells (Borie et al. 2005). Prevention of graft rejection and prolongation of kidney allograft survival in cynomolgus monkeys has been suggested to be related to the decrease of NK cells and T cells and lower production levels of IFN-γ into facitinib-treated animals (Paniagua et al. 2005; Rousvoal et al. 2006). JAK3 inhibition with this compound was also effective in a transplantation setup in rodents and prevented allograft vasculopathy in rats (Busque et al. 2009). Hence, the JAK3 inhibitor tofacitinib has shown promise as a CNI substitute in immunosuppressive protocols. In humans, favorable nonhuman primate and Phase I results, a Phase IIa pilot study comparing two doses of tofacitinib (15 and 30 mg b.i.d.) as group 1 and 2 to tacrolimus as group 3 in de novo kidney allograft recipients was undertaken with results published in 2009 (Vincenti et al. 2012). All subjects received IL-2 receptor antagonist induction, MMF, and corticosteroids. The 6-month incidence of biopsy-proven acute rejection (BPAR) was 5.3%, 21.1%, and 4.8% for 15 mg b.i.d. (T-15), 30 mg b.i.d. (T-30), and tacrolimus, respectively. The cardiometabolic data for tofacitinib were mixed. The 6-month eGFR was similar across the three groups: however, the eGFR of the 12-month extension study were 83.6, 77.6, and 73.3 mL/min for T-15, T-30, and tacrolimus, respectively. By month 12, there was no difference between the three groups in regard to hemoglobin concentration or in the incidence of new onset diabetes mellitus after transplant (NODAT). Compared to tacrolimus at 12 months, subjects enrolled in the high dose tofacitinib group (T-30) experienced a significantly higher rate of BK virus nephropathy (20%) as well as cytomegalovirus (CMV) disease (21.1%). In comparison, the 12-month rate of BK virus nephropathy and CMV disease in tacrolimus-treated patients was 0%. The conclusion from this study was that the preferred dose of tofacitinib was 15 mg b.i.d. as the 30 mg b.i.d. dose regimen was associated with overimmunosuppression without improved efficacy. In the next stage, a large Phase IIb trial was performed to evaluate the effectiveness of two different lower dose strategies of tofacitinib compared to a CsA-based regimen. Group 1 received the active comparator CsA. Group 2 received T-15 for months 1–6, then 10 mg b.i.d. (T-10) for months 7–12, while group 3 received T-15 for months 1–3 followed by T-10 for months 4–12. All three groups also received MMF and steroids. The 12-month BPAR rate were 18.8%, 17.4%, and 15.4% for groups 1, 2, and 3, respectively. Both tofacitinib-dosing strategies were non-inferior to CsA. The 12-month mean measured eGFR was significantly better for both tofacitinib groups: 53.9, 64.6, and 64.7 mL/min for groups 1, 2 and 3, respectively. The incidence of CAN in protocol biopsies was also lower for tofacitinib-treated patients: 48.3%, 25%, and 23.9% for groups 1, 2, and 3, respectively. The incidence of NODAT was also lower for tofacitinib-treated patients at 9.9% and 9.3% for groups 2 and 3, respectively. This is in comparison to CsA with an incidence of NODAT of 20.8%. Overall, both tofacitinib groups demonstrated noninferiority in 12-month BPAR rate and statistically higher measured eGFR compared to CsA. There was an additional benefit in tofacitinib-treated patients of a lower rate of CAN and NODAT. It is clear that tofacitinib results in potent immunosuppression and further tweaking of the regimen may be required to minimize the risk of immune deficiency. Given the promising results with a large Phase IIb trial, a long-term extension (LTE) was performed to evaluate the patients who completed 12 months of CsA or tofacitinib treatment in the phase IIb parent study were enrolled into the LTE study. Patients were analyzed by tofacitinib less-intensive (LI) or more-intensive (MI) regimens received in the parent study. For both groups, tofacitinib dose was reduced from 10 to 5 mg twice daily by 6 months into the LTE. Patients were followed up through month 72 post-transplant, with a focus on month 36 results. Tofacitinib demonstrated similar 36-month patient and graft survival rates to CsA. Biopsy-proven acute rejection rates at month 36 were 11.2% for CsA, versus 10.0% and 7.4% (both $p > 0.05$) for tofacitinib LI and MI, respectively. Least GFR were 9 to 15 mL/min per 1.73m^2 higher for tofacitinib versus CsA at month 36. The proportions of patients with grade 2/3 interstitial fibrosis and tubular atrophy in month 36 protocol biopsies were 20.0% for LI and 18.2% for MI (both $p >$ 0.05) versus 33.3% for CsA. Kaplan-Meier cumulative serious infection rates at month 36 were numerically higher for tofacitinib LI (43.9%: $p = 0.45$) and significantly higher for MI (55.9%: $p <$ 0.05) versus CsA (37.1%). Long-term tofacitinib continued to be effective in preventing renal allograft acute rejection and preserving renal function. However, long-term tofacitinib and MMF product combination was associated with persistent serious infection risk. Further studies have been needed to evaluate the benefit of anti-reject therapy and serious infection by the blockade of JAK-STAT signaling pathway.

10.5 Conclusion

Current therapeutic strategy in diabetic nephropathy, lupus nephritis, and kidney transplantation is based on the blockade of ANG II-activation by the ACE/ARB and inflammatory cytokine cascade by the steroids and immunosuppressive agents such as cyclophosphamide, MMF, and CNI, which confers the inefficient and the risk of the serious infection and renal toxicity. The JAK-STAT pathway transmits signals from extracellular ligands, including many cytokines and chemokines as well as growth factors and hormones, directly to the nucleus to induce a variety of cellular responses. Since the JAK-STAT pathways play major activating roles in a variety of disease processes, there has been a robust effort to develop specific inhibitors of this pathway. JAK-STAT pathway and its components are good candidates for immunological intervention for various disease controls.

The growing evidence for JAK-STAT activation in the pathogenesis of renal disorders establishes a new set of targets for potential intervention in this disease. Whether JAK inhibitors will show a positive effect on the progression of renal diseases in terms of loss of kidney function and development of ESRD and other complications remains uncertain, though the preclinical and early clinical data are quite encouraging. However, we should consider whether the long-term administration of JAK inhibitors in the treatment of renal diseases will be safe. These agents were developed for the treatment of autoimmune conditions as well as myelodysplastic syndrome, and treatment to date with any of the inhibitors has been of significantly shorter duration than the many years of treatment that effective renal diseases therapy would be likely to entail. For example, the long-term effect of JAK2 inhibition may be problematic if the treatment potentially aggravates the anemia that many renal disorders patients have. Despite these concerns, there is good reason to be optimistic about the outcomes of long-term JAK inhibition as a treatment that could substantially slow the progression of renal disorders and make outcomes better for the millions of individuals worldwide who suffer from this often-fatal disease.

REFERENCES

Amiri, F., S. Shaw, X. Wang et al. 2002. Angiotensin II activation of the JAK/STAT pathway in mesangial cells is altered by high glucose. *Kidney Int* 61(5):1605–1616.

Andrews, B. S., R. A. Eisenberg, A. N. Theofilopoulos et al. 1978. Spontaneous murine lupus-like syndromes. Clinical and immunopathological manifestations in several strains. *J Exp Med* 148 (5):1198–1215.

Aringer, M., J. S. Smolen. 2004. Tumour necrosis factor and other pro-inflammatory cytokines in systemic lupus erythematosus: a rationale for therapeutic intervention. *Lupus* 13(5):344–347.

Bagavant, H., S. M. Fu. 2009. Pathogenesis of kidney disease in systemic lupus erythematosus. *Curr Opin Rheumatol* 21(5):489–494.

Banes, A. K., S. Shaw, J. Jenkins et al. 2004. Angiotensin II blockade prevents hyperglycemia-induced activation of JAK and STAT proteins in diabetic rat kidney glomeruli. *Am J Physiol Renal Physiol* 286(4):653–659.

Baxter, E. J., L. M. Scott, P. J. Campbell et al. 2005. Acquired mutation of the tyrosine kinase JAK2 in human myeloproliferative disorders. *Lancet* 365:1054–1061.

Berthier, C. C., H. Zhang, M. Schin et al. 2009. Enhanced expression of Janus kinase-signal transducer and activator of transcription pathway members in human diabetic nephropathy. *Diabetes* 58 (2):469–477.

Bhat, G. J., T. J. Thekkumkara, W. G. Thomas et al. 1994. Angiotensin II stimulates sis-inducing factor-like DNA binding activity: evidence that the AT1A receptor activates transcription factor-Stat91 and/or a related protein. *J Biol Chem* 269(50):31443–31449.

Bhat, G. J., T. J. Thekkumkara, W. G. Thomas et al. 1995. Activation of the STAT pathway by angiotensin II in T3CHO/AT1A cells: cross-talk between angiotensin II and interleukin-6 nuclear signaling. *J Biol Chem* 270(32):19059–19065.

Blanco, P., V. Pitard, J. F. Viallard et al. 2005. Increase in activated CD8+ T lymphocytes expressing perforin and granzyme B correlates with disease activity in patients with systemic lupus erythematosus. *Arthritis Rheum* 52(1):201–211.

Boers, M., M. T. Nurmohamed, C. J. Doelman et al. 2003. Influence of glucocorticoids and disease activity on total and high density lipoprotein cholesterol in patients with rheumatoid arthritis. *Ann Rheum Dis* 62(9):842–845.

Boggio, E., N. Clemente, A. Mondino et al. 2004. IL-17 protects T cells from apoptosis and contributes to development of ALPS-like phenotypes. *Blood* 123(8):1178–1186.

Borie, D. C., P. S. Changelian, M. J. Larson et al. 2005. Immunosuppression by the JAK3 inhibitor CP-690,550 delays rejection and significantly prolongs kidney allograft survival in nonhuman primates. *Transplantation* 79(7):791–801.

Borie, D. C., M. S. Si, R. E. Morris et al. 2003. JAK3 inhibition as a new concept for immune suppression. *Curr Opin Investig Drugs* 4(11):1297–1303.

Boyle, D. L., K. Soma, J. Hodge et al. 2015. The JAK inhibitor tofacitinib suppresses synovial JAK1-STAT signalling in rheumatoid arthritis. *Ann Rheum Dis* 74(6):1311–1316.

Brosius, F. C., 3rd, J. C. He. 2015. JAK inhibition and progressive kidney disease. *Curr Opin Nephrol Hypertens* 24(1):88–95.

Brosius, F. C., K. R. Tuttle, M. Kretzler. 2016. JAK inhibition in the treatment of diabetic kidney disease. *Diabetologia* 59(8):1624–1627.

Buckley, R. H. 2004. The multiple causes of human SCID. *J Clin Invest* 114(10):1409–1411.

Burmester, G. R., J. M. Kremer, F. Van Den Bosch et al. 2018. Safety and efficacy of upadacitinib in patients with rheumatoid arthritis and inadequate response to conventional synthetic disease-modifying anti-rheumatic drugs (SELECT-NEXT): a randomised, double-blind, placebo-controlled phase 3 trial. *Lancet* 391(10139):2503–2512.

Busque, S., J. Leventhal, D. C. Brennan et al. 2009. Calcineurin-inhibitor-free immunosuppression based on the JAK inhibitor CP-690,550: a pilot study in de novo kidney allograft recipients. *Am J Transplant* 9(8):1936–1945.

Cash, H., M. Relle, J. Menke et al. 2010. Interleukin 6 (IL-6) deficiency delays lupus nephritis in MRL-Faslpr mice: the IL-6 pathway as a new therapeutic target in treatment of autoimmune kidney disease in systemic lupus erythematosus. *J Rheumatol* 37(1):60–70.

Changelian, P. S., M. E. Flanagan, D. J. Ball et al. 2003. Prevention of organ allograft rejection by a specific Janus kinase 3 inhibitor. *Science* 302(5646):875–878.
Chappey, O., C. Dosquet, M. P. Wautier et al. 1997. Advanced glycation end products, oxidant stress and vascular lesions. *Eur J Clin Invest* 27(2):97–108.
Charles-Schoeman, C., R. Fleischmann, J. Davignon et al. 2015. Potential mechanisms leading to the abnormal lipid profile in patients with rheumatoid arthritis versus healthy volunteers and reversal by tofacitinib. *Arthritis Rheumatol* 67(3):616–625.
Chen, J., J. K. Chen, R. C. Harris. 2012. Angiotensin II induces epithelial-to-mesenchymal transition in renal epithelial cells through reactive oxygen species/Src/caveolin-mediated activation of an epidermal growth factor receptor-extracellular signal-regulated kinase signaling pathway. *Mol Cell Biol* 32 (5):981–991.
Choudhury, G. G., N. Ghosh-Choudhury, H. E. Abboud. 1998. Association and direct activation of signal transducer and activator of transcription1alpha by platelet-derived growth factor receptor. *J Clin Invest* 101(12):2751–2760.
Chun, H. Y., J. W. Chung, H. A. Kim et al. 2007. Cytokine IL-6 and IL-10 as biomarkers in systemic lupus erythematosus. *J Clin Immunol* 27(5):461–466.
Clark, J. D., M. E. Flanagan, J. B. Telliez. 2014. Discovery and development of Janus kinase (JAK) inhibitors for inflammatory diseases. *J Med Chem* 57(12):5023–5038.
Cohen, P. L., R. A. Eisenberg. 1991. Lpr and gld: single gene models of systemic autoimmunity and lymphoproliferative disease. *Annu Rev Immunol* 9:243–269.
Conklyn, M., C. Andresen, P. Changelian et al. 2004. The JAK3 inhibitor CP-690550 selectively reduces NK and CD8+ cell numbers in cynomolgus monkey blood following chronic oral dosing. *J Leukoc Biol* 76(6):1248–1255.
Craiglow, B. G., B. A. King. 2015. Tofacitinib citrate for the treatment of vitiligo: a pathogenesis-directed therapy. *JAMA Dermatol* 151(10):1110–1112.
Crispin, J. C., M. Oukka, G. Bayliss et al. 2008. Expanded double negative T cells in patients with systemic lupus erythematosus produce IL-17 and infiltrate the kidneys. *J Immunol* 181(12):8761–8766.
Croker, B. A., H. Kiu, S. E. Nicholson. 2008. SOCS regulation of the JAK/STAT signaling pathway. *Semin Cell Dev Biol* 19(4):414–422.
Denny, M. F., S. Yalavarthi, W. Zhao et al. 2010. A distinct subset of pro-inflammatory neutrophils isolated from patients with systemic lupus erythematosus induces vascular damage and synthesizes type I IFNs. *J Immunol* 184(6):3284–3297.
Ducloux, D., G. Motte, M. Kribs et al. 2002. Hypertension in renal transplantation: donor and recipient risk factors. *Clin Nephrol* 57(6):409–413.
Edgerton, C., J. C. Crispin, C. M. Moratz et al. 2009. IL-17 producing CD4+ T cells mediate accelerated ischemia/reperfusion-induced injury in autoimmunity-prone mice. *Clin Immunol* 130(3):313–321.
Febbraio, M. A., S. Rose-John, B. K. Pedersen. 2010. Is interleukin-6 receptor blockade the Holy Grail for inflammatory diseases? *Clin Pharmacol Ther* 87(4):396–398.
Fridman, J. S., P. A. Scherle, R. Collins et al. 2010. Selective inhibition of JAK1 and JAK2 is efficacious in rodent models of arthritis: preclinical characterization of INCB028050. *J Immunol* 184(9):5298–5307.
Fujimoto, M., H. Tsutsui, O. Xinshou et al. 2004. Inadequate induction of suppressor of cytokine signaling-1 causes systemic autoimmune diseases. *Int Immunol* 16(2):303–314.
Furumoto, Y., M. Gadina. 2013. The arrival of JAK inhibitors: advancing the treatment of immune and hematologic disorders. *Bio Drugs* 27(5):431–438.
Furumoto, Y., C. K. Smith, L. Blanco et al. 2017. Tofacitinib ameliorates murine lupus and its associated vascular dysfunction. *Arthritis Rheumatol* 69(1):148–160.
Gaston, R. S., J. M. Cecka, B. L. Kasiske et al. 2010. Evidence for antibody-mediated injury as a major determinant of late kidney allograft failure. *Transplantation* 90(1):68–74.
Genovese, M. C., M. Greenwald, C. Codding et al. 2017. Peficitinib, a JAK inhibitor, in combination with limited conventional synthetic disease-modifying antirheumatic drugs in the treatment of moderate-to-severe rheumatoid arthritis. *Arthritis Rheumatol* 69(5):932–942.
Genovese, M. C., F. Yang, M. Østergaard et al. 2016. Efficacy of VX-509 (decernotinib) in combination with a disease-modifying antirheumatic drug in patients with rheumatoid arthritis: clinical and MRI findings. *Ann Rheum Dis* 75(11):1979–1983.

Ghoreschi, K., M. I. Jesson, X. Li et al. 2011. Modulation of innate and adaptive immune responses by tofacitinib (CP-690,550). *J Immunol* 186(7):4234–4243.

Ghoreschi, K., A. Laurence, J. J. O'Shea. 2009. Janus kinases in immune cell signaling. *Immunol Rev* 228 (1):273–287.

Gohda, T., M. A. Niewczas, L. H. Ficociello et al. 2012. Circulating TNF receptors 1 and 2 predict stage 3 CKD in type 1 diabetes. *J Am Soc Nephrol* 23(3):516–524.

Grisouard, J., H. Hao-Shen, S. Dirnhofer et al. 2014. Selective deletion of Jak2 in adult mouse hematopoietic cells leads to lethal anemia and thrombocytopenia. *Haematologica* 99(4):52–54.

Ha, H., H. B. Lee. 2000. Reactive oxygen species as glucose signaling molecules in mesangial cells cultured under high glucose. *Kidney Int Supp* 177:S19–S25.

Harrison, C., J. J. Kiladjian, H. K. Al-Ali et al. 2012. JAK inhibition with ruxolitinib versus best available therapy for myelofibrosis. *N Engl J Med* 366(9):787–798.

Holland, S. M., F. R. DeLeo, H. Z. Elloumi. 2007. STAT3 mutations in the hyper-IgE syndrome. *N Engl J Med* 357(16):1608–1619.

Illei, G. G., Y. Shirota, C. H. Yarboro et al. 2010. Tocilizumab in systemic lupus erythematosus: data on safety. preliminary efficacy, and impact on circulating plasma cells from an open-label phase I dosage-escalation study. *Arthritis Rheum* 62(2):542–552.

James, C., V. Ugo, J. P. Le Couédic et al. 2005. A unique clonal JAK2 mutation leading to constitutive signalling causes polycythaemia vera. *Nature* 434(7037):1144–1148.

Jedidi, A., C. Marty, C. Oligo et al. 2009. Selective reduction of JAK2V617F-dependent cell growth by siRNA/shRNA and its reversal by cytokines. *Blood* 114(9):1842–1851.

Jeon, J. Y., K. Y. Kim, H. A. Kim et al. 2010. Interleukin 6 gene polymorphisms are associated with systemic lupus erythematosus in Koreans. *J Rheumatol* 37(11):2251–2258.

Jevnikar, A. M., R. B. Mannon. 2008. Late kidney allograft loss: what we know about it and what we can do about it. *Clin J Am Soc Nephrol* 3(Suppl 2):S56–567.

Johnson, R. J., J. Floege, W. G. Couser et al. 1993. Role of platelet-derived growth factor in glomerular disease. *J Am Soc Nephrol* 4(2):119–128.

Kaplan, M. J. 2011. Neutrophils in the pathogenesis and manifestations of SLE. *Nat Rev Rheumatol* 7 (12):691–699.

Karaman, M. W., S. Herrgard, D. K. Treiber et al. 2008. A quantitative analysis of kinase inhibitor selectivity. *Nat Biotechnol* 26(1):127–132.

Kavanaugh, A., J. Kremer, L. Ponce et al. 2017. Filgotinib (GLPG0634/GS-6034), an oral selective JAK1 inhibitor, is effective as monotherapy in patients with active rheumatoid arthritis: results from a randomised, dose-finding study (DARWIN 2). *Ann Rheum Dis* 76(6):1009–1019.

Kirou, K. A., E. Gkrouzman. 2013. Anti-interferon alpha treatment in SLE. *Clin Immunol* 148(3):303–312.

Knight, J. S., W. Zhao, W. Luo et al. 2013. Peptidylarginine deiminase inhibition is immunomodulatory and vasculoprotective in murine lupus. *J Clin Invest* 123(7):2981–2993.

Kralovics, R., F. Passamonti, A. S. Buser et al. 2005. A gain-of-function mutation of JAK2 in myeloproliferative disorders. *N Engl J Med* 352(17):1779–1790.

Krempler, A., Y. Qi, A. A. Triplett et al. 2004. Generation of a conditional knockout allele for the Janus kinase 2 (Jak2) gene in mice. *Genesis* 40(1):52–57.

Kudlacz, E., B. Perry, P. Sawyer et al. 2004. The novel JAK–3 inhibitor CP-690550 is a potent immunosuppressive agent in various murine models. *Am J Transplant* 4(1):51–57.

Kuhrt, D., D. M. Wojchowski. 2015. Emerging EPO and EPO receptor regulators and signal transducers. *Blood* 125(23):3536–3541.

Lan, H. Y., A. C. Chung. 2012. TGF-β/Smad signaling in kidney disease. *Semin Nephrol* 32(3):236–243.

Lannutti, B. J., A. Epp, J. Roy et al. 2009. Incomplete restoration of Mpl expression in the mpl-/- mouse produces partial correction of the stem cell-repopulating defect and paradoxical thrombocytosis. *Blood* 113(8):1778–1785.

Levine, R. L., M. Wadleigh, J. Cools et al. 2005. Activating mutation in the tyrosine kinase JAK2 in polycythemia vera, essential thrombocythemia, and myeloid metaplasia with myelofibrosis. *Cancer Cell* 7(4):387–397.

Levy, L. L., J. Urban, B. A. King. 2015. Treatment of recalcitrant atopic dermatitis with the oral Janus kinase inhibitor tofacitinib citrate. *J Am Acad Dermatol* 73(3):395–399.

Li, J., G. Gobe. 2006. Protein kinase C activation and its role in kidney disease. *Nephrology (Carlton)* 11 (5):428–434.
Li, X., Y. Wang, K. Wang, Y. Wu et al. 2018. Renal protective effect of Paeoniflorin by inhibition of JAK2/STAT3 signaling pathway in diabeticmice. *Biosci Trends* 12(2):168–176.
Ling, B. N., E. E. Seal, D. C. Eaton. 1993. Regulation of mesangial cell ion channels by insulin and angiotensin II. Possible role in diabetic glomerular hyperfiltration. *J Clin Invest* 92(5):2141–2151.
Liu, R., Y. Zhong, X. Li et al. 2014. Role of transcription factor acetylation in diabetic kidney disease. *Diabetes* 63(7):2440–2453.
Lodhi, S. A., K. E. Lamb, H. U. Meier-Kriesche. 2011. Solid organ allograft survival improvement in the United States: the long-term does not mirror the dramatic short-term success. *Am J Transplant* 11 (6):1226–1235.
López-Pedrera, C., M. A. Aguirre, N. Barbarroja et al. 2010. Accelerated atherosclerosis in systemic lupus erythematosus: role of pro-inflammatory cytokines and therapeutic approaches. *J Biomed Biotechnol* 2010:607084.
Lu, T. C., Z. H. Wang, X. Feng et al. 2009. Knockdown of Stat3 activity in vivo prevents diabetic glomerulopathy. *Kidney Int* 76(1):63–71.
Mahajan, S., J. K. Hogan, D. Shlyakhter et al. 2015. VX-509 (decernotinib) is a potent and selective janus kinase 3 inhibitor that attenuates inflammation in animal models of autoimmune disease. *J Pharmacol Exp Ther* 353(2):405–414.
Manshouri, T., A. Quintás-Cardama, R. H. Nussenzveig et al. 2008. The JAK kinase inhibitor CP-690,550 suppresses the growth of human polycythemia vera cells carrying the JAK2V617F mutation. *Cancer Science* 99(6):1265–1273.
Markham, A. 2017. Baricitinib: first global approval. *Drugs* 77(6):697–704.
Marrero, M. B., A. K. Banes-Berceli, D. M. Stern et al. 2006. Role of the JAK/STAT signaling pathway in diabetic nephropathy. *Am J Physiol Renal Physiol* 290(4):F762–7768.
Mathis, A. S., N. Dave, G. T. Knipp et al. 2004. Drug-related dyslipidemia after renal transplantation. *Am J Health Syst Pharm* 61(6):565–585.
Mazière, C., M. A. Conte, J. C. Mazière. 2001. Activation of JAK2 by the oxidative stress generated with oxidized low-density lipoprotein. *Free Radic Biol Med* 31(11):1334–1340.
McInnes, I. B., R. Higgs, J. Lee. 2017. Ex vivo comparison of baricitinib, upadacitinib, filgotinib, and tofacitinib for cytokine signaling in human leukocyte subpopulations. *Arthritis Rheumatol* 69(Suppl 10):Abstract 2870.
Meier-Kriesche, H. U., J. D. Schold, T. R. Srinivas et al. 2004. Lack of improvement in renal allograft survival despite a marked decrease in acute rejection rates over the most recent era. *Am J Transplant* 4(3):378–383.
Mesa, R. A., J. Cortes. 2013. Optimizing management of ruxolitinib in patients with myelofibrosis: the need for individualized dosing. *J Hematol Oncol* 6:79.
Migita, K., Y. Izumi, T. Torigoshi et al. 2013. Inhibition of Janus kinase/signal transducer and activator of transcription (JAK/STAT) signaling pathway in rheumatoid synovial fibroblasts using small molecule compounds. *Clin Exp Immunol* 174(3):356–363.
Mohamed, M. F., H. S. Camp, P. Jiang et al. 2016. Pharmacokinetics, safety and tolerability of ABT-494, a novel selective JAK 1 inhibitor, in healthy volunteers and subjects with rheumatoid arthritis. *Clin Pharmacokinet* 55(12):1547–1558.
Navarro-González, J. F., C. Mora-Fernández, M. Muros de Fuentes et al. 2011. Inflammatory molecules and pathways in the pathogenesis of diabetic nephropathy. *Nat Rev Nephrol* 7(6):327–340.
Neubauer, H., A. Cumano, M. Müller et al. 1998. Jak2 deficiency defines an essential developmental checkpoint in definitive hematopoiesis. *Cell* 93(3):397–409.
Ng, A. P., M. Kauppi, D. Metcalf et al. 2014. Mpl expression on megakaryocytes and platelets is dispensable for thrombopoiesis but essential to prevent myeloproliferation. *Proc Natl Acad Sci U S A* 111(16):5884–5889.
Niewczas, M. A., T. Gohda, J. Skupien et al. 2012. Circulating TNF receptors 1 and 2 predict ESRD in type 2 diabetes. *J Am Soc Nephrol* 23(3):507–515.
Niewold, T. B., J. Hua, T. J. Lehman et al. 2007. High serum IFN-alpha activity is a heritable risk factor for systemic lupus erythematosus. *Genes Immun* 8(6):492–502.

O'Shea, J. J., R. Plenge. 2012. JAK and STAT signaling molecules in immunoregulation and immune-mediated disease. *Immunity* 36(4):542–550.

O'Shea, J. J., D. M. Schwartz, A. V. Villarino et al. 2015. The JAK–STAT pathway: impact on human disease and therapeutic intervention. *Annu Rev Med* 66:311–328.

Okiyama, N., Y. Furumoto, V. A. Villarroel et al. 2014. Reversal of CD8 T-cell-mediated mucocutaneous graft-versus-host-like disease by the JAK inhibitor tofacitinib. *J Invest Dermatol* 134(4):992–1000.

Olivera, P., S. Danese, L. Peyrin-Biroulet. 2017. JAK inhibition in inflammatory bowel disease. *Expert Rev Clin Immunol* 13(7):693–703.

Õrtiz-Munoz, G., V. Lopez-Parra, O. Lopez-Franco et al. 2010. Suppressors of cytokine signaling abrogate diabetic nephropathy. *J Am Soc Nephrol* 21(5):763–772.

Paniagua, R., M. S. Si, M. G. Flores et al. 2005. Effects of JAK3 inhibition with CP-690,550 on immune cell populations and their functions in nonhuman primate recipients of kidney allografts. *Transplantation* 80(9):1283–1292.

Parganas, E., D. Wang, D. Stravopodis et al. 1998. Jak2 is essential for signaling through a variety of cytokine receptors. *Cell* 93(3):385–395.

Park, S. O., H. L. Wamsley, K. Bae et al. 2013. Conditional deletion of Jak2 reveals an essential role in hematopoiesis throughout mouse ontogeny: implications for Jak2 inhibition in humans. *PLoS One* 8(3):e59675.

Perry, D., A. Sang, Y. Yin et al. 2011. Murine models of systemic lupus erythematosus. *J Biomed Biotechnol* 2011:271694.

Pikman, Y., B. H. Lee, T. Mercher et al. 2006. MPLW515L is a novel somatic activating mutation in myelofibrosis with myeloid metaplasia. *PLoS Med* 3(7):270.

Podder, H., B. D. Kahan. 2004. Janus kinase 3: a novel target for selective transplant immunosupression. *Expert Opin Ther Targets* 8(6):613–629.

Quintás-Cardama, A., K. Vaddi, P. Liu, T. Manshouri et al. 2010. Preclinical characterization of the selective JAK1/2 inhibitor INCB018424: therapeutic implications for the treatment of myeloproliferative neoplasms. *Blood* 115(15):3109–3117.

Richez, C., M. E. Truchetet, M. Kostine et al. 2017. Efficacy of baricitinib in the treatment of rheumatoid arthritis. *Expert Opin Pharmacother* 18(13):1399–1407.

Ripley, B. J., B. Goncalves, D. A. Isenberg et al. 2005. Raised levels of interleukin 6 in systemic lupus erythematosus correlate with anaemia. *Ann Rheum Dis* 64(6):849–853.

Rousvoal, G., M. S. Si, M. Lau et al. 2006. Janus kinase 3 inhibition with CP-690,550 prevents allograft vasculopathy. *Transpl Int* 19(12):1014–1021.

Saran, R., Y. Li, B. Robinson, J. Ayanian et al. 2015. US Renal Data System 2014 annual data report: epidemiology of kidney disease in the United States. *Am J Kidney Dis* 66(1 Suppl):1–305.

Schwartz, D. M., M. Bonelli, M. Gadina. 2016. Type I/II cytokines. JAKs, and new strategies for treating autoimmune diseases. *Nat Rev Rheumatol* 12(1):25–36.

Scott, L. M., W. Tong, R. L. Levine et al. 2007. JAK2 exon 12 mutations in polycythemia vera and idiopathic erythrocytosis. *N Engl J Med* 356(5):459–468.

Shaw, S., X. D. Wang, H. Redd et al. 2003. High glucose augments the angiotensin II-induced activation of JAK2 in vascular smooth muscle cells via the polyol pathway. *J Biol Chem* 278 (33):30634–30641.

Shi, J. G., X. Chen, F. Lee et al. 2014. The pharmacokinetics, pharmacodynamics, and safety of baricitinib, an oral JAK 1/2 inhibitor, in healthy volunteers. *J Clin Pharmacol* 54(12):1354–1361.

Silver, J. S., C. A. Hunter. 2010. gp130 at the nexus of inflammation, autoimmunity, and cancer. *J Leukoc Biol* 88(6):1145–1156.

Simon, A. R., U. Rai, B. L. Fanburg et al. 1998. Activation of the JAK–STAT pathway by reactive oxygen species. *Am J Physiol Cell Physiol* 275(6):1640–1652.

Siragy, H. M., R. M. Carey. 2010. Role of the intrarenal renin-angiotensin-aldosterone system in chronic kidney disease. *Am J Nephrol* 31(6):541–550.

Tam, C. S., S. Verstovsek. 2013. Investigational Janus kinase inhibitors. *Expert Opin Investig Drugs* 22 (6):687–699.

Thacker, S. G., C. C. Berthier, D. Mattinzoli et al. 2010. The detrimental effects of IFN-alpha on vasculogenesis in lupus are mediated by repression of IL-1 pathways: potential role in atherogenesis and renal vascular rarefaction. *J Immunol* 185(7):4457–4469.

Thacker, S. G., X. Rousset, S. Esmail et al. 2015. Increased plasma cholesterol esterification by LCAT reduces diet-induced atherosclerosis in SR-BI knockout mice. *Journal of Lipid Research* 56(7):1282–1295.
Thacker, S. G., W. Zhao, C. K. Smith et al. 2012. Type I interferons modulate vascular function, repair, thrombosis, and plaque progression in murine models of lupus and atherosclerosis. *Arthritis Rheum* 64(9):2975–2985.
Tiedt, R., J. Coers, S. Ziegler et al. 2009. Pronounced thrombocytosis in transgenic mice expressing reduced levels of Mpl in platelets and terminally differentiated megakaryocytes. *Blood* 113(8):1768–1777.
Townsend, M. J., J. G. Monroe, A. C. Chan. 2010. B-cell targeted therapies in human autoimmune diseases: an updated perspective. *Immunol Rev* 237(1):264–283.
Tsokos, G. C. 2011. Systemic lupus erythematosus. *N Engl J Med* 365(22):2110–2121.
Tucci, M., L. Lombardi, H. B. Richards et al. 2008. Overexpression of interleukin-12 and T helper 1 predominance in lupus nephritis. *Clin Exp Immunol* 154(2):247–254.
Tuttle, K. R., F. C. Brosius, 3rd, S. G. Adler et al. 2018. JAK1/JAK2 inhibition by baricitinib in diabetic kidney disease: results from a Phase 2 randomized controlled clinical trial. *Nephrol Dial Transplant* 33(11):1950–1959.
Ushio-Fukai, M., R. W. Alexander, M. Akers et al. 1999. Reactive oxygen species mediate the activation of Akt/protein kinase B by angiotensin II in vascular smooth muscle cells. *J Biol Chem* 274 (32):22699–22704.
Usui, F., H. Kimura, T. Ohshiro et al. 2012. Interleukin-17 deficiency reduced vascular inflammation and development of atherosclerosis in Western diet-induced apoE-deficient mice. *Biochemical Biophys Res Commun* 420(1):72–77.
Van, R. L., R. Galien, E. M. van der Aar et al. 2013. Preclinical characterization of GLPG0634, a selective inhibitor of JAK1, for the treatment of inflammatory diseases. *J Immunol* 191(7):3568–3577.
Verstovsek, S., H. Kantarjian, R. A. Mesa et al. 2010. Safety and efficacy of INCB018424, a JAK1 and JAK2 inhibitor, in myelofibrosis. *N Engl J Med* 363(12):1117–1127.
Verstovsek, S., R. A. Mesa, J. Gotlib et al. 2012. A double-blind, placebo-controlled trial of ruxolitinib for myelofibrosis. *N Engl J Med* 366(9):799–807.
Vincenti, F., S. Friman, E. Scheuermann et al. 2007. Results of an international, randomized trial comparing glucose metabolism disorders and outcome with cyclosporine versus tacrolimus. *Am J Transplant* 7(6):1506–1514.
Vincenti, F., A. D. Kirk. 2008. What's next in the pipeline. *Am J Transplant* 8(10):1972–1981.
Vincenti, F., S. H. Tedesco, S. Busque et al. 2012. Randomized phase 2b trial of tofacitinib (CP-690,550) in de novo kidney transplant patients: efficacy, renal function and safety at 1 year. *Am J Transplant* 12(9):2446–2456.
Wada, J., H. Makino. 2013. Inflammation and the pathogenesis of diabetic nephropathy. *Clin Sci (Lond)* 124(3):139–152.
Wang, S., N. Yang, L. Zhang et al. 2010. Jak/STAT signaling is involved in the inflammatory infiltration of the kidneys in MRL/lpr mice. *Lupus* 19(10):1171–1180.
Wang, X., S. Shaw, F. Amiri et al. 2002. Inhibition of the Jak/STAT signaling pathway prevents the high glucose-induced increase in TGF-beta and fibronectin synthesis in mesangial cells. *Diabetes* 51 (12):3505–3509.
Wenzel, U. O., B. Fouqueray, P. Biswas et al. 1995. Activation of mesangial cells by the phosphatase inhibitor vanadate. Potential implications for diabetic nephropathy. *J Clin Invest* 95(3):1244–1252.
Westhovens, R., P. C. Taylor, R. Alten et al. 2017. Filgotinib (GLPG0634/GS-6034), an oral JAK1 selective inhibitor, is effective in combination with methotrexate (MTX) in patients with active rheumatoid arthritis and insufficient response to MTX: results from a randomised, dose-finding study (DARWIN 1). *Ann Rheum Dis* 76(6):998–1008.
Wiik, A. C., C. Wade, T. Biagi et al. 1994. A deletion in nephronophthisis 4 (NPHP4) is associated with recessive cone-rod dystrophy in standard wire-haired dachshund. *Genome Res* 18(9):1415–1421.
Winter, O., K. Moser, E. Mohr et al. 2010. Megakaryocytes constitute a functional component of a plasma cell niche in the bone marrow. *Blood* 116(11):1867–1875.
Witthuhn, B. A., F. W. Quelle, O. Silvennoinen et al. 1993. JAK2 associates with the erythropoietin receptor and is tyrosine phosphorylated and activated following stimulation with erythropoietin. *Cell* 74(2):227–236.

Witthuhn, B. A., O. Silvennoinen, O. Miura et al. 1994. Involvement of the JAK–3 Janus kinase in signalling by interleukins 2 and 4 in lymphoid and myeloid cells. *Nature* 370(6485):153–157.
Wolfe, R. A., V. B. Ashby, E. L. Milford et al. 1999. Comparison of mortality in all patients on dialysis, patients on dialysis awaiting transplantation, and recipients of a first cadaveric transplant. *N Engl J Med* 341(23):1725–1730.
Wong, C. K., L. C. Lit, L. S. Tam et al. 2008. Hyperproduction of IL-23 and IL-17 in patients with systemic lupus erythematosus: implications for Th17-mediated inflammation in auto-immunity. *Clin Immunol* 127(3):385–393.
Woroniecka, K. I., A. S. Park, D. Mohtat et al. 2011. Transcriptome analysis of human diabetic kidney disease. *Diabetes* 60(9):2354–2369.
Xing, L., Z. Dai, A. Jabbari et al. 2014. Alopecia areata is driven by cytotoxic T lymphocytes and is reversed by JAK inhibition. *Nature Med* 20(9):1043–1049.
Yu, C. C., A. A. Mamchak, A. L. DeFranco. 2003. Signaling mutations and autoimmunity. *Curr Dir Autoimmun* 6:61–88.
Zhang, H., V. Nair, J. Saha, K. B. Atkins et al. 2017. Podocyte-specific JAK2 overexpression worsens diabetic kidney disease in mice. *Kidney Int* 92(4):909–921.
Zhuang, S. 2013. Regulation of STAT signaling by acetylation. *Cell Signal* 25(9):1924–1931.

11

JAK-STAT Signaling in Hematologic Malignancies

Thomas Pincez
Division of Pediatric Hematology-Oncology
Charles-Bruneau Cancer Center, Ste-Justine Hospital
Quebec, Canada

Thai Hoa Tran

Division of Pediatric Hematology-Oncology
Charles-Bruneau Cancer Center, Ste-Justine Hospital
Quebec, Canada

Department of Pediatrics
University of Montreal
Quebec, Canada

11.1 Introduction

Hematopoiesis leads to the production of different types of mature blood cells derived from hematopoietic stem cells (HSC). This process is highly regulated by cytokines, which govern cell fate choices by inducing intracellular signals. Throughout hematopoiesis, cytokine signaling is under tight control to regulate cellular proliferation and differentiation. As most cytokine receptors lack intrinsic kinase activity, they are coupled to other signaling intermediates to mediate signal transduction. JAK-STAT proteins are a major system of downstream signaling in hematopoietic ontogeny. In physiology, its critical role has been demonstrated in mice models and the loss-of-function of JAK1, JAK3, TYK2, STAT1, STAT2, STAT4, or STAT6 impair lymphopoiesis (Khwaja 2006). Moreover, JAK2-, STAT3-, and STAT5-deficient mice are not viable (Khwaja 2006). In humans, several primary immunodeficiencies resulting from JAK-STAT deficiency have been described (Bousfiha et al. 2018).

Hematological malignancies encompass neoplasms arising from blood cells at any differentiation stage (Figure 11.1). The genomic aberrations underlying the malignant transformation vary upon different entities. Some diseases share disrupted signaling pathways; but the molecular mechanisms, functional consequences, and therapeutic relevance may differ. Therefore, the JAK-STAT pathway alterations within each hematological malignancy will be discussed separately in this chapter. Three major categories of hematological malignancies harbor recurrent alterations of this pathway: acute leukemias, myeloproliferative neoplasms (MPN), and lymphomas. The 2016 revised classification of the World Health Organization will be used in this chapter (Arber et al. 2016; Swerdlow et al. 2016).

Aside from acute lymphoblastic leukemia (ALL), the remaining diseases are more frequently encountered in adults than children. Consequently, numerous studies are performed on adult samples and the knowledge has to be extrapolated from these data. However, the clinical presentation, evolution, and the underpinning biology of each malignancy may differ and data interpretation needs to be nuanced. The relevant studies from the adult population will be discussed here and adjusted to pediatric patients.

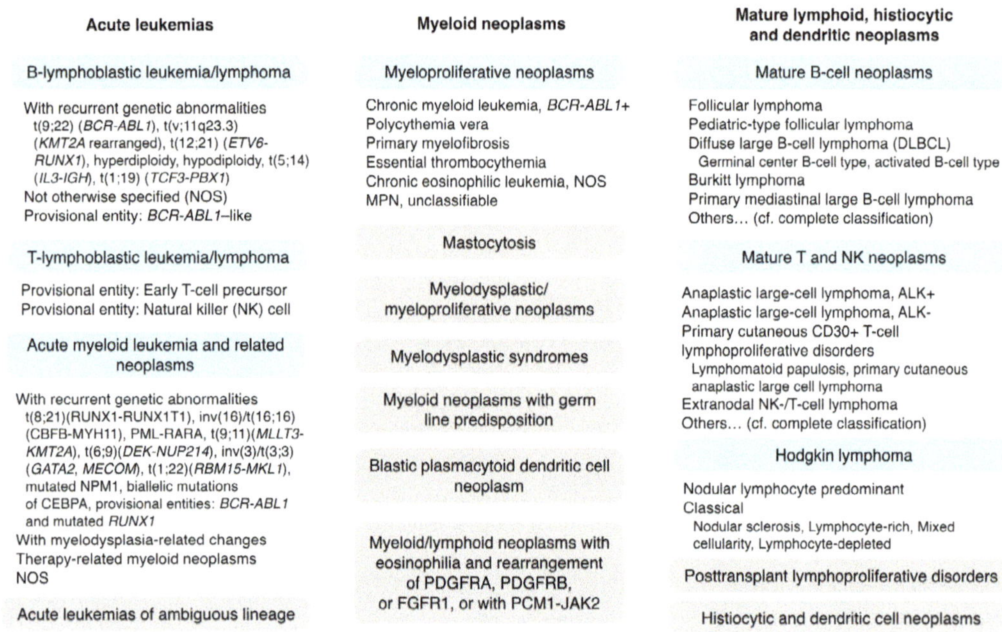

FIGURE 11.1 Schematic representation of classification of hematological malignancies.

Classification according to the World Health Organization 2016 revision. In blue, the malignancies described in this chapter with their subtypes. In grey, the other hematological malignancies.

JAK-STAT pathway involvement in several diseases led to the development of several pharmacologic inhibitors, which can be divided in two categories (Vainchenker et al. 2018). Type I inhibitors target the ATP-binding site of the JAK proteins under the active conformation of the kinase domain. Type II inhibitors bind to the ATP-binding pocket of kinase domains in inactive conformation and provide not only stronger inhibition, but also more side effects than type I drugs. The preclinical and clinical data of both will be discussed for each malignancy.

11.2 Acute Leukemias

Acute leukemia is a malignant hematopoietic proliferation of immature blood cells from the bone marrow. The involved lineage can be lymphoid or myeloid, defining ALL or acute myeloid leukemia (AML).

ALL is markedly more frequent in the pediatric than adult population. In the Unites States of America (USA), 35.9 cases per million of person younger than 20 years old are diagnosed each year with ALL (Noone et al. 2018). The peak incidence lies from 3–5 years of age with a slight male predominance (55%) (Noone et al. 2018). Over the past decades, long-term survival rates have reached nearly 90% with modern chemotherapy regimens (Hunger and Mullighan 2015a). ALL may arise from B- or T-cell progenitors; hence called B-ALL and T-ALL, respectively. Advances in genomic profiling and next-generation sequencing (NGS) technologies have provided new insights into biology and defined the genomic landscape of these different leukemia subtypes. Mature B-ALL (Burkitt leukemia) is a specific entity with distinct treatment and prognosis. The JAK-STAT pathway has not been shown to be disrupted and this leukemia will not be discussed here. B-ALL can develop from different stages of maturation of B-cell precursors and harbors recurrent genomic alterations associated with prognosis. High hyperdiploidy (>50 chromosomes) and the t(12;21) translocation encoding *ETV6-RUNX1*, representing each ~25% of B-ALL, are associated with favorable outcome (Hunger and Mullighan 2015a). *KMT2A* rearrangements (occurring mainly in children <1 year) and hypodiploidy (<44 chromosomes) are both poor prognostic factors. The t(9;22), also named Philadelphia chromosome, encodes the constitutively activated tyrosine kinase BCR-ABL1 and is present in 3–5% of children with ALL (the incidence increases with age). Philadelphia chromosome-positive ALL (Ph+ ALL) was historically associated with poor prognosis but targeted therapy in combination with intensive chemotherapy backbone drastically improved survival rates (Aricò et al. 2010; Biondi et al. 2012; Schultz et al. 2014; Slayton et al. 2018). T-ALL used to be an adverse prognostic factor, especially in the case of early precursor T-ALL. However, outcome differences between B-ALL and T-ALL become increasingly smaller in the era of risk-stratified intensive chemotherapy regimens (Winter et al. 2018). In addition, recurrent cytogenetic features do not confer prognosis and are not incorporated in the risk stratification of T-ALL treatment. In this chapter, B- and T-ALL will be discussed separately. The relative high frequency of ALL in children allows an in-depth investigation from exclusively pediatric samples.

The annual incidence of AML is much lower in pediatric (<20 years) than in the adult population, 8.7 cases compared to 42.6 cases per million in the USA, respectively (Noone et al. 2018). The highest incidence in childhood AML is found before the first year of life (18.8 cases per million), while the lowest occurs among children between 5 and 9 years old. AML remains a clinical challenge with lower cure rates than ALL and an overall survival around 65% despite modern chemotherapy regimens (Cooper et al. 2012; Gamis et al. 2013, 2014). If AML classification is heavily based on morphology, its prognosis is, however, greatly influenced by the underlying biology (Bolouri et al. 2018). A great diversity of cytogenetic and/or genomic alterations is encountered in pediatric AML and varies according to age. Moreover, in some cases, such as *KMT2A* rearrangements, the prognostic significance of a given gene rearrangement depends on the involved gene partner (Creutzig et al. 2012). To allow for a clinically relevant stratification in this heterogeneous group, subtypes of AML are defined by distinct genomic alterations, which confer prognosis. The presence of t(8;21), inv(16), or t(16;16) are classically favorable prognostic factors. Conversely, monosomy 7, monosomy 5/5q deletion, and high allelic ratio *FLT3* internal tandem duplication are associated with poor outcome (Rooij et al. 2015). The genomic landscape of AML differs greatly between adults and children; therefore, the clinical relevance of the JAK-STAT pathway within the pediatric AML population will be highlighted in this chapter.

11.2.1 JAK-STAT Pathway Activation Is a Recurrent Feature of Ph-Like and Down-Syndrome Associated B-ALL

11.2.1.1 Ph-like ALL

In 2009, genome-wide profiling studies identified a new subtype of B-ALL that exhibits a gene expression signature similar to those found in Ph+ ALL but lacking the canonical *BCR-ABL1* gene fusion (Den Boer et al. 2009; Mullighan et al. 2009). Thus, this new subtype is known as BCR-ABL1-like or Ph-like ALL, and represents approximately 20% of B-ALL. In contrast to Ph+ ALL, in which the prevalence augments with increasing age, Ph-like ALL occurs in 14% among children with National Cancer Institute (NCI) standard-risk B-ALL between 1 and 9 years of age, 20% in NCI high-risk B-ALL between 10 and 21 years old, peaks at 28% among young adults between 21 and 39 years old, and decreases to 22% in older adults from 40 to 86 years old (Roberts et al. 2014, 2016, 2018). The Ph-like signature is associated with high frequency of *IKZF1* alterations and confers poor outcomes despite intensive chemotherapy regimens (Jain et al. 2017; Roberts et al. 2014; Tran et al. 2018). Ph-like ALL is particularly prevalent in patients of Hispanic ethnicity (Harvey et al. 2010; Reshmi et al. 2017). Genetic susceptibility to Ph-like ALL has been linked to some *GATA3* polymorphisms, which are more frequent in individuals of US Hispanics and Native

American genetic ancestry than in those of European descent (Perez-Andreu et al. 2013, 2015). Large-scale comprehensive NGS studies have depicted the genomic landscape of Ph-like ALL. It is now known that over 90% of Ph-like ALL harbor a diverse spectrum of genomic alterations activating cytokine receptor genes and signaling kinase pathways, most of which converges to the JAK-STAT signaling pathways in more than 70% of Ph-like ALL cases.

11.2.1.2 Down Syndrome-Associated ALL

Children with Down Syndrome (DS) carry a ~20-fold higher risk of developing ALL compared to those without DS (Hasle et al. 2000). Infant ALL is extremely rare among the DS population and the peak of incidence is slightly higher than in children without DS (Buitenkamp et al. 2014; Dördelmann et al. 1998). Down syndrome-associated ALL (DS-ALL) carries a higher risk of relapse compared to ALL patients without DS (Buitenkamp et al. 2014; Goto et al. 2011; Whitlock et al. 2005; Zeller et al. 2005). In a retrospective analysis of 653 DS-ALL patients enrolled in 16 international trials from 1995 to 2004, the 8-year cumulative incidence of relapse was 26% ± 2%, compared to 15% ± 1% in those without DS ($p < 0.001$). The treatment-related mortality is notoriously higher and mainly due to bacterial and viral infections (Buitenkamp et al. 2014; Shah et al. 2009; Whitlock et al. 2005; Zeller et al. 2005). Both these elements contribute to the poor overall survival associated with DS-ALL (Buitenkamp et al. 2014; Chessells et al. 2001; Whitlock et al. 2005). Precursor B-ALL is almost the exclusive ALL phenotype seen in DS-ALL and is characterized by a distinct biology. The favorable cytogenetic features (such as *ETV6-RUNX1* translocation and high hyperdiplody) are less frequent and 40–45% of DS-ALL has a normal karyotype (Buitenkamp et al. 2014; Forestier et al. 2008). As detailed below, DS-ALL is associated with frequent alterations of the JAK-STAT signaling pathway.

11.2.1.3 CRLF2 Rearrangements and Concomitant JAK Mutations

Approximately half of Ph-like ALL and 60% of DS-ALL patients harbor *CRLF2* rearrangements ($CRLF2_r$, Table 11.1) (Den Boer et al. 2009; Hertzberg et al. 2010; Mullighan et al. 2009b; Reshmi et al. 2017; Roberts et al. 2014; Yoda et al. 2010). Two *CRLF2* partners have been described in these populations: *P2RY8* and *IGH*, the resulting aberrant gene fusion leads to CRLF2 overexpression (Den Boer et al. 2009; Mullighan et al. 2009b; Roberts et al. 2014). Rarely, *P2RY8-CRLF2* fusions can also be found in non-Ph-like ALL (Harvey et al. 2010; Reshmi et al. 2017). A more rare *CRLF2* alteration leading to CRLF2 overexpression is the $CRLF2^{F232C}$ activating point mutation (Chapiro et al. 2010; Yoda et al. 2010). In about half of $CRLF2_r$ patients, concomitant *JAK1* or *JAK2* mutations are found (Den Boer et al. 2009; Hertzberg et al. 2010; Reshmi et al. 2017; Roberts et al. 2014; Yoda et al. 2010). They mostly affect the *JAK2* Janus homology domain (JH2) pseudokinase domain (which negatively regulates the JH1 kinase domain), especially at the R683 residue, either as a point mutation (the most frequent being $JAK2^{R683G}$) or a 5-amino acid deletion removing R683 (JAK2ΔIREED) (Gaikwad et al. 2009; Kearney et al. 2009; Malinge et al. 2007). *JAK1* mutations are less common, the most frequent is $JAK1^{V658F}$, which is analogous to the $JAK2^{V617F}$ mutation found in MPN (Staerk et al. 2005). Interestingly, both $CRLF2_r$ and *JAK* mutations are particularly prevalent in patients with DS-ALL as they may occur in up to 30–50% of cases (Gaikwad et al. 2009; Hertzberg et al. 2010; Kearney et al. 2009).

CRLF2 gene is located at the pseudoautosomal region 1 of Xp22/Yp11 and encodes for the thymic stromal lymphopoeitin (TSLP) receptor, which is formed by the heterodimerization of CRLF2 with interleukin-7 receptor-α (IL7R-α) on the cell membrane, and plays an important role in early B-cell development (Scheeren et al. 2010). Upon its ligand binding, the physiological downstream cascade involves STAT5 phosphorylation through JAK2 activation, leading to dysregulated JAK-STAT signaling (Isaksen et al. 1999, 2002). In ALL patients' cells, phospho-JAK2 (pJAK2) and pSTAT5 are more elevated in the presence of *CRLF2* alterations, regardless

TABLE 11.1

JAK-STAT Alterations and Targeted Therapies in Acute Leukemias

Acute leukemias	JAK-STAT alterations	Frequency	Treatment options: Pre-clinical	Treatment options: Clinical
Ph-like ALL				
	CRLF2 rearrangement (*CRLF2*$_r$) ± *CRLF2*^{F232C}	50% of Ph-like ALL, 60% of DS-ALL	Type I and II JAK inhibitors ± mTOR or MEK inhibitor. Givinostat, CART-cells	Ruxolitinib (NCT02723994) Ruxolitinib (NCT03117751 and NCT02420717
	± *JAK1/2* alterations	50% of *CRLF2*$_r$		
	± *IL7R-α* alterations	Around 5% of Ph-like ALL		
	± *SH2B3* alterations	Individual cases		
	JAK2 fusion	5-10% of Ph-like ALL		
	EPOR rearrangement	Around 5% of Ph-like ALL	Type I JAK inhibitors	
DS-ALL				
	CRLF2$_r$	60% of DS-ALL	Type I and II JAK inhibitors ± mTOR or MEK inhibitor. Givinostat, CART-cells	None
T-ALL				
	ETV6-JAK2	Individual cases	Licochalcone A	None
	JAK1, JAK2, JAK3, IL7R-α, PTPN2, FLT3, STATB5, SH2B3 and *IL2RB* point mutations	Around 25% of T-ALL	Ruxolitinib + venetoclax (*IL7R-α*)	
AML				
	JAK mutations	Individual cases of AMKL	Ruxolitinib + venetoclax (AML regardless of JAK alterations)	Ruxolitinib (adults with relapse AML regardless of JAK alterations, NCT00674479)
	STAT3 activation	Around 50% of adult AML		

ALL: acute lymphoblastic leukemia, AML: acute myeloid leukemia, DS: Down Syndrome

of *JAK* mutation (Tasian et al. 2012). However, CRLF2 overexpression is probably not sufficient on its own to drive leukemogenesis. In murine fetal liver cells, *CRLF2* RNA-transduction is associated with constitutive JAK2 and STAT5 phosphorylation as well as enhanced lymphoid progenitor proliferation in the presence of growth factors (Russell et al. 2009). However, *CRLF2*$_r$-transduced Ba/F3 cells result neither in constitutive activation nor cytokine-independent growth (Russell et al. 2009; Tasian et al. 2012). In addition, transgenic-mediated CRLF2 overexpression in mice result neither in JAK-STAT pathway activation nor ALL development (Kim et al. 2018).

Leukemogenesis is potentially driven by the cooperation between *CRLF2*$_r$ and *JAK* mutations, as suggested by their co-occurrence. *JAK* mutations are usually not able to transform Ba/F3 cells but can lead to cytokine-independent proliferation when co-expressed with erythropoietin (EPO) or thrombopoietin (TPO) (Bercovich et al. 2008; Mullighan et al. 2009a). The co-expression of *P2RY8-CRLF2* and *JAK2* mutations (*JAK2*^{R683G} or *JAK2*^{P933R}) in these cells result in constitutive JAK2 and STAT5 activation as well as cytokine-independent proliferation (Hertzberg et al.

2010; Mullighan et al. 2009; Yoda et al. 2010). In mouse models, transduction of CRLF2-overexpressing/*JAK2*$^{R683G\text{ or }P933R}$ fetal liver cells induce an aggressive leukemia with hepatosplenic infiltration and circulating mutant progenitor B-cell blasts (Kim et al. 2018). The same cooperation was also demonstrated between *CRLF2*^{F232C} and *JAK2* mutations (Bodegom et al. 2012). JAK pharmacologic inhibition is effective in these preclinical models to abolish hyperactive STAT signaling, cell proliferation, and *in vivo* tumor growth (Maude et al. 2012; Mullighan et al. 2009b; Roberts et al. 2017; Tasian et al. 2012, 2017; Yoda et al. 2010). Interestingly, in addition to JAK-STAT signaling, PI3K/mTOR pathway was also hyperactivated in cells harboring *CRLF2*$_r$ and *JAK* mutation, providing another potential therapeutic avenue (Tasian et al. 2012).

However, two studies challenge the driver effect of CRLF2$_r$. In fact, *P2RY8-CRLF2*-positive clone only represents a minor population at diagnosis and 30–50% of these patients lose this clone at relapse (Morak et al. 2012; Vesely et al. 2017). Other genetic alterations are persistent at relapse, particularly *IKZF1* deletions, which become enriched in the re-emerging clones. The above observations suggest that *CRLF2*$_r$ may be dispensable for mediating relapse.

11.2.1.4 Other Mechanisms of JAK-STAT Activation in Ph-Like ALL

In a subset of Ph-like ALL patients without *CRLF2*$_r$ or *JAK2* mutations, *JAK2* fusions were identified in 6% of children with NCI high-risk B-ALL, 7% of adolescents between 16 and 21 years old, 6% in young adults from 21 to 39 years old, and 8% in older adults from 40 to 86 years old (Jain et al. 2017; Pui et al. 2017; Roberts et al. 2014, 2016). At least 19 partner genes have been reported, sharing the same mechanism consisting of in-frame fusions at the N-terminus extremity of *JAK2*, which conserve the kinase ± the pseudokinase domains (Kawamura et al. 2015; Reshmi et al. 2017; Roberts et al. 2012, 2014, 2016; Tirado et al. 2010; Yano et al. 2015). *In vitro*, the expression of *JAK2* fusions results in constitutive STAT5 activation and cytokine-independent growth (Maude et al. 2012; Roberts et al. 2012, 2014; Schinnerl et al. 2015).

In approximately 5% of patients with Ph-like ALL, rearrangements involving the EPO receptor (EPOR) have been reported with four different partners: *IGH*, *IGK*, *LAIR1*, *THADA* (Iacobucci et al. 2016; Reshmi et al. 2017; Roberts et al. 2012, 2014). The fusion of the first three genes have been well described; these *EPOR* rearrangements result in the truncation of EPOR's C-terminus, which contains distal regulatory residues (Iacobucci et al. 2016). This leads to stabilized expression of EPOR and hyperactive JAK-STAT, PI3K/mTOR, and MAPK signaling. In mouse fetal liver cells, these rearrangements induce B-ALL development (Iacobucci et al. 2016). Interestingly, analogous-truncated EPOR proteins are found in autosomal benign erythrocytosis, which is caused by a germline mutation (de la Chapelle et al. 1993).

IL7R-α gain-of-function is mediated by either amino acid 185 substitution or in-frame insertion/deletion in less than 10% of Ph-like ALL (Roberts et al. 2014; Shochat et al. 2011). *IL7R-α* mutations are often found in patients with *CRLF2* fusions without concomitant *JAK* mutations. Paired with *CRLF2*, the mutant *IL7R-α* leads to STAT5 hyperactivation and cytokine-independent growth of progenitor lymphoid cells (Shochat et al. 2011).

Loss-of-function point mutations and small deletions of *SH2B3* (*Lnk*) have been reported in children with Ph-like ALL harboring *CRLF2* alterations but without *JAK* or *IL7R-α* mutations (Maude et al. 2012; Roberts et al. 2014). A recent report observed that *SH2B3* deletions are more enriched among B-ALL cases with intrachromosomal amplification of chromosome 21 (iAMP21) (Baughn et al. 2018). Lnk protein, encoded by *SH2B3*, is a negative regulator of JAK-STAT; therefore, *SH2B3* deletion leads to hyperactive JAK-STAT signaling. In a mouse model, Lnk deficiency synergizes with *TP53* deletion to promote leukemogenesis via STAT5 activation (Cheng et al. 2016). In xenograft models, *IL7R-α* and *SH2B3*-mutant ALL are sensitive to ruxolitinib (Maude et al. 2012).

11.2.2 JAK-STAT Pathway Alterations Are Commonly Found in Pediatric T-ALL

11.2.2.1 JAK2 Fusions

Although *JAK2* fusions are more prevalent in Ph-like ALL, the former has been reported but occurs in low frequency in T-lineage ALL. *ETV6-JAK2* was initially described in a patient with T-ALL and several other cases have since been published, which although are uncommon (Lacronique et al. 1997). In T-ALL, the translocation fuses exon 5 of *TEL* and exon 19 of *JAK2*, whereas in B-ALL, exon 4 of TEL is juxtaposed to exon 17 of *JAK2* (Lacronique et al. 1997; Peeters et al. 1997). The aberrant transcript exhibits constitutive tyrosine kinase activity and promotes STAT1, STAT5, PI3K, RAS/ERK, and NF-κB activation (Nguyen et al. 2001). In several *TEL-JAK2* transgenic mice models, the mice develop an invasive T-cell leukemia (Carron et al. 2000; Santos et al. 2007; Schwaller et al. 1998).

11.2.2.2 JAK-STAT Pathway Mutations

Somatic mutations involving the JAK-STAT pathway are recurrent events in childhood T-ALL, but only one cohort has investigated the prevalence of such alterations (Liu et al. 2017). In this study, almost 25% of 264 patients harbor a point mutation involving this pathway, most commonly *JAK3* (n = 20), *IL7R* (n = 18), *PTPN2* (n = 13), *FLT3* (n = 12), *STATB5* (n = 11), *JAK1* (n = 7), *SH2B3* (n = 3), *JAK2* (n = 3), and *IL2RB* (n = 1) mutations. Interestingly, 19 patients have several concomitant mutations implicating this pathway, while 17 patients involve other pathways (Ras or PI3K/AKT). The nature and functional impact of these alterations, particularly in the setting of T-ALL, are largely unknown with a few exceptions. For example, 10 different *JAK3* mutations have been reported in childhood T-ALL, mainly missense, the most frequent being $JAK3^{M511I}$. Most of *JAK3* mutations require both binding to the common γ-chain and the presence of JAK1 for their transforming capacities (Degryse et al. 2014). In particular, $JAK3^{M511I}$ mutation alone may have limited oncogenic potential in Ba/F3 cells as it is competing with wild-type *JAK3* in one model. Interestingly, a second *JAK3* mutation was found in 30% of the adults studied and the double *JAK3* mutants ($JAK3^{M511I}$ and $JAK3^{A572T}$ or $JAK3^{A573V}$ mutations) induce increased STAT5 phosphorylation and cytokine-independent leukemia transformation. In the pediatric cohort cited earlier, only 3 of the 30 patients contain a double *JAK3* mutant but 50% harbors another potentially oncogenic mutation (Liu et al. 2017). *HOXA* cluster rearrangements (leading to HOXA overexpression) frequently co-occur with *JAK3* mutations. *In vivo*, HOXA overexpression cooperates with $JAK3^{M511I}$ to develop leukemia (Bock et al. 2014). Taken together, these data suggest that *JAK3* mutations are not "drivers" and require an additional event to promote leukemogenesis.

IL7R alterations are different to those seen in B-ALL. In T-ALL, they represent gain-of-function indels within exon 6 (Zenatti et al. 2011). In most cases, these alterations introduce an unpaired cysteine in the juxtamembrane/transmembrane extracellular region, which lead to dimerization of *IL7R* mutants. In cell lines, they drive constitutive JAK1 signaling, cytokine-independent growth, and promote viability; a phenotype which is reversed through type I JAK inhibitor and STAT5 inhibitor (Zenatti et al. 2011).

FLT3 mutations have been reported mainly in hyperdiploid pediatric T-ALL (Armstrong et al. 2004). These activating mutations were mostly present in the receptor-activation loop, a different mechanism from the internal tandem duplication seen in AML (Armstrong et al. 2004). These *FLT3* mutants were constitutively phosphorylated and transformed 32D cells (Yamamoto et al. 2001).

JAK1 mutations are missense substitutions located in the FERM domain (band 4.1, ezrin, radixin, and moesin) or pseudokinase domain. In adults, the most frequent is $JAK1^{V658F}$, which is analogous to the $JAK2^{V617F}$ described earlier in B-ALL (Staerk et al. 2005). Among the seven children reported within the largest cohort, this mutation was present in only one case and four harbored the $JAK1^{R724H}$ mutation (Liu et al. 2017). Both mutations induced cytokine-independent growth in Ba/F3 cell lines (Flex et al. 2008; Porcu et al. 2009), but these mutants need IL-9Rα or IL-2Rβ expression for constitutive activation (Hornakova et al. 2009).

The emergence of mutations can evolve under the selective pressure of therapy. In *JAK3*-mutated murine cell lines, *JAK1* mutations appear following treatment with type I JAK inhibitors as a mechanism of resistance and, *vice versa*, *JAK3* mutations emerge when *JAK1*-mutated cell lines are treated with pan-JAK inhibitors (Springuel et al. 2014). This example suggests that acquired *JAK* mutations may represent a mechanism of resistance to type I JAK inhibitors.

11.2.3 JAK-STAT Alterations Are Rarely Found in AML Despite Frequent STAT3 Phosphorylation

11.2.3.1 JAK Mutations

The current data suggest that the implication of the JAK-STAT pathway is less frequent in AML than ALL. The genomic landscape of pediatric and adult AML did not reveal frequent recurrent alterations of the JAK-STAT pathway genes (Bolouri et al. 2018; Ley et al. 2013). The canonical MPN-associated $JAK2^{V617F}$ point mutation is uncommon in AML and may be present in around 5% of adult AML (Hidalgo-López et al. 2017; Jelinek et al. 2005; Steensma et al. 2005, 2006). However, at least in some cases, it may be due to leukemic transformation of undiagnosed pre-existing MPN.

Two *JAK1* mutations (T478S, V623A) have been reported in adult but not in pediatric AML patients (Tomasson et al. 2008). Although they lead to STAT1 phosphorylation, both are unable to transform primary murine cells, suggesting that they act more as disease-modifier than as driver mutations (Xiang et al. 2008).

Several *JAK3* non-recurrent mutations have been reported in acute megakaryoblastic leukemia (AMKL) occurring in patients with or without DS, including children (Hama et al. 2008; Kiyoi et al. 2007; Malinge et al. 2008; Norton et al. 2007; Sato et al. 2008; Vita et al. 2007). Not all of these mutations have been functionally characterized, but, in those that have been described, they commonly lead to constitutive JAK3 activation, drive growth, and promote viability of Ba/F3 cell lines (Malinge et al. 2008; Sato et al. 2008).

11.2.3.2 STAT3 Phosphorylation

Despite the paucity of JAK-STAT pathway gene alterations, constitutive STAT3 phosphorylation in adult AML blasts has been found in approximately 50% of patients and confers a poor prognosis (Benekli et al. 2002; Xia et al. 1998). In children, the constitutive phosphorylation of STAT3 and STAT5 on tyrosine 705 has been reported to be recurrent in a single study (Redell et al. 2013). In this paper, STAT3 phosphorylation at basal state is not associated with outcome, but a subgroup (11%) of patients, who displayed hyperphosphorylated STAT3 in response to moderate doses of both granulocyte-colony stimulating factor and IL-6, has an extremely favorable outcome (Redell et al. 2013). The pattern of STAT3 sensitivity changes at relapse and represents a prognostic factor (Stevens et al. 2015). Furthermore, in pediatric AML cell lines, pharmacologic inhibition of STAT3 led to apoptosis (Redell et al. 2011). Collectively, these data highlight the important role of STAT3 activation in a subgroup of pediatric AML even though the molecular mechanisms underpinning this dysregulation remain unknown.

11.2.4 Therapeutic Implications of JAK-STAT Hyperactivation in Acute Leukemias

As JAK-STAT signaling pathways are constitutively dysregulated in a subset of Ph-like ALL patients described above, abrogating its signal represents the logical therapeutic avenue. Several approaches have been studied in preclinical models but only ruxolitinib has moved into clinical trials for further investigation. The use of JAK inhibitors in T-ALL and AML is less explored but will be briefly discussed.

11.2.4.1 Type I JAK Inhibitor in Ph-like ALL

In vitro, type I JAK inhibitor ruxolitinib allows to limit, if not inhibit, cell proliferation (Mullighan et al. 2009; Tasian et al. 2012; Yoda et al. 2010). Its efficacy has been also demonstrated in Ph-like ALL patient-derived xenograft (PDX) models (Maude et al. 2012; Roberts et al. 2017; Tasian et al. 2017). Nevertheless, these preclinical studies also highlight potential challenges, which may limit the benefits of ruxolitinib. First, ruxolitinib response in PDX models can be variable depending on the underlying JAK-STAT pathway alterations, with *JAK2* rearrangements exhibiting a more robust sensitivity to ruxolitinib than *CRLF2* rearrangements (Maude et al. 2012). Secondly, type I JAK inhibitor may induce paradoxical JAK2 hyperphosphorylation conferring resistance to ruxolitinib (Weigert et al. 2012). Thirdly, in genetically engineered mouse models, after a transient response to JAK2 inhibition, mutant B-cells remain viable by upregulating c-Myc expression (Kim et al. 2018). However, the combination of BET bromodomain inhibitors with ruxolitinib can abrogate this effect (Kim et al. 2018). Lately, JAK-STAT hypersignaling may be toxic to cells and some patients have a therapeutic-emergent loss-of-function of USP9X, an adaptive mechanism to limit this effect (Schwartzman et al. 2017). In this study, tempering JAK-STAT signaling paradoxically improved proliferation; hence, if present *in vivo*, this effect could limit the efficacy of suboptimal JAK inhibition. The clinical relevance of all these phenomena remains to be demonstrated and ongoing clinical studies will address the efficacy of ruxolitinib in this population.

Three clinical trials are currently investigating the safety and efficacy of the combination of ruxolitinib and conventional chemotherapy backbone. The Children's Oncology Group (COG) phase 2 trial AALL1521 (NCT02723994) is active and enrolls NCI high-risk Ph-like ALL patients with JAK-STAT pathway lesions who are stratified according to the underlying genetic alterations (*$CRLF2_r$*, *JAK* mutations, *JAK2* fusions or *EPOR* rearrangements) and end-induction MRD status. The first phase of this trial has determined that continuous dosing of ruxolitinib at 40mg/m^2/dose for 28 days per cycle in combination with intensive multi-agent chemotherapy regimen is safe and tolerable in children and young adults with eligible Ph-like genetic lesions (Tasian et al. 2018). St. Jude Children's Research Hospital's Total therapy XVII protocol (NCT03117751) adds ruxolitinib to chemotherapy in children with JAK–mutant Ph-like ALL. For adults, a phase 2 MD Anderson Cancer Center trial (NCT02420717) is evaluating ruxolitinib for relapsed/refractory JAK–mutant ALL. Patients are initially treated with up to 3 weeks of ruxolitinib as monotherapy and multiagent chemotherapy will be eventually added for those with incomplete response.

11.2.4.2 Others Approaches in Ph-Like ALL

Type II JAK inhibitors may represent an alternative to allow more profound inhibition of JAK-STAT signaling. The compound CHZ868 was able to abrogate JAK-STAT signaling and prolonged JAK–mutant PDX mice survival. However, similar to ruxolitinib, the emergence of resistance mediated by acquired point mutations was also documented (Wu et al. 2015). The combination of JAK inhibitors with HSP90 or BET bromodomain inhibitors has been tested in preclinical models and has shown to be effective in inhibiting JAK-STAT signaling in Ph-like ALL. In addition to JAK-STAT pathway, PI3K/mTOR and Ras-MAPK pathways are also activated in Ph-like ALL cell lines with *$CRLF2_r$* and *JAK* mutations, the addition of a pharmacological inhibition of the mTOR pathway to ruxolitinib was investigated. It resulted in improved inhibition of leukemia growth in Ph-like ALL PDX models (Tasian et al. 2017). Furthermore, the addition of selumitinib, a MEK inhibitor, to AZD1480, a JAK1/2 inhibitor, markedly improved the inhibition in cell lines but not in PDX mice (Suryani et al. 2015). The histone deacetylase inhibitor givinostat (ITF2357) has shown potent efficacy as monotherapy to treat *$CRLF2_r$* Ph-like ALL *in vitro* and in PDX mice (Savino et al. 2017). Finally, chimeric antigen receptor T-cells targeting TSLP receptor has demonstrated efficacy to eradicate CRLF2-overexpressing ALL in PDX mice without off-target effect (Qin et al. 2015).

11.2.4.3 JAK Inhibitors in T-ALL and AML

In a preclinical study, the combination of ruxolitinib to the BCL2 inhibitor venetoclax has shown to control cell proliferation of *IL7R-α*-mutated T-ALL *in vivo* and *in vitro*, in contrast to ruxolitinib alone. This therapeutic combination has not yet been investigated in patients (Senkevitch et al. 2018). Another study has found the same combination to be efficacious in AML, regardless of the underlying genomic alteration, which could be explained by effective targeting of the AML niche (Karjalainen et al. 2017).

A phase II study (NCT00674479) investigated ruxolitinib as a monotherapy for adult patients with relapsed/refractory acute leukemias, including 28 with AML (which was secondary to MPN in 10 cases), irrespective of JAK-STAT mutation status. Despite being well tolerated, the drug has only modest effect; among three patients (all with secondary AML) who achieved a complete response, two had $JAK2^{V617F}$ mutation (Eghtedar et al. 2012). A phase I/II in adult AML found similar results and no data are available for pediatric AML (Pemmaraju et al. 2015).

11.2.5 Conclusion

JAK-STAT signaling is involved in a significant proportion of the three leukemia types presented here although the mechanisms differ between each disease entity. Moreover, the clinical impact of JAK-STAT involvement may vary between these diseases. In AML, genetic alterations involving this pathway are rare but functional dysregulation is frequently associated. The underlying mechanism remains largely unknown and if this element represents the driver event or only secondary to other alterations is yet to be determined. Nevertheless, the implication of JAK-STAT pathway has been associated with prognosis in AML, highlighting its contribution to leukemogenesis.

In T-ALL, JAK-STAT alterations comprise a quarter of cases through different mechanisms and genes. Interestingly, these alterations are frequently associated with each other as well as with other pathways. This illustrates the cooperation between different oncogenes to foster cell proliferation or survival.

Ph-like ALL is the most characterized and studied disease entity, harboring frequent and heterogeneous JAK-STAT pathway alterations. Its discovery presents a unique opportunity for molecularly targeted therapy to improve the unfavorable outcomes of this patient population. The therapeutic incorporation of JAK inhibitors in different preclinical studies has shown promising responses despite several challenges have since arisen. The ongoing clinical studies will allow confirming the safety and efficacy of ruxolitinib as a clinically active JAK inhibitor in Ph-like ALL.

11.3 Myeloproliferative Neoplasms

Myeloproliferative neoplasms (MPN) are clonal hematopoietic stem cell disorders resulting in the excessive production of cytokine-independent mature myeloid blood cells (Tefferi 2006). Under the WHO 2016-revised classification (Arber et al. 2016), they are divided into four distinct "classic" MPN: chronic myelogenous leukemia (CML), polycythemia vera (PV), essential thrombocythemia (ET), and primary myelofibrosis (PMF) (Swerdlow et al. 2008). CML is a distinct entity within the MPN spectrum; since *BCR-ABL* fusion has no involvement with the JAK-STAT pathway, it will thereby not be discussed. Three other entities belong to MPN in this classification: chronic eosinophilic leukemia not otherwise specified, chronic neutrophilic leukemia, and unclassifiable MPN, which are all exceptionally rare in childhood. This chapter will focus on PV, ET, and PMF in children.

Although being the more prevalent MPN in this age group, PV and ET remain extremely rare in children. The annual incidences of PV and ET are estimated around 0.2 and between 0.1 and 0.4 per million of person under 20 years old, respectively (Hasle 2000; Hasle et al. 1995, 1999; Jensen et al. 2000; McNally et al. 1997), which correspond to an incidence of 40–90-fold lower than in adults (Teofili et al. 2008). There is a female predominance in both adults and children with PV (Chim et al. 2005; Giona et al. 2012). PMF is uncommon in children with less than 50 cases published in the literature to date (DeLario et al. 2010, 2012). With regards to the clinical picture, the disease presentation in children is distinct from adult patients. PV is symptomatic in the majority of adult patients and thrombotic events occur frequently (Marchioli et al. 2005; Tefferi et al. 2013), whereas, in children, less than half will present with clinical symptoms and thrombosis is rare (Giona et al. 2012). The prognosis also differs according to age and death is estimated in around 8% of adults but none in children (Giona et al. 2012; Marchioli et al. 2005; Tefferi et al. 2013). ET is diagnosed mainly as an incidental finding on blood counts in children (Giona et al. 2012), but in adults, 50–70% of patients present with symptoms at diagnosis such as asthenia, vasomotor symptoms, bleeding, and thrombosis (Cervantes et al. 2002; Chim et al. 2005; Cortelazzo et al. 1990; Fenaux et al. 1990). PMF occurs mainly in toddlers with frequent transfusion support. The progression to leukemia is a major concern in adults with PMF but has never been reported in children (Saksena et al. 2014). Spontaneous resolution of the disease in children has been described in single case reports but the only published cohort reports 26% of spontaneous remission among 19 patients (DeLario et al. 2012). Due to these important differences in the disease evolution, data from adult studies should be carefully interpreted and should not be extrapolated to children.

Cases of polycythemia, thrombocytosis, and myelofibrosis occurring in childhood comprise a higher proportion of hereditary disorders due to a constitutional defect. Notably, germline mutations in *MPL*, *THPO*, *VHL*, *PHD2*, and *HIF-2α* have been reported in patients with hereditary polycythemia or thrombocytosis (Ang et al. 2002; Ding et al. 2004; Lenglet et al. 2018; Semenza 2009; Skoda 2010; Teofili et al. 2010). These rare monogenic defects lead to polyclonal proliferation with a high penetrance and early disease onset in childhood. As they do not meet the classical definition of MPN, they will not be discussed here. These conditions are distinct from the "hereditary" MPN, but the boundary between these two entities are becoming more porous as clonal hematopoiesis may also be found in hereditary polycythemia (Kralovics et al. 2003); but for didactical purpose, the distinction, even if somewhat artificial, will be kept here.

11.3.1 MPN-Associated JAK2 Mutations Are Less Frequent in Children than in Adults

11.3.1.1 Epidemiology of JAK2^{V617F} Mutation

The first molecular defect identified in PV was the loss-of-heterozygosity (LOH) at 9p in approximately a third of adults (Kralovics et al. 2002). The minimal amplified region contained several genes, including *JAK2*, of which the specific implication was initially not demonstrated.

A recurrent somatic guanine-to-thymidine substitution of nucleotide 1849 within the exon 14 of *JAK2* was simultaneously identified in 2005 by several teams. This missense heterozygous mutation results in the substitution of valine to phenylalanine at codon 617 (*JAK2*^{V617F}) and was found in more than 85% of adults with PV and 50% of those with ET and PMF (Baxter et al. 2005; James et al. 2005; Kralovics et al. 2005; Levine et al. 2005; Lippert et al. 2006). It preferentially occurs in a common germline 46/1 (also called GGCC) *JAK2* haplotype, present in about half of the population (Jones et al. 2010; Pietra et al. 2012).

In children, due to the smaller cohort studied, the frequency of this mutation is less precisely known but, as in adults, it frequency varies among the different MPN (Table 11.2). In PV, four studies identified *JAK2*^{V617F} to be present in 25–37% of children with sporadic PV (maximum population studied: 11 patients), while another study reported a frequency as high as 75% among eight cases (Cario et al. 2008; Giona et al. 2012; Ismael et al. 2012; Teofili et al. 2007a) Interestingly, this mutation can be detected in congenital PV but not in familial PV (Kelly et al. 2008). This fact highlights the difficulty to assess the real etiology of numerous cases of polycythemia with subnormal EPO levels and a significant proportion of these cases may not represent a real MPN. In fact, not all the cases studied (with the exception of the last paper) had documented evidence of

TABLE 11.2

JAK-STAT Alterations and Targeted Therapies in Myeloproliferative Neoplasms

Myeloproliferative neoplasms	JAK-STAT alterations	Frequency in children	Treatment options: Pre-clinical	Treatment options: Clinical	FDA-approved therapy
Polycythemia vera					
	JAK2^{V617F} mutation	25-75% of cases	Type I JAK inhibitors	Ruxolitinib in adults, 2 children reported	Ruxolitinib
	Exon 12 *JAK2* mutations	Individual cases (n=2)			
	SH2B3 mutations	2 cases among 320 MPN			
Essential thrombocythemia					
	JAK2^{V617F} mutation	22-56% of cases	Type I JAK inhibitors	Ruxolitinib in adults, no experience in children	
	MPL mutations	Individual cases (n=2)			
	CALR mutations	23.5% of cases (n=8/34)			
	SH2B3 mutations	6 cases among 320 MPN			
Primary myelofibrosis					
	JAK2^{V617F} mutation	Not reported (n=0/17)	Type I JAK inhibitors	Ruxolitinib and other type I JAK inhibitors (pacritinib, memelotinib and fedratinib) in adults, no experience in children	Ruxolitinib
	CALR mutations	50% of cases (n=7/14)			
	SH2B3 mutations	15 cases among 320 MPN			

MPN: Myeloproliferative neoplasms

clonal disease with clonal hematopoiesis and/or endogenous erythroid colonies and/or *PRV-1* RNA overexpression (Cario et al. 2008; Teofili et al. 2007b). Despite these, the cases fulfilled the WHO 2016-revised criteria for PV (Arber et al. 2016). Due to important differences between pediatric and adult PV, these criteria may not apply accurately in children than in adults. It has been suggested that, in the absence of *JAK2*^{V617F} mutation in a child with sporadic PV, the initial WHO 2001 criteria (made before the identification of *JAK2*^{V617F} and incorporating other indirect evidence of MPN) may be more appropriate to accurately make the diagnosis of PV (Barbui et al. 2014; Tefferi et al. 2007; Teofili et al. 2008; Vardiman et al. 2002).

In pediatric ET, the prevalence of *JAK2*^{V617F} mutation is less different than that of adults, from 22% to 56% gathered from the eight published studies, the largest enrolling 63 patients (Fu et al. 2016; Giona et al. 2012; Ismael et al. 2012; Randi et al. 2015; Sekiya et al. 2016; Teofili et al. 2007a, 200, 2009). This mutation has not been reported in hereditary ET but, interestingly, a dominant germline *JAK2*^{V617I} mutation was found in a family with hereditary thrombocytosis (Mead et al. 2012). Sporadic ET represents a similar challenge than PV with regards to the diagnostic criteria. Some patients with persistent thrombocytosis may have other underlying mechanisms than clonal proliferation, which point to the MPN (Fu et al. 2013; Randi et al. 2015). As with PV, it should be emphasized to rule out a hereditary cause of thrombocythemia (Teofili et al. 2008).

In summary, a significant proportion of cases of polycythemia and thrombocythemia in childhood may not be related to MPN despite biological findings suggest no other causes, which render the interpretation of cohort and biological studies more difficult. Among cases of polycythemia and thrombocythemia with markers of MPN, the *JAK2*^{V617F} mutation is not as frequent as in adults.

Due to the extreme rarity of PMF among children, very few data are available. A single study found no *JAK2*^{V617F} mutation among 17 children (DeLario et al. 2012).

11.3.1.2 JAK2^{V617F} Effect on Downstream Signaling

JAK2^{V617F} is located in the JH2 domain (Ungureanu et al. 2011). The point mutation seems to destabilize the JH1-JH2 interaction, and acts as a gain-of-function mutation resulting in constitutive activation of not only JAK2 and STAT, but also of MAPK and PI3K pathways (James et al. 2005; Silvennoinen and Hubbard 2015; Wernig et al. 2006; Zhao et al. 2005). Transduced in cell lines, *JAK2*^{V617F} promotes cytokine-independent growth and excess proliferation in the presence of EPO, IL-3, and IGF-1 (James et al. 2005; Levine et al. 2005; Staerk et al. 2005; Zhao et al. 2005). In murine models, this mutation induces a myeloproliferative PV-like disorder manifested by splenomegaly, bone marrow hyperplasia, and EPO-independent erythrocytosis within the first 3 months (James et al. 2005; Lacout et al. 2006; Wernig et al. 2006). Subsequently, an important myelofibrosis develops, causing further increase in splenomegaly, followed by regression of polycythemia and development of cytopenias (Lacout et al. 2006).

The clinical phenotype of MPN probably depends on the intensity of the JAK-STAT signaling, linked to the number of *JAK2*^{V617F} copies. Transgenic mice with a higher JAK-STAT signal intensity develop a PV phenocopy, whereas lower intensity results in an ET-like disease with mainly thrombocytosis (Tiedt et al. 2008). These *in vivo* data support the observation that most patients with PV harbor a 9p LOH leading to homozygous *JAK2*^{V617F} in contrast to the heterozygous mutation in patients with ET (Dupont et al. 2007; Scott et al. 2006).

In addition, the mechanism contributing to phenotype variation may be related to the activation of a specific downstream effector. PV, ET, and PMF are associated with preferential phosphorylation of STAT5, STAT1, and STAT3, respectively (Chen et al. 2010; Teofili et al. 2007c). Consistently, in a murine model, *STAT1* knockout results in decreased thrombocytosis and favors erythropoiesis (Duek et al. 2014). The fact that the same nucleotide substitution is found in three different diseases with different downstream effects is currently not completely understood. Other modifying factors, such as host polymorphisms or other cooperating mutations, may contribute to the phenotype diversity and are currently under investigation.

Despite these phenotype variations, STAT5 probably plays a central role in all MPN. pSTAT5 is elevated in each phenotype and even in the absence of *JAK2*^{V617} mutation (Teofili et al. 2007c). *In vitro*, both STAT5 and PI3K activation are required for the EPO-independent growth induced by *JAK2*^{V617F} (Garçon et al. 2006; Ugo et al. 2004). In mice models, STAT5 activation is required for the development of *JAK2*^{V617F}-induced MPN (Walz et al. 2012; Yan et al. 2012). The high pSTAT5 in PV is partially linked to its capacity to upregulate and cooperate with the anti-apoptotic protein Bcl-xL, which is also upregulated in PV (Garçon et al. 2006; Silva et al. 1998).

PI3K/Akt/mTOR pathway is activated downstream of *JAK2* activation. In *JAK2*^{V617F} cell lines, the pharmacologic inhibition of AKT hampered cell growth (Khan et al. 2013). The combination of PI3K and JAK1/JAK2 inhibition reduce spleen volume in mice inoculated with *JAK2*^{V617F} Ba/F3 cells (Bartalucci et al. 2013; Khan et al. 2013).

Whether *JAK2*^{V617F} is sufficient by itself to initiate MPN is controversial. This mutation has been identified in primitive $CD34^+$ $CD38^-$ HSC and skews the maturation toward erythroid differentiation (Jamieson et al. 2006). However, a study found no expansion of this compartment in the sole presence of *JAK2*^{V617F} mutation (Anand et al. 2011). In a mouse model, the presence of the thrombopoietin receptor (MPL) (but not thrombopoietin) was mandatory to induce the MPN phenotype in a *JAK2*^{V617F} background (Sangkhae et al. 2014). This highlights the fact that this mutation may not be a driver and requires additional cooperative event(s). Furthermore,

MPN clonality may not correlate with *JAK2*^{V617F} allele burden (Nussenzveig et al. 2007). A study of families with more than one person affected by MPN showed that in some families, a relative may have a *JAK2*^{V617F}-positive MPN, while another family member is afflicted with a *JAK2*^{V617F}-negative MPN (Bellanné-Chantelot et al. 2006). In the same line, *JAK2*^{V617F} allele may be detected at a low frequency in the general population without evidence of MPN (Martinaud et al. 2010; Nielsen et al. 2011; Sidon et al. 2006; Xu et al. 2007). In conclusion, the contribution of this mutation to the pathogenesis of MPN has been demonstrated in several *in vitro* and *in vivo* models. Nevertheless, this molecular defect probably does not explain entirely the pathophysiology of MPN. Further mechanistic studies are still needed to better define its functional role.

11.3.1.3 Exon 12 JAK2 Mutations

Among the subset of patients with *JAK2*^{V617F}-negative PV, some present with mutations in the exon 12 of *JAK2* (Tefferi 2010). Initially described in 2007 (Scott et al. 2007), more than 15 heterozygous gain-of-function mutations have been reported to date in adults, including point mutations, in-frame deletions, and duplications, mainly affecting seven highly conserved amino acid residues (F537–E543) within exon 12 (Butcher et al. 2008; Pardanani et al. 2007; Passamonti et al. 2011; Pietra et al. 2008). These mutation carriers tend to be adult patients with a familial history of PV and mostly present at a younger age with an isolated erythrocytosis, a subnormal serum erythropoietin level, and a similar prognosis to that of *JAK2*-mutant PV (Martínez-Avilés et al. 2007; Pardanani et al. 2007, 12; Passamonti et al. 2011). To date, only two children have been reported with an exon 12 *JAK2* mutation (Cario et al. 2008). In Ba/F3 cell lines, among the four mutations tested, all resulted in IL-3-independent growth (Scott et al. 2007). In a retroviral transfer murine model, these mutations led to a myeloproliferative disease with marked erythrocytosis and modest thrombocytosis (Scott et al. 2007).

11.3.2 Other Genomic Alterations Activating JAK-STAT in PMF and ET

11.3.2.1 MPL Mutations

Among adults with ET and PMF, less than 10% harbor a *MPL* mutation with a slightly different clinical phenotype: older age, higher platelet counts but lower hemoglobin level, and microvascular symptoms associated with arterial thrombosis over the course of disease, and similar survival rates (Beer et al. 2008; Pardanani et al. 2006; Pikman et al. 2006; Vannucchi et al. 2008). Several somatic *MPL* mutations have been reported since 2006 in ET and PMF, the most frequent being *MPL*^{W515L}, which has been reported in two children with ET (Beer et al. 2008; Farruggia et al. 2013; Pardanani et al. 2006; Putti et al. 2017; Teofili et al. 2010). These patients display functional similarities with patients carrying a germline *MPL*^{S505N} mutation but follow a different disease course; the latter may evolve towards myelofibrosis but no leukemic transformation has been reported and hematopoiesis remains polyclonal (DeLario et al. 2012; Ding et al. 2004; Teofili et al. 2007a, 2010).

MPL encodes for the thrombopoietin receptor and plays a key role in megakaryocyte proliferation and survival. Both somatic and germline mutations are located in the exon 10 and result in a gain-of-function of the protein. The mutations are mainly heterozygous but acquire homozygosity through uniparental isodisomy as previously described (Szpurka et al. 2009). In case of a somatic mutation, the origin of clonal mutations was set at the multipotent HSC level (Chaligné et al. 2007). Exon 10 is located in the juxtamembrane domain of MPL, which has an auto-inhibitory role, limiting MPL activation (Staerk et al. 2006). Activating mutations result in loss of the inhibitory signal and constitutive activation, despite the absence of thrombopoietin binding (Staerk et al. 2006). In several cell lines (Ba/F3, 32D and UT7), transduction of *MPL*^{W515L}, but not of all mutations found in patients, led to cytokine-independent growth (Chaligné et al. 2008; Pikman et al. 2006). In Ba/F3 cell lines, *MPL*^{S505N} results in cytokine-independent proliferation (Ding et al. 2004). As in normal physiologic states, MPL dimerization induced by TPO leads to JAK2 phosphorylation (Drachman et al. 1995; Tortolani et al. 1995), in *MPL*^{W515L} Ba/F3 cells, JAK-STAT pathway was activated not only with JAK2, STAT3, STAT5, but also AKT and ERK being phosphorylated (Pikman et al. 2006). Moreover, *MPL*^{S505N} Ba/F3 cells result in constitutive MEK1/2 and STAT5B phosphorylation in the absence of IL-3 (Ding et al. 2004). In a mouse model, transplantation of *MPL*^{W515L} bone marrow leads to a MPN-like phenotype characterized by splenomegaly, splenic infarct, marked thrombocytosis, and reduced life expectancy (Pikman et al. 2006). The contribution in the pathogenesis of these somatic mutations is yet not known in patients. Some cases harbor several *MPL* mutations or concomitant *JAK2*^{V617F} within the same clone (Lasho et al. 2006). The clinical impact of these two mutations differs as patients with germline *MPL*^{S505N} may evolve toward myelofibrosis without leukemic transformation and hematopoiesis remains polyclonal (Teofili et al. 2010).

11.3.2.2 CALR Mutations

CALR gene encodes calreticulin, a Ca^{2+} binding protein associated with multiple functions, including proper protein folding in the endoplasmic reticulum (ER) and calcium homeostasis (Eggleton and Michalak 2013). Through whole exome sequencing, somatic *CALR* mutations have been identified in 2006 and may be present in up to 70% of adults with ET or PMF without *JAK2* or *MPL* mutations, but not in patients with PV (Klampfl et al. 2013; Nangalia et al. 2013). In children, they were found in 23.5% among 34 ET cases and 50% in 14 PMF cases (An et al. 2014; Giona et al. 2014). Despite the rarity of these mutations, they still occur more frequently than *MPL* mutations in ET and PMF for both children and adults (Tefferi et al. 2014a). More than 50 mutations have been described; they consist of insertions and deletions located in exon 9 of *CALR* (Klampfl et al. 2013; Nangalia et al. 2013). The most frequent *CALR* mutations in adults are the 52-base pair (bp) deletion (type 1) and the 5-bp insertion (type 2 mutation), encompassing more than 80% of *CALR* mutations (Rumi et al. 2014a, 2014b). In children with PMF, only type 2 mutations were reported whereas both types were described in pediatric ET (An et al. 2014; Giona et al. 2014).

The pathophysiological relationship between CALR, a housekeeping chaperone protein, and MPN was not obvious. Studies have demonstrated that mutant CALR resulted in abnormal physical interaction with the MPL protein inside the ER (Araki et al. 2016; Chachoua et al. 2016). The coupled proteins are exported on the cell surface and result in constitutive TPO activation leading to JAK2 dimerization and activation (Araki et al. 2016; Chachoua et al. 2016). In a mouse model, type 1 and type 2 mutants induce thrombocytosis with frequent progression to myelofibrosis (Marty et al. 2016). This phenotype was more frequently and rapidly induced by type 1 than type 2 mutations. MPL, but not TPO, is mandatory in the disease progression of these mice (Marty et al. 2016). In adults with ET, *CALR* mutations (especially type 2) are associated with higher platelet counts, lower hemoglobin and leukocyte counts than those with *JAK2*^{V617F}, suggesting that differences in JAK-STAT dysregulation depends on the involved gene and the type of *CALR* mutation (Tefferi et al. 2014b). Coexistence of *CALR* and *JAK2*^{V617F} mutations is rare but has been reported (Lundberg et al. 2014).

11.3.2.3 SH2B3 Mutations

Five to seven percent of MPN patients harbor *SH2B3* mutations (McMullin and Cario 2016). The *SH2B3* gene is located on chromosome 12q24.12, encodes for the protein LNK, which belongs to the LNK/SH2B family of adaptor proteins, which plays an important role in early hematopoiesis. Most of *SH2B3* mutations are missense mutations that predominantly target exon 2 coding for the Pleckstrin homology (PH) domain, resulting in a reduced level of protein activity. The other reported *SH2B3* alterations consist of deletions of exons 2 and 5, targeting once again the PH domain (Maslah et al. 2017). Functional analyses demonstrate that these *SH2B3* mutants *in vitro* exhibit aberrant JAK2/STAT3/STAT5 activation as a result of decreased LNK activity (Oh et al. 2010). Therefore, LNK acts a negative regulator of the JAK-STAT pathway. *SH2B3* mutations can be found in all MPN subtypes and can co-exist with other MPN driver genes (*JAK2, MPL, CALR*) or other genes (*TET2, CBL*) (Maslah et al. 2017). Interestingly, it has also been observed that MPN patients may develop *SH2B3* mutations during leukemic transformation, during which their frequency increases to 13% (Maslah et al. 2017; Pardanani et al. 2010). This suggests that *SH2B3* mutations may contribute to the leukemic transformation of MPN. However, the method used in these studies has low sensitivity and may not detect subclonal *SH2B3* mutations during the chronic phase of MPN.

11.3.2.4 Mutations in Other Genes and Interaction between Co-existing Mutations

Adult patients without *JAK2*, *MPL* and *CALR* mutations represent less than 10% of PMF and around 15% of ET; they are often referred as triple-negative MPN. Alterations in other genes have been described but without functional impact on the JAK-STAT pathway (Cabagnols et al. 2016; Feenstra et al. 2016). For example, mutations in *TET2* have been reported in both *JAK2*^{V617F}-positive and *JAK2*^{V617F}-negative patients, in addition to mutations in *CBL*, *ASXL1*, *IDH*, *IKZF1* and *EZH2* in triple-negative patients (Guglielmelli et al. 2014; Tefferi and Vainchenker 2011). They represent a small subset of adults and their functional consequences remain under investigation. In children, the potential implication of these genes stays unknown.

Lastly, the mutations described herein appear at various stages of clonal evolution and may co-exist with each other (Lundberg et al. 2014). This has a potential impact on the disease course and the risk of leukemic transformation in adults (Klampfl et al. 2011; Lundberg et al. 2014). The impact on the JAK-STAT pathway for each specific alteration or combination is not entirely known. However, genomic analyses showed that for all adults with MPN, regardless of the underlying mutation, a transcriptional signature consistent with activated JAK2 signaling is present (Rampal et al. 2014).

11.3.3 Therapeutic Opportunities of JAK-STAT Inhibition in MPN

MPN in adults was the first field leading to the development of type 1 JAK inhibitors. Currently, ruxolitinib is the only drug with the Food and Drug Administration (FDA) and European Medicine Agency (EMA) approval for PMF as first-line therapy since 2011 and then hydroxyurea (HU)-resistant or -intolerant PV since 2014.

In HU-resistant or -intolerant PV in adults, ruxolitinib has shown to be superior to placebo in improving symptoms, splenomegaly and hematocrit (Passamonti et al. 2017; Vannucchi et al. 2015). Data in children is limited to a 26-month old patient with a *JAK2*^{V617F} PV-associated Budd-Chiari syndrome and another isolated patient enrolled on the COG ADVL1011 phase 1 trial (NCT01164163) (Coskun et al. 2017; Loh et al. 2015).

In adults with HU-resistant or -intolerant ET, ruxolitinib allows clinical and biological improvement but the impact on hemorrhage, thrombosis and leukemic transformation is yet unclear (Gunawan et al. 2018; Harrison et al. 2017a; Verstovsek et al. 2017). No experience has been published in children with HU-resistant or –intolerant ET.

Ruxolitinib has shown efficacy in adults with PMF, allowing to reduce clinical symptoms and spleen volume (Al-Ali et al. 2016; Verstovsek et al. 2010). As previously evocated, PMF is rare and has a markedly more benign evolution in the pediatric population. Therefore, no pediatric experience of ruxolitinib in PMF is yet published.

More recently, other type I JAK inhibitors have demonstrated significant clinical benefits in adult PMF (such as pacritinib, memelotinib and fedratinib) (Harrison et al. 2017b; Mascarenhas et al. 2018; Mesa et al. 2017; Pardanani et al. 2015), but there is no such experience for these therapies in children.

In conclusion, JAK inhibitors have become the standard treatment for adult MPN, as first or second-line therapies. However, it is not clear whether these therapies are truly disease-modifying and improve outcomes (Cervantes and Pereira 2017). The symptoms' improvement may also be linked to the anti-inflammatory effect of ruxolitinib (Hasselbalch 2013). Moreover, MPN clones persist during therapy and adaptive mechanisms of resistance through paradoxical JAK-STAT activation have been described (Koppikar et al. 2012). To overcome these pitfalls, others strategies including combinatorial therapy between ruxolitinib and novel agents (eg. histone deacetylase inhibitors) are being developed for adult patients but their benefits have not been established in children (Bose and Verstovsek 2017).

11.3.4 Conclusion

The JAK-STAT pathway plays an important role in the pathogenesis of the three classical MPN discussed above. Functional analyses showed the central role of this pathway in the disease evolution. The identification of the *JAK2*^{V617F} mutation was an important breakthrough deepening our understanding of the underlying causes of the disease. However, the molecular mechanisms and the *primum movens* remain not known precisely. Other genetic factors seem to contribute to the disease onset and evolution. In children, the knowledge of these studies pertains only to a subgroup of patients and a significant proportion of children do not have a molecular or mechanistic explanation of their MPN. Investigations are needed in this specific population to address this question. Finally, the paucity of MPN in children, the distinct evolution and the diverse genomic spectrum have halted the development of targeted therapies in this population. To date, the experience is limited; international, multicentric collaborations will be needed to allow a comprehensive pediatric-tailored management for MPN.

11.4 Lymphomas

The lymphoproliferative neoplasms encompass a large category of malignant lymphoid proliferations arising from B- or T-lymphocytes at various maturation stages. The 2016 WHO revised classification distinguishes more than 40 entities. Lymphomas are classically divided into two broad categories: Hodgkin lymphomas (HL) and non-Hodgkin lymphomas (NHL).

HL are the most frequent lymphomas among children with a peak incidence during adolescence (30.1 cases per million between 15 and 19 years in the USA) (Noone et al. 2018). The most frequent histologic subtype is classical HL (cHL). Longterm survival rates in pediatric HL now reaches above 90% with contemporary treatment regimens but are negatively affected by treatment-related toxicities, especially in high-risk disease being exposed to intensified combinatorial chemo-radiotherapy (Smith et al. 2010; Thomson and Wallace 2002). The dysregulated pathways of lymphomatous cells are partly derived from genetic influences and partly as the result of the important interaction with the microenvironment surrounding the tumoral cells (Mathas, Hartmann, and Küppers 2016). The impact of age on the biology of HL is poorly understood. Clinical observations often indicate that older children present with more extensive disease, despite the comparable survival rates among all age subgroups (Englund et al. 2018). However, this difference may be related to the difference in growth and development rather than the underlying disease biology. Thus, the majority of biological studies in HL are not age-based and the discussed literature herein includes representation from the adult population including adolescents and young adults.

NHL are less frequent in childhood than adulthood as the incidence rises with increasing age, from 7.4 per million between 1 and 4 years to 20 per million between 15 and 19 years in the USA (Noone et al. 2018). Pediatric NHL is markedly different from their adult counterpart, being almost exclusively high-grade lymphomas and mostly of B-cell origin (Burkhardt et al. 2005). Almost all NHL subtypes are less common during childhood with the exception of sporadic Burkitt lymphoma (BL). No JAK-STAT pathway implication has been demonstrated in BL; hence, this entity will not be discussed here. Diffuse large B-cell lymphoma (DLBCL) is the second most frequent NHL in childhood. Gene expression profiling defined two expression signatures in DLBCL: Germ cell B (GCB-DLBCL) or activated B-cell (ABC-DLBCL) type, the latter conferring a worse prognosis. In children, the GCB-DLBCL subtype is more common and is not associated with prognosis. A third NHL subtype is the primary mediastinal large B-cell lymphoma (PMBCL), mainly occurring in young adult women with a distinct gene expression pattern, treatment and outcome. Anaplastic large-cell lymphoma (ALCL) is the main peripheral T-cell lymphoma in the pediatric age group and comprises 10–15% of NHL. In over 95% of pediatric ALCL, the *ALK*, Anaplastic Lymphoma Kinase, gene rearrangement is identified and drives lymphomagenesis. The *ALK*-negative (ALK-) ALCL has a different gene expression profile and a worse prognosis.

Due to the higher incidence of NHL in adults, most of molecular studies presented here were performed in adults only or using mixed pediatric/adult samples. We will also discuss the relevance of these findings in the pediatric population when applicable.

11.4.1 JAK-STAT Pathway Gene Alterations Are Frequent in Different Lymphoma Subtypes

11.4.1.1 JAK1 Alterations

JAK1 mutations occur mostly in T-cell lymphomas (Table 11.3). They represent 15% of 88 ALCL, 6% of 110 natural killer-cell/T-cell lymphoma (NKTCL) and 11% of 9 cutaneous T-cell lymphoma (CTCL) cases (samples include both children and adults) (Crescenzo et al. 2015; Song et al. 2018). In ALCL, *JAK1* mutations are located in the JH1 kinase domain and involve almost exclusively the G1097 residue with several missense mutations substituting to aspartic acid, asparagine or serine (Crescenzo et al. 2015). *In vivo*, mutant *JAK1* leads to increased levels of pSTAT3 and cell proliferation. This oncogenic mutation is synergistic with *STAT3* mutations and exhibits sensitivity to JAK inhibition (Crescenzo et al. 2015).

11.4.1.2 JAK2 Alterations

JAK2 sequence mutations have not been described in lymphomas. In contrast to MPN, no canonical *JAK2*^{V617F} mutations were found in cell lines derived from lymphomas, PMBCL and cHL (Melzner et al. 2006; Scott et al. 2005; Wu et al. 2009). Moreover, activating *JAK2* mutations in exon 12, described in MPN, were not found in PBMCL (Wu et al. 2009). *JAK2* fusions have been identified in 4 adult patients with cHL (Roosbroeck et al. 2011). Two patients carry a *SEC31A-JAK2* fusion resulting from a reciprocal t(4;9)(q21;p24) translocation. Its expression in hematopoietic cell lines results in cytokine-independent growth and can be abolished by pharmacological JAK inhibition. *SEC31A-JAK2*–transduced murine cells lead to a fatal T-ALL in 6 out of 16 mice. Two additional patients harbor *JAK2* fusions but the respective partners could not be identified.

11.4.1.3 JAK3 Alterations

The identification of somatic *JAK3* mutations has been made in NKTCL, representing 20-35% of the studied cohort (adult/pediatric) and in approximately 20% of CTCL samples (Almeida et al. 2015; Bouchekioua et al. 2014; Koo et al. 2012; McGirt et al. 2015). NKTCL has the highest prevalence in Asia and Latin America but remains extremely rare in children among whom only a few case reports have been reported (Lai and Dunleavy 2013; Pillai et al. 2016).

Two recurrent missense cytosine-for-thymine substitutions (A572V and A573V) were described in NHL. Both are located in exon 12 of *JAK3*, within the JH2 domain. These mutations are mostly detected at the heterozygous state but some lymphomas harbor compound heterozygous or homozygous mutations. A single patient was reported with a V722I mutation, located in exon 16, also belonging to the JH2 domain (Bouchekioua et al. 2014). The *JAK3*^{A572V} mutation has also

TABLE 11.3

JAK-STAT Alterations and Targeted Therapies in Lymphomas

Lymphomas	JAK-STAT alterations	Frequency	Treatment options	
			Pre-clinical	**Clinical**
Hodgkin Lymphoma				
	JAK2 fusion	Individual cases (n=4)	Type I JAK inhibitors	Ruxolitinib in adult patients, phase II completed (NCT01877005) and ongoing (NCT01965119). Pacritinib in adults, phase I completed (NCT00741871)
	STAT6 mutations	Around 30% of cases		
	9p24 amplification	Up to 90% of cases		
	SOCS1 mutation	Around 55% of cases		
	PTPN1 mutation	Around 20% of cases		
B-cell non-Hodgkin lymphoma				
DLBCL	*STAT6* mutations	Around 15% of adults GCB-DLBCL	Type I JAK inhibitors	Pacritinib in adults, phase I completed (NCT00741871)
	SOCS1 mutation	Around 27% of adults		
	MYD88 mutations	Around 30% of ABC-DLBCL		
PMBCL	9p24 amplification	Around 35% of cases	Type I JAK inhibitors	
	SOCS1 mutation	Around 55% of cases		
	PTPN1 mutation	Around 20% of cases		
	STAT6 mutations	36-72% of cases		
T-cell non-Hodgkin lymphoma				
ALCL	*JAK1* mutation	Around 15% of cases	Type I JAK inhibitors	Pacritinib in adults, phase I completed (NCT00741871). Cerdulatinib in adult patients, phase IIa (NCT01994382)
	TYK2 activation	Most of ALCL	TYK2 inhibitors	
	STAT3 mutation	Around 10% of ALK- ALCL	Type I JAK inhibitors	
NKTCL	*JAK1* mutation	Around 6% of cases	Type I JAK inhibitors	
	JAK3 mutation	20-35% of cases	JAK3 inhibitors	
	STAT3 mutation	Around 20% of cases	Type I JAK inhibitors	
PTCL	*PTPN2* mutation	Individual cases (n=2)	Type I JAK inhibitors	
CTCL	*JAK1* mutation	Around 10% of cases	Type I JAK inhibitors	
	JAK3 mutation	Around 20% of cases	JAK3 inhibitors	

DLBCL: Diffuse large B-cell lymphoma, PMBCL: Primary mediastinal large B-cell lymphoma, ALCL: Anaplastic large-cell lymphoma, NKTCL: natural killer-cell/T-cell lymphoma, PTCL: Peripheral T-cell lymphoma, CTCL: cutaneous T-cell lymphoma

been reported in adult patients with CTCL (Cornejo et al. 2009; McGirt et al. 2015). In NKTCL cell lines, *JAK3*^{A572V} leads to constitutive activation of both JAK3 and STAT5, resulting in cytokine-independent growth (Koo et al. 2012). *JAK3*^{A572V}-transduced cells in a murine model lead to a fatal TCRαβ+ T-lymphoproliferative disease in all mice (Cornejo et al. 2009). Mutations in the FERM domain of *JAK3* have been reported in adult T-cell leukemia/lymphoma, a very rare disease not occurring during childhood.

11.4.1.4 TYK2 Alterations

To date, *TYK2* mutations have never been reported in lymphomas. However, activation of TYK2 has been identified as a central event in ALK+ and ALK- ALCL (Prutsch et al. 2018). TYK2 is highly expressed and its knockdown or inhibition in human ALCL cells induces death. In transgenic mice models of ALCL, the suppression of TYK2 reduces tumoral growth. The effect on cell growth depends on STAT1 and STAT3 phosphorylation and is strengthened by an autocrine mechanism involving IL-10 and IL-22: TYK2 overexpression results in production of these cytokines which bind to their cognate receptor on tumor cells, activating STAT1 and STAT3 (Prutsch et al. 2018). In ALK- ALCL with a JAK-STAT mutation, the presence of cytokine signaling is also necessary for cell survival (Chen et al. 2017). These data suggest that cytokine signaling plays a major role in sustaining cell viability in spite of constitutive activation of the JAK-STAT pathway. The molecular mechanism underlying TYK2 activation in ALK+ ALCL is not known; however, in some patients with ALK- ALCL, transcriptomic analysis unravels recurrent *NPM1-TYK2* or *NF-κB2-TYK2* fusions (Crescenzo et al. 2015; Velusamy et al. 2014). *NPM1-TYK2* fusion is found in cutaneous ALK- ALCL and is also present in lymphomatoid papulosis. A single case with ALK- ALCL harboring a *PABPC4-TYK2* fusion was reported (Crescenzo et al. 2015). The *NPM1-TYK2* fusion gene induces constitutive activation of TYK2 and leads to phosphorylation of STAT1, STAT3 and STAT5. TYK2 knockdown in *NPM1-TYK2* MyLa cell lines results in a significant decrease in cell proliferation (Velusamy et al. 2014). As for the *NF-κB2-TYK2* fusion, it constitutes a chimeric transcription factor (Crescenzo et al. 2015). Its transduction in cell lines results in tumorigenesis and induces JAK and STAT3 phosphorylation. Another fusion of NF-κB2 within the intracytoplasmic domain of ROS1 (*NF-κB2-ROS1*) has been identified in the same work (Crescenzo et al. 2015). ROS1 is a JAK-STAT-independent tyrosine kinase with structural similarities to the ALK protein. Interestingly, the effect of the fusion phenocopies that of NF-κB2-TYK2 in regards to STAT3 dependency.

11.4.1.5 STAT3 Alterations

Somatic *STAT3* mutations were primarily described in granular lymphocytic leukemia but have been found in 21% of 101 patients with NKTCL, 10% of 88 patients with ALK- ALCL and 3% of 44 patients with cutaneous ALCL (mixed adult/pediatric samples) (Crescenzo et al. 2015; Hu et al. 2013; Koskela et al. 2012; Lohr et al. 2012; Morin et al. 2011; Song et al. 2018). These point mutations are not recurrent but occur predominantly within the SH2 domain. The pathogenicity of each mutation is not equivalent. In murine lymphoid Ba/F3 cells, among 4 mutations predicted to be pathogenic (p.D427H, p.E616G, p.E616K and p.E696K), all four lead to STAT3 phosphorylation, but only the p.E616K mutation can confer cytokine-independent growth (Song et al. 2018). *In vitro*, *STAT3*^{Y640F} cell lines are constitutively activated and can be abrogated by STAT3 inhibition. *STAT3*^{Y640F}-transplanted hematopoietic cells in mice lead to the development of a myeloproliferative-like disease (Couronné et al. 2013).

In NKTCL cell lines, pSTAT3 levels correlate with PD-L1 expression on the tumor cell surface (Song et al. 2018). This effect is mediated by STAT3 direct regulation of *PD-L1* transcriptional activity. Remarkably, STAT3 silencing and inhibition decrease PD-L1 expression, providing a potential therapeutic approach combining STAT3 inhibitors to anti-PD-1/PD-L1 antibodies (Song et al. 2018).

In T-cell lymphoma, *STAT3* mutations co-exist with other JAK-STAT mutations in 62% of 171 cases (Song et al. 2018). As previously mentioned, the co-existence of *STAT3* and *JAK1* mutations in ALK- ALCL exerts synergistic effect on cell proliferation *in vitro* (Crescenzo et al. 2015).

11.4.1.6 STAT6 Alterations

Heterozygous *STAT6* mutations have been found in 36-72% of PMBCL, 14% of adult GCB-DLBCL and 32% of adult/pediatric cHL (Dubois et al. 2016; Ritz et al. 2009; Tiacci et al. 2018).

These mutations occur in the hotspot DNA-binding domain within the exon 12 and overlap between the different disease subtypes. Interestingly, a large proportion of patients harbor concomitant alterations in other genes regulating the JAK-STAT pathway, such as *SOCS1* and *JAK2*, providing additional evidence to support that alterations affecting this pathway are not mutually exclusive. Nevertheless, the downstream effect of *STAT6* mutations is not fully understood. They could lead to constitutive phosphorylation resulting in tumor growth (Ritz et al. 2009; Tiacci et al. 2018); however, mutant-transduced cell lines exhibit an impaired rather than enhanced DNA binding activity of STAT6 (Ritz et al. 2009). Moreover, in a limited gene expression study, the transcription of STAT6 target genes was not modulated by these mutations (Ritz et al. 2009).

STAT6 mutations have also been described in adult follicular lymphoma (Yildiz et al. 2015). However, this entity is quite uncommon in children and exhibits distinct pathological and genetic features from the adult counterpart (Lorsbach et al. 2002; Oschlies et al. 2006). Notably, MAPK pathway mutations are more prevalent and the involvement of *STAT6* mutations in these pediatric cases has not been reported (Louissaint et al. 2016). The clinical relevance of these findings is currently not known in children.

11.4.2 JAK-STAT Pathway Collaborates in Lymphomagenesis through the Recurrent 9p24 Amplification in cHL and PMBCL

PMBCL gene expression studies have revealed that it displays a distinct profile compared to that of DLBCL and cHL (Rosenwald et al. 2003; Savage et al. 2003). Recurrent gene alterations are shared between these different B-cell lymphomas, especially the classical focal 9p24.1 amplification (Borchmann and Engert 2017; Lenz et al. 2008; Meier et al. 2009). Up to 94% of patients with cHL may harbor a 9p24.1 copy gain (> 2 copies) or amplification (>5 copies) (Roemer et al. 2016). The intensity and the spanning of the amplified region vary, but the minimal common region of copy number gain seems to encompass 7–10 genes among which figure *PD-L1*, *PD-L2*, *KDM4C*, *RANBP6* and *JAK2*. Being a part of the amplicon, *JAK2* may contribute to the pathogenesis and mediate several effects in lymphomas. In cHL and PMBCL cell lines, *JAK2* knockdown or drug-mediated inhibition induces apoptosis but does not inhibit proliferation (Rui et al. 2010). The effect of *JAK2* on cell viability is dependent on IL-13, of which its secretion is upregulated by an autocrine loop (Rui et al. 2010). Moreover, *JAK2* has a direct effect on increasing PD-L1 expression (Green et al. 2010). Since *JAK2* is not the only upregulated gene within the amplicon, the other genes located in the amplicon probably share an important role in tumorigenesis. In fact, in contrast to *JAK2*, *KDM4C* and *RANBP6* knockdown reduces proliferation without inducing apoptosis (Rui et al. 2010). Moreover, the cooperation within the amplified region fosters cell survival and proliferation. Functional analyses showed that *KDM4C*, a histone demethylase, cooperates with *JAK2* in the epigenome modification (Rui et al. 2010). The inhibition of both *KDM4C* and *JAK2* increases heterochromatin and kills PMBCL and HL cell lines (Rui et al. 2010). Upon action of these genes in cHL, transcription–repression of hundred genes is reduced, several of which are likely to be involved in lymphomagenesis. One of them is *MYC*, which does not allow cell proliferation by itself but remains important in promoting survival (Rui et al. 2010). In conclusion, this amplicon provides multiple cooperative key genes involving the JAK-STAT pathway and providing a potential therapeutic action, notably through JAK inhibition in combination with immune checkpoint inhibitors, as demonstrated in NKTCL (Song et al. 2018).

11.4.3 Loss-of-Function of JAK-STAT Regulators Is Frequent in Lymphomas

11.4.3.1 SOCS1 Mutations

SOCS1 mutations have been reported as a mechanism of JAK-STAT pathway activation in some lymphomas. Suppressor-of-cytokine-signaling (SOCS) family proteins constitute major regulators of JAK activity and are involved in cellular homeostasis. These proteins contain a SH2 domain, which binds to JAK, and ultimately lead to proteosomal degradation through poly-ubiquitination (Ungureanu et al. 2002). STAT5 activation induces *SOCS1* transcription, which limits JAK activity (Hennighausen and Robinson 2008). Therefore, loss-of-function mutations in *SOCS1* result in the abolition of this negative-feedback loop and hyperactivation of the JAK-STAT pathway. *SOCS1* mutations have been identified in several types of B-cell lymphomas in pediatric and adult patients (Melzner et al. 2005; Mottok et al. 2009; Weniger et al. 2006). They are present in up to 56% of cHL (pediatric/adult samples), 55% of PMBCL and 27% of adult DLBCL (Dubois et al. 2016; Mottok et al. 2009; Tiacci et al. 2018; Weniger et al. 2006). Although less frequent, 4% of 171 T-cell lymphomas harbor a *SOCS1* mutation (Song et al. 2018). In B-cell lymphomas, these mutations result from the somatic hypermutation process occurring during mature B-cell development upon B-cell receptor engagement (Mottok et al. 2009, 2007). Consequently, there is no specific hotspot in *SOCS1* and *SOCS1* alterations translate in most cases in partial or complete deletion of the protein. Most frequent types of alterations are non-sense mutations resulting in a truncated protein followed by interstitial deletions (Mottok et al. 2007). Both heterozygous and homozygous mutations have been reported (Melzner et al. 2005; Mottok et al. 2009). In PMBCL-derived MedB-1 cell lines, impaired SOCS1 leads to high levels of pSTAT5 and pJAK2 (Melzner et al. 2005). The constitutive JAK2 activation was due to its defective degradation, which can be corrected by ectopic *SOCS1* expression. In this cell line, *SOCS1* loss-of-function also mediates the loss of *RB1* expression, another important tumor-suppressor gene (Melzner et al. 2005). *SOCS1* mutations and 9p24.1 amplification are not mutually exclusive, since a subset of patients with PBMCL harbors both alterations (Melzner et al. 2005; Mottok et al. 2007). Furthermore, some cHL patients have been reported with both *SOCS1* and *STAT6* mutations (Tiacci et al. 2018).

11.4.3.2 PTPN1 Mutations

Another mechanism of regulation of JAK-STAT activation is the dephosphorylation of activated JAK/TYK2 proteins. PTPN1 and PTPN2 are phosphatases removing the phosphorylation of activated JAK2/TYK2 and JAK1/JAK3, respectively. Whole-genome and whole-transcriptome sequencing identified somatic *PTPN1* mutations in 20% of 6 cHL and 22% of 77 PMBCL cases (Gunawardana et al. 2014). Mutation types include predominantly missense, then nonsense and frameshift mutations. Most mutations are located in the catalytic domain and the impairment impact varies according to the mutation. In cHL cell line, these mutations decrease PTPN1 phosphatase activity, leading to hyperphosphorylation of JAK-STAT proteins. Co-occurrence of *PTPN1* and *SOCS1* mutations were frequent in this cohort (Gunawardana et al. 2014). Functional analysis of PTP1BΔ6, a PTPN1 variant lacking exon 6, demonstrates that its expression in cHL cell lines induces STAT activity and cell proliferation. This dominant negative variant protects against cytotoxic agents and impedes the normal PTP1B function by heterodimerization (Zahn et al. 2017).

11.4.3.3 PTPN2 Mutations

Bi-allelic inactivation of *PTPN2* has been found in 2 out of 39 patients with peripheral T-cell lymphoma not otherwise specified (Kleppe et al. 2011). One patient carries two frameshift mutations (p.G41fs and p.A45fs) and the second patient has a missense mutation (c.133G>A) in one allele and a 18p deletion in the other allele. One variant was found in a patient with cHL but without functional consequence. It has been shown that cHL SUP-HD1 cell lines have very low PTPN2 protein levels. These cells carry a missense mutation (p.L249P) in one allele and a gene deletion in the other. In a similar fashion to PTPN1 loss-of-function, JAK-STAT proteins are constitutively activated when PTPN2 is inactivated.

11.4.4 MYD88 Mutations Activate the JAK-STAT Pathway in ABC-DLBCL

ABC-DLBCL is characterized by a constitutive activation of the anti-apoptotic NF-κB signaling pathway (Compagno et al. 2009; Davis et al. 2001). Mutations in *MYD88* have been reported in approximately 30% of ABC-DLBCL and 10% of mucosa-associated lymphoid tissue lymphoma (Ngo et al. 2011; Schmitz et al. 2018). They are also present in more than 90% of Waldenstrom's macroglobulinemia, a disease restricted only to adult patients (Treon et al. 2012).

MYD88 is a signaling adaptor protein downstream of IL-1, IL-18 and all toll-like receptors that activate the NF-κB pathway upon ligand binding. The large majority of mutations identified consists of L265P point mutations, located in the Toll-like/IL-1 receptor domain (Ngo et al. 2011). In ABC-DLBCL cell line, $MYD88^{L265P}$ results in a gain-of-function and activates NF-κB signaling pathway. In addition, these cells exhibit activated STAT3, which synergizes with NF-κB to promote cell proliferation and survival (Ding et al. 2008; Lam et al. 2008). These two pathways interact through an autocrine loop involving IL-6 and IL-10 (Gupta et al. 2012). Through IRAK1 and IRAK4 signaling, NF-κB stimulates IL-6, IL-10 and interferon-β synthesis. These cytokines activate their receptors on the surface of lymphoma cell, causing STAT3 activation, which will enhance NF-κB activation (Ngo et al. 2011).

11.4.5 Therapeutic Implications of the JAK-STAT Pathway in Lymphomas

Due to the frequent activation of the JAK-STAT pathway in lymphomas and its central role in lymphomagenesis, the use of JAK-STAT pathway inhibitors seems like a promising therapeutic approach but its translation into the clinic was not as successful as expected. Only three drugs have moved to early phase trials for adult lymphoma patients.

First, ruxolitinib has demonstrated *in vitro* and *in vivo* anti-proliferative activity (Lee et al. 2018). A phase II study (NCT01877005) of ruxolitinib in patients with relapsed or refractory HL reported modest responses, while another multicentric study (NCT01965119) is ongoing (Kim et al. 2014; Neste et al. 2018). Pacritinib (SB1518) demonstrated some response in patients with cHL and various NHL in a phase I study (NCT00741871) (Hart et al. 2011; Younes et al. 2012). Cerdulatinib has shown *in vitro* activity in DLBCL cell lines and allowed responses in around 40% of T-cell lymphoma in the preliminary analysis of the phase IIa trial (NCT01994382) (Hamlin et al. 2018; Ishikawa et al. 2018; Ma et al. 2015).

Other JAK–family inhibitors have been tested clinically but their development has since halted as in the examples of fedratinib because of Wernicke encephalopathy and lestaurtinib for lack of efficacy (Diaz et al. 2011; Hao et al. 2014; Marshall et al. 2005). Some agents have only been used in preclinical studies (tofacitinib, PRN371 and AZD1480) (Bouchekioua et al. 2014; Derenzini et al. 2011; Koo et al. 2012; Nairismägi et al. 2018).

As discussed above, there is biologic rationale for the combination of a JAK-STAT inhibitor with other novel agents, notably with immune-checkpoint inhibitors, but no such experience has been reported in lymphoma yet (only a case report related the efficacy of the combination between nivolumab and ruxolitinib in a patient with a Merkel cell carcinoma) (Debureaux et al. 2018). Nevertheless, the combination of ruxolitinib with the MMAE-conjugated anti-CD30 monoclonal antibody brentuximab-vedotin has demonstrated an increased efficacy *in vitro* and *in vivo* (Ju et al. 2016).

11.4.6 Conclusion

JAK-STAT hyperactivation is common in lymphomas and can be dysregulated in numerous subtypes by a diverse array of genomic alterations. This pathway plays a central role in lymphomagenesis by promoting cell survival and proliferation. Furthermore, several examples in different lymphoma subtypes illustrate that the JAK-STAT pathway was not the only deregulated signal but interacts with other signaling pathways to promote cell proliferation and/or immune-escape.

Due to frequency of the JAK-STAT hyperactivation and its important effect in lymphomas, its inhibition seems like a logical strategy. Despite a strong rationale derived from robust *in vitro* and *ex vivo* data, clinical development of JAK inhibitors in this patient population is lagging behind because of the lack of durable response in early phase trials. Perhaps a more effective approach would be the combination of JAK-STAT pathway inhibition with cytotoxic chemotherapy or immunotherapeutic agents such as brentuximab vedotin and anti-PD1/PD-L1 antibodies, given that JAK-STAT pathway is not the only driver present in lymphoma cells residing within a rich immune microenvironment.

11.5 Summary

JAK-STAT dysregulation is the recurrent central event upon the different hematologic malignancies reviewed herein. The underlying mechanisms activating JAK-STAT signaling are diverse and differ between disease types. JAK-STAT family of genes may be directed altered but other genes regulating this pathway might be also affected. Not all of the genomic alterations identified are functionally characterized, but those which have been studied all converge to the central theme of reinforcing JAK-STAT activity. Altogether, these data attribute this pathway as the cornerstone of tumorigenesis. The collaborating events foster JAK-STAT signaling which drives cell proliferation in the different models presented. In most of *in vitro* and *in vivo* experiments, this proliferation could have been attenuated or even abolished with JAK inhibitors.

Based on these compelling data, the clinical use of type I JAK inhibitor ruxolitinib has been swiftly increased in the last 10 years; first in MPN then expanding to almost all hematologic malignancies. In most cases, ruxolitinib results in symptomatic improvement but survival benefits have not been convincingly demonstrated. In fact, some studies suggest that ruxolitinib is unable to eliminate the primitive clone and acquired mechanisms of resistance have emerged from longterm drug exposure. These phenomena may be in part explained by the fact that the JAK-STAT pathway is undoubtedly involved in a large number of hematologic malignancies to promote cell proliferation, but it remains unclear if it constitutes the driver event. Hence, even in MPN where JAK-STAT involvement has been the most studied, the role of *JAK2*^{V617F} in initiating the disease remains controversial.

The selection of the ideal JAK inhibitor to use in the clinic is yet unknown. Currently, only type I JAK inhibitors are being clinically investigated. They have an acceptable safety profile with moderate immunosuppression as the main side effect. The use of type II JAK inhibitors leads to stronger JAK inhibition but potentially more adverse toxicities as well. In particular, since JAK-STAT signaling plays a key role in the immune system, its inhibition represents an effective therapeutic approach in several inflammatory and auto-immune diseases (Banerjee et al. 2017). This pathway inhibition affects T-cells, NK-cells and dendritic cells' function (Assi et al. 2014; Curran et al. 2017; Keohane et al. 2015; Parampalli Yajnanarayana et al. 2015). It might be of benefit to limit the cancer-associated inflammation, as it seems to be in MPN. (Hasselbalch 2012; Tefferi et al. 2011) However, it also dampens the tumor immunosurveillance and the potential action of the immune system upon cell proliferation (Groner and von Manstein 2017). Therefore, the overall effect of JAK inhibition depends on the balance between its tumoral and immunological effects. The benefits are probably variable depending on the specific tumor type and the given host. The current data on JAK-STAT implication in the micro-environment and immune reaction surrounding tumor cells is yet largely unknown. Future research should focus on an integrative approach to study genomic alterations and their functional impact on the tumor cells and their microenvironment. This will allow a deepened understanding of the disease biology to design more effective therapeutic strategies.

REFERENCES

Al-Ali, Haifa, Martin Griesshammer Kathrin, Philipp le Coutre, Cornelius F. Waller, Anna Marina Liberati, Philippe Schafhausen, Renato Tavares, et al. 2016. "Safety and Efficacy of Ruxolitinib in an Open-Label, Multicenter, Single-Arm Phase 3b Expanded-Access Study in Patients with Myelofibrosis: A Snapshot of 1144 Patients in the JUMP Trial." *Haematologica* 101 (9): 1065–73. doi: 10.3324/haematol.2016.143677.

Almeida, Silva, Ana Carolina Da, Francesco Abate, Hossein Khiabanian, Estela Martinez-Escala, Joan Guitart, Cornelis P. Tensen, Maarten H. Vermeer, Raul Rabadan, Adolfo Ferrando, and Teresa Palomero. 2015. "The Mutational Landscape of Cutaneous T Cell Lymphoma and Sézary Syndrome." *Nature Genetics* 47 (12): 1465–70. doi: 10.1038/ng.3442.

An, Wenbin, Yang Wan, Ye Guo, Xiaojuan Chen, Yuanyuan Ren, Jingliao Zhang, Lixian Chang, Wei Wei, Peihong Zhang, and Xiaofan Zhu. 2014. "CALR Mutation Screening in Pediatric Primary Myelofibrosis." *Pediatric Blood & Cancer* 61 (12): 2256–62. doi: 10.1002/pbc.25211.

Anand, Shubha, Frances Stedham, Philip Beer, Emma Gudgin, Christina A. Ortmann, Anthony Bench, Wendy Erber, Anthony R. Green, and Brian J. P. Huntly. 2011. "Effects of the JAK2 Mutation on the Hematopoietic Stem and Progenitor Compartment in Human Myeloproliferative Neoplasms." *Blood* 118 (1): 177–81. doi: 10.1182/blood-2010-12-327593.

Ang, Sonny O., Hua Chen, Victor R. Kiichi Hirota, Jaroslav Jelinek Gordeuk, Yongli Guan, Enli Liu, et al. 2002. "Disruption of Oxygen Homeostasis Underlies Congenital Chuvash Polycythemia." *Nature Genetics* 32 (4): 614–21. doi: 10.1038/ng1019.

Araki, Marito, Yinjie Yang, Nami Masubuchi, Yumi Hironaka, Hiraku Takei, Soji Morishita, Yoshihisa Mizukami, et al. 2016. "Activation of the Thrombopoietin Receptor by Mutant Calreticulin in CALR-Mutant Myeloproliferative Neoplasms." *Blood* 127 (10): 1307–16. doi: 10.1182/blood-2015-09-671172.

Arber, Daniel A., Attilio Orazi, Robert Hasserjian, Michael J. Jürgen Thiele, Michelle M. Borowitz, Clara D. Le Beau, Mario Cazzola Bloomfield, and James W. Vardiman. 2016. "The 2016 Revision to the World Health Organization Classification of Myeloid Neoplasms and Acute Leukemia." *Blood* 127 (20): 2391–405. doi: 10.1182/blood-2016-03-643544.

Aricò, Maurizio, Stephen P. Martin Schrappe, William L. Hunger, Valentino Conter Carroll, Stefania Galimberti, Atsushi Manabe, et al. 2010. "Clinical Outcome of Children with Newly Diagnosed Philadelphia Chromosome–Positive Acute Lymphoblastic Leukemia Treated between 1995 and 2005." *Journal of Clinical Oncology* 28 (31): 4755–61. doi: 10.1200/JCO.2010.30.1325.

Armstrong, Scott A., Meghann E. Mabon, Lewis B. Silverman, Aihong Li, John G. Gribben, Edward A. Fox, Stephen E. Sallan, and Stanley J. Korsmeyer. 2004. "FLT3 Mutations in Childhood Acute Lymphoblastic Leukemia." *Blood* 103 (9): 3544–6. doi: 10.1182/blood-2003-07-2441.

Assi, Hikmat, Jaclyn Espinosa, Sarah Suprise, Michael Sofroniew, Robert Doherty, Daniel Zamler, Pedro R. Lowenstein, and Maria G. Castro. 2014. "Assessing the Role of STAT3 in DC Differentiation and Autologous DC Immunotherapy in Mouse Models of GBM." *PLoS One* 9 (5): e96318. doi: 10.1371/journal.pone.0096318.

Banerjee, Shubhasree, Ann Biehl, Massimo Gadina, Sarfaraz Hasni, and Daniella M. Schwartz. 2017. "JAK–STAT Signaling as a Target for Inflammatory and Autoimmune Diseases: Current and Future Prospects." *Drugs* 77 (5): 521–46. doi: 10.1007/s40265-017-0701-9.

Barbui, T., J. Thiele, A. M. Vannucchi, and A. Tefferi. 2014. "Rethinking the Diagnostic Criteria of Polycythemia Vera." *Leukemia* 28 (6): 1191–5. doi: 10.1038/leu.2013.380.

Bartalucci, Niccolò, Lorenzo Tozzi, Costanza Bogani, Serena Martinelli, Giada Rotunno, Jean-Luc Villeval, and Alessandro M. Vannucchi. 2013. "Co-Targeting the PI3K/MTOR and JAK2 Signalling Pathways Produces Synergistic Activity against Myeloproliferative Neoplasms." *Journal of Cellular and Molecular Medicine* 17 (11): 1385–96. doi: 10.1111/jcmm.12162.

Baughn, L. B., M. M. Meredith, L. Oseth, T. A. Smolarek, and B. Hirsch. 2018. "SH2B3 Aberrations Enriched in IAMP21 B Lymphoblastic Leukemia." *Cancer Genetics* 226–227 (October): 30–5. doi: 10.1016/j.cancergen.2018.05.004.

Baxter, E. Joanna, Linda M Scott, Peter J Campbell, Clare East, Nasios Fourouclas, Soheila Swanton, George S Vassiliou, et al. 2005. "Acquired Mutation of the Tyrosine Kinase JAK2 in Human Myeloproliferative Disorders." *The Lancet* 365 (9464): 1054–61. doi: 10.1016/S0140-6736(05)71142-9.

Beer, Philip A., Peter J. Campbell, Linda M. Scott, Anthony J. Bench, Wendy N. Erber, David Bareford, Bridget S. Wilkins, et al. 2008. "MPL Mutations in Myeloproliferative Disorders: Analysis of the PT-1 Cohort." *Blood* 112 (1): 141–9. doi: 10.1182/blood-2008-01-131664.

Ding, B. Belinda, J. Jessica Yu, Raymond Y.-L. Yu, Lourdes M. Mendez, Rita Shaknovich, Yonghui Zhang, Giorgio Cattoretti, and B. Hilda Ye. 2008. "Constitutively Activated STAT3 Promotes Cell

Proliferation and Survival in the Activated B-Cell Subtype of Diffuse Large B-Cell Lymphomas." *Blood* 111 (3): 1515–23. doi: 10.1182/blood-2007-04-087734.

Bellanné-Chantelot, Christine, Isabelle Chaumarel, Myriam Labopin, Florence Bellanger, Véronique Barbu, Claudia De Toma, François Delhommeau, et al. 2006. "Genetic and Clinical Implications of the Val617Phe JAK2 Mutation in 72 Families with Myeloproliferative Disorders." *Blood* 108 (1): 346–52. doi: 10.1182/blood-2005-12-4852.

Benekli, Mustafa, Kathleen A. Zheng Xia, Laurie A. Donohue, Lynda A. Ford, Maria R. Pixley, Heinz Baumann Baer, and Meir Wetzler. 2002. "Constitutive Activity of Signal Transducer and Activator of Transcription 3 Protein in Acute Myeloid Leukemia Blasts Is Associated with Short Disease-Free Survival." *Blood* 99 (1): 252–7. doi: 10.1182/blood.V99.1.252.

Bercovich, Dani, Ithamar Ganmore, Linda M Scott, Gilad Wainreb, Yehudit Birger, Arava Elimelech, Chen Shochat, et al. 2008. "Mutations of JAK2 in Acute Lymphoblastic Leukaemias Associated with Down's Syndrome." *The Lancet* 372 (9648): 1484–92. doi: 10.1016/S0140-6736(08)61341-0.

Biondi, Andrea, Martin Schrappe, Paola De Lorenzo, Anders Castor, Giovanna Lucchini, Virginie Gandemer, Rob Pieters, et al. 2012. "Imatinib after Induction for Treatment of Children and Adolescents with Philadelphia-Chromosome-Positive Acute Lymphoblastic Leukaemia (Esphall): A Randomised, Open-Label, Intergroup Study." *The Lancet Oncology* 13 (9): 936–45. doi: 10.1016/S1470-2045(12)70377-7.

Bock, Charles E. de, Sandrine Degryse, Sofie Demeyer, Bram Sweron, Olga Gielen, Nicole Mentens, Ellen Geerdens, and Jan Cools. 2014. "Synergism between HOXA9 and Mutant JAK3 (M511I) Leads to Rapid Leukemia Development within an In Vivo Murine Bone Marrow Transplant Model." *Blood* 124 (21): 1078. www.bloodjournal.org/content/124/21/1078.

Bodegom, Diederik van, Jun Zhong, Nadja Kopp, Chaitali Dutta, Min-Sik Kim, Liat Bird, Oliver Weigert, et al. 2012. "Differences in Signaling through the B-Cell Leukemia Oncoprotein CRLF2 in Response to TSLP and through Mutant JAK2." *Blood* 120 (14): 2853–63. doi: 10.1182/blood-2012-02-413252.

Bolouri, Hamid, Jason E. Farrar, Rhonda E. Timothy Triche, Emilia L. Ries, Todd A. Lim, Yussanne Ma Alonzo, et al. 2018. "The Molecular Landscape of Pediatric Acute Myeloid Leukemia Reveals Recurrent Structural Alterations and Age-Specific Mutational Interactions." *Nature Medicine* 24 (1): 103–12. doi: 10.1038/nm.4439.

Borchmann, Sven, and Andreas Engert. 2017. "The Genetics of Hodgkin Lymphoma: An Overview and Clinical Implications." *Current Opinion in Oncology* 29 (5): 307–14. doi: 10.1097/CCO.0000000000000396.

Bose, Prithviraj, and Srdan Verstovsek. 2017. "JAK2 Inhibitors for Myeloproliferative Neoplasms: What Is Next?" *Blood* 130 (2): 115–25. doi: 10.1182/blood-2017-04-742288.

Bouchekioua, A., L. Scourzic, O. de Wever, Y. Zhang, P. Cervera, A. Aline-Fardin, T. Mercher, et al. 2014. "JAK3 Deregulation by Activating Mutations Confers Invasive Growth Advantage in Extranodal Nasal-Type Natural Killer Cell Lymphoma." *Leukemia* 28 (2): 338–48. doi: 10.1038/leu.2013.157.

Bousfiha, Aziz, Leïla Jeddane, Capucine Picard, H. Fatima Ailal, Bobby Gaspar, Waleed Al-Herz, Talal Chatila, et al. 2018. "The 2017 IUIS Phenotypic Classification for Primary Immunodeficiencies." *Journal of Clinical Immunology* 38 (1): 129–43. doi: 10.1007/s10875-017-0465-8.

Buitenkamp, Trudy D., Shai Izraeli, Martin Zimmermann, Erik Forestier, Nyla A. Heerema, Marry M. van Den Heuvel-Eibrink, Rob Pieters, et al. 2014. "Acute Lymphoblastic Leukemia in Children with down Syndrome: A Retrospective Analysis from the Ponte Di Legno Study Group." *Blood* 123 (1): 70–7. doi: 10.1182/blood-2013-06-509463.

Burkhardt, Birgit, Martin Zimmermann, Ilske Oschlies, Felix Niggli, Georg Mann, Reza Parwaresch, Hansjoerg Riehm, Martin Schrappe, and Alfred Reiter. 2005. "The Impact of Age and Gender on Biology, Clinical Features and Treatment Outcome of Non-Hodgkin Lymphoma in Childhood and Adolescence." *British Journal of Haematology* 131 (1): 39–49. doi: 10.1111/j.1365-2141.2005.05735.x.

Butcher, C. M., U. Hahn, L. B. To, J. Gecz, E. J. Wilkins, H. S. Scott, P. G. Bardy, and R. J. D'Andrea. 2008. "Two Novel *JAK2* Exon 12 Mutations in JAK2V617F-Negative Polycythaemia Vera Patients." *Leukemia* 22 (4): 870–3. doi: 10.1038/sj.leu.2404971.

Cabagnols, Xénia, Fabrizia Favale, Florence Pasquier, Kahia Messaoudi, Jean Philippe Defour, Jean Christophe Ianotto, Christophe Marzac, et al. 2016. "Presence of Atypical Thrombopoietin Receptor (MPL) Mutations in Triple-Negative Essential Thrombocythemia Patients." *Blood* 127 (3): 333–42. doi: 10.1182/blood-2015-07-661983.

Cancer Genome Atlas Research Network. Ley, Timothy J., Christopher Miller, Li Ding, Benjamin J. Raphael, Andrew J. Mungall, A. Gordon Robertson, et al. 2013. "Genomic and Epigenomic Landscapes of Adult De Novo Acute Myeloid Leukemia." *The New England Journal of Medicine* 368 (22): 2059–74. doi: 10.1056/NEJMoa1301689.

Cario, Holger, Jan M. Klaus Schwarz, Vladimir Komrska Herter, Mary F. McMullin, Milen Minkov, Charlotte Niemeyer, et al. 2008. "Clinical and Molecular Characterisation of a Prospectively Collected Cohort of Children and Adolescents with Polycythemia Vera." *British Journal of Haematology* 142 (4): 622–6. doi: 10.1111/j.1365-2141.2008.07220.x.

Carron, Clémence, Françoise Cormier, Anne Janin, Virginie Lacronique, Marco Giovannini, Marie-Thérèse Daniel, Olivier Bernard, and Jacques Ghysdael. 2000. "TEL-JAK2 Transgenic Mice Develop T-Cell Leukemia." *Blood* 95 (12): 3891–9. www.bloodjournal.org/content/95/12/3891.

Cervantes, Francisco, Alberto Alvarez-Larrán, Carme Talarn, Marta Gómez, and Emilio Montserrat. 2002. "Myelofibrosis with Myeloid Metaplasia following Essential Thrombocythaemia: Actuarial Probability, Presenting Characteristics and Evolution in a Series of 195 Patients." *British Journal of Haematology* 118 (3): 786–90.

Cervantes, Francisco, and Arturo Pereira. 2017. "Does Ruxolitinib Prolong the Survival of Patients with Myelofibrosis?" *Blood* 129 (7): 832–7. doi: 10.1182/blood-2016-11-731604.

Chachoua, Ilyas, Christian Pecquet, Mira El-Khoury, Harini Nivarthi, Roxana-Irina Albu, Caroline Marty, Vitalina Gryshkova, et al. 2016. "Thrombopoietin Receptor Activation by Myeloproliferative Neoplasm Associated Calreticulin Mutants." *Blood* 127 (10): 1325–35. doi: 10.1182/blood-2015-11-681932.

Chaligné, R., C. Tonetti, R. Besancenot, L. Roy, C. Marty, P. Mossuz, J.-J. Kiladjian, et al. 2008. "New Mutations of MPL in Primitive Myelofibrosis: Only the MPL W515 Mutations Promote a G_1/S-Phase Transition." *Leukemia* 22 (8): 1557–66. doi: 10.1038/leu.2008.137.

Chaligné, Ronan, Chloé James, Carole Tonetti, Rodolphe Besancenot, Jean Pierre Le Couédic, Fanny Fava, Fréderic Mazurier, et al. 2007. "Evidence for MPL W515L/K Mutations in Hematopoietic Stem Cells in Primitive Myelofibrosis." *Blood* 110 (10): 3735–43. doi: 10.1182/blood-2007-05-089003.

Chapelle, A. de la, A. L. Träskelin, and E. Juvonen. 1993. "Truncated Erythropoietin Receptor Causes Dominantly Inherited Benign Human Erythrocytosis." *Proceedings of the National Academy of Sciences of the United States of America* 90 (10): 4495–9.

Chapiro, E., L. Russell, E. Lainey, S. Kaltenbach, C. Ragu, V. Della-Valle, K. Hanssens, et al. 2010. "Activating Mutation in the TSLPR Gene in B-Cell Precursor Lymphoblastic Leukemia." *Leukemia* 24 (3): 642–5. doi: 10.1038/leu.2009.231.

Chen, Edwin, Philip A. Beer, Anna L. Godfrey, Christina A. Ortmann, Juan Li, Ana P. Costa-Pereira, Catherine E. Ingle, Emmanouil T. Dermitzakis, Peter J. Campbell, and Anthony R. Green, 2010. "Distinct Clinical Phenotypes Associated with JAK2V617F Reflect Differential STAT1 Signaling." *Cancer Cell* 18 (5): 524–35. doi: 10.1016/j.ccr.2010.10.013.

Chen, Jing, Yong Zhang, N. Michael, Wenming Petrus, Alina Xiao, Mark Nicolae, Stefania Pittaluga Raffeld, et al. 2017. "Cytokine Receptor Signaling Is Required for the Survival of ALK− Anaplastic Large Cell Lymphoma, Even in the Presence of JAK1/STAT3 Mutations." *Proceedings of the National Academy of Sciences* 114 (15): 3975–80. doi: 10.1073/pnas.1700682114.

Cheng, Ying, Kudakwashe Chikwava, Chao Wu, Haibing Zhang, Anchit Bhagat, Dehua Pei, John K. Choi, and Wei Tong. 2016. "LNK/SH2B3 Regulates IL-7 Receptor Signaling in Normal and Malignant B-Progenitors." *The Journal of Clinical Investigation* 126 (4): 1267–81. doi: 10.1172/JCI81468.

Chessells, J. M., G. Harrison, S. M. Richards, C. C. Bailey, F. G. H. Hill, B. E. Gibson, and I. M. Hann. 2001. "Down's Syndrome and Acute Lymphoblastic Leukaemia: Clinical Features and Response to Treatment." *Archives of Disease in Childhood* 85 (4): 321–5. doi: 10.1136/adc.85.4.321.

Chim, Chor-Sang, Yok-Lam Kwong, Albert Kwok-Wei Lie, Siu-Kwan Ma, Chi-Chung Chan, Lap-Gate Wong, Bonnie Chi San Kho, et al. 2005. "Long-Term Outcome of 231 Patients with Essential

Thrombocythemia: Prognostic Factors for Thrombosis, Bleeding, Myelofibrosis, and Leukemia." *Archives of Internal Medicine* 165 (22): 2651–8. doi: 10.1001/archinte.165.22.2651.
Compagno, Mara, Wei Keat Lim, Subhadra V. Adina Grunn, Manisha Brahmachary Nandula, Qiong Shen, Francesco Bertoni, et al. 2009. "Mutations of Multiple Genes Cause Deregulation of NF-KB in Diffuse Large B-Cell Lymphoma." *Nature* 459 (7247): 717–21. doi: 10.1038/nature07968.
Cooper, Todd M., Robert B. Janet Franklin, Todd A. Gerbing, Craig Hurwitz Alonzo, Susana C. Raimondi, Betsy Hirsch, et al. 2012. "AAML03P1, A Pilot Study of the Safety of Gemtuzumab Ozogamicin in Combination with Chemotherapy for Newly Diagnosed Childhood Acute Myeloid Leukemia: A Report from the Children's Oncology Group." *Cancer* 118 (3): 761–9. doi: 10.1002/cncr.26190.
Cornejo, Melanie G., Michael G. Kharas, Miriam B. Werneck, Séverine Le Bras, Sandra A. Moore, Brian Ball, Marie Beylot-Barry, et al. 2009. "Constitutive JAK3 Activation Induces Lymphoproliferative Syndromes in Murine Bone Marrow Transplantation Models." *Blood* 113 (12): 2746–54. doi: 10.1182/blood-2008-06-164368.
Cortelazzo, S., P. Viero, G. Finazzi, A. D'Emilio, F. Rodeghiero, and T. Barbui. 1990. "Incidence and Risk Factors for Thrombotic Complications in a Historical Cohort of 100 Patients with Essential Thrombocythemia." *Journal of Clinical Oncology: Official Journal of the American Society of Clinical Oncology* 8 (3): 556–62. doi: 10.1200/JCO.1990.8.3.556.
Coskun, Mehmet Enes, Sue Height, Anil Dhawan, and Nedim Hadzic. 2017. "Ruxolitinib Treatment in an Infant with JAK2+ Polycythaemia Vera-Associated Budd-Chiari Syndrome." *BMJ Case Reports* 2017 (July): bcr-2017-220377. doi: 10.1136/bcr-2017-220377.
Couronné, Lucile, Laurianne Scourzic, Camilla Pilati, Véronique Della Valle, Yannis Duffourd, Eric Solary, William Vainchenker, et al. 2013. "STAT3 Mutations Identified in Human Hematologic Neoplasms Induce Myeloid Malignancies in a Mouse Bone Marrow Transplantation Model." *Haematologica* 98 (11): 1748–52. doi: 10.3324/haematol.2013.085068.
Crescenzo, Ramona, Francesco Abate, Elena Lasorsa, Fabrizio Tabbo', Marcello Gaudiano, Nicoletta Chiesa, Filomena Di Giacomo, et al. 2015. "Convergent Mutations and Kinase Fusions Lead to Oncogenic STAT3 Activation in Anaplastic Large Cell Lymphoma." *Cancer Cell* 27 (4): 516–32. doi: 10.1016/j.ccell.2015.03.006.
Creutzig, Ursula, Marry M. van Den Heuvel-Eibrink, Michael N. Brenda Gibson, Souichi Adachi Dworzak, Eveline de Bont, Jochen Harbott, et al. 2012. "Diagnosis and Management of Acute Myeloid Leukemia in Children and Adolescents: Recommendations from an International Expert Panel." *Blood* 120 (16): 3187–205. doi: 10.1182/blood-2012-03-362608.
Curran, Shane A., Justin A. Shyer, Erin T. St Angelo, Lillian R. Talbot, David J. Sneh Sharma, Glenn Heller Chung, Katharine C. Hsu, Brian C. Betts, and James W. Young. 2017. "Human Dendritic Cells Mitigate NK-Cell Dysfunction Mediated by Nonselective JAK1/2 Blockade." *Cancer Immunology Research* 5 (1): 52–60. doi: 10.1158/2326-6066.CIR-16-0233.
Debureaux, P. E., J. Arrondeau, D. Bouscary, and F. Goldwasser. 2018. "Nivolumab Combined with Ruxolitinib: Antagonism or Synergy?" *Annals of Oncology* 29 (5): 1334–5. doi: 10.1093/annonc/mdy077.
Degryse, Sandrine, Charles E. de Bock, Luk Cox, Sofie Demeyer, Olga Gielen, Nicole Mentens, Kris Jacobs, et al. 2014. "JAK3 Mutants Transform Hematopoietic Cells through JAK1 Activation, Causing T-Cell Acute Lymphoblastic Leukemia in a Mouse Model." *Blood* 124 (20): 3092–100. doi: 10.1182/blood-2014-04-566687.
DeLario, Melissa R., Andrea Sheehan, Ramona Ataya, Alison A. Bertuch, Carlos Vega, C. Renee Webb, Dolores lopez-Terrada, and Lakshmi Venkateswaran. 2010. "Primary Myelofibrosis in Children." *Blood* 116 (21): 3079. www.bloodjournal.org/content/116/21/3079.
DeLario, Melissa R., Andrea M. Sheehan, Ramona Ataya, Alison A. Bertuch, Carlos Vega, C. Renee Webb, Dolores Lopez-Terrada, and Lakshmi Venkateswaran. 2012. "Clinical, Histopathologic, and Genetic Features of Pediatric Primary Myelofibrosis—An Entity Different from Adults." *American Journal of Hematology* 87 (5): 461–4. doi: 10.1002/ajh.23140.
Den Boer, Monique L., Marjon van Slegtenhorst, Renée X. De Menezes, Meyling H. Cheok, Jessica G. C. A. M. Buijs-Gladdines, Susan T. C. J. M. Peters, Laura J. C. M. Van Zutven, et al. 2009. "A Subtype of Childhood Acute Lymphoblastic Leukaemia with Poor Treatment Outcome:

A Genome-Wide Classification Study." *The Lancet Oncology* 10 (2): 125–34. doi: 10.1016/S1470-2045(08)70339-5.

Derenzini, E., M. Lemoine, D. Buglio, H. Katayama, Y. Ji, R. E. Davis, S. Sen, and A. Younes, 2011. "The JAK Inhibitor AZD1480 Regulates Proliferation and Immunity in Hodgkin Lymphoma." *Blood Cancer Journal* 1 (12): e46. doi: 10.1038/bcj.2011.46.

Diaz, Tania, Alfons Navarro, Gerardo Ferrer, Bernat Gel, Anna Gaya, Rosa Artells, Beatriz Bellosillo, et al. 2011. "Lestaurtinib Inhibition of the JAK/STAT Signaling Pathway in Hodgkin Lymphoma Inhibits Proliferation and Induces Apoptosis." *PLoS One* 6 (4): e18856. doi: 10.1371/journal.pone.0018856.

Ding, Jianmin, Hirokazu Komatsu, Atsushi Wakita, Miyuki Kato-Uranishi, Masato Ito, Atsushi Satoh, Kazuya Tsuboi, et al. 2004. "Familial Essential Thrombocythemia Associated with a Dominant-Positive Activating Mutation of the c-MPL Gene, Which Encodes for the Receptor for Thrombopoietin." *Blood* 103 (11): 4198–200. doi: 10.1182/blood-2003-10-3471.

Dördelmann, M., M. Schrappe, A. Reiter, M. Zimmermann, N. Graf, G. Schott, F. Lampert, et al. 1998. "Down's Syndrome in Childhood Acute Lymphoblastic Leukemia: Clinical Characteristics and Treatment Outcome in Four Consecutive BFM Trials." *Leukemia* 12 (5): 645. doi: 10.1038/sj.leu.2400989.

Drachman, J. G., J. D. Griffin, and K. Kaushansky. 1995. "The C-Mpl Ligand (Thrombopoietin) Stimulates Tyrosine Phosphorylation of Jak2, Shc, and c-Mpl." *The Journal of Biological Chemistry* 270 (10): 4979–82.

Dubois, Sydney, Pierre-Julien Viailly, Sylvain Mareschal, Elodie Bohers, Philippe Bertrand, Philippe Ruminy, Catherine Maingonnat, et al. 2016. "Next-Generation Sequencing in Diffuse Large B-Cell Lymphoma Highlights Molecular Divergence and Therapeutic Opportunities: A LYSA Study." *Clinical Cancer Research* 22 (12): 2919–28. doi: 10.1158/1078-0432.CCR-15-2305.

Duek, Adrian, Pontus Lundberg, Takafumi Shimizu, Jean Grisouard, Axel Karow, Lucia Kubovcakova, Hui Hao-Shen, Stephan Dirnhofer, and Radek C. Skoda. 2014. "Loss of Stat1 Decreases Megakaryopoiesis and Favors Erythropoiesis in a JAK2-V617F–Driven Mouse Model of MPNs." *Blood* 123 (25): 3943–50. doi: 10.1182/blood-2013-07-514208.

Dupont, Sabrina, Aline Massé, Chloé James, Irène Teyssandier, Yann Lécluse, Frédéric Larbret, Valérie Ugo, et al. 2007. "The JAK2 617V>F Mutation Triggers Erythropoietin Hypersensitivity and Terminal Erythroid Amplification in Primary Cells from Patients with Polycythemia Vera." *Blood* 110 (3): 1013–21. doi: 10.1182/blood-2006-10-054940.

Eggleton, Paul, and Marek Michalak. 2013. "Calreticulin for Better or for Worse, in Sickness and in Health, until Death Do Us Part." *Cell Calcium* 54 (2): 126–31. doi: 10.1016/j.ceca.2013.05.006.

Eghtedar, Alireza, Srdan Verstovsek, Zeev Estrov, Jan Burger, Jorge Cortes, Carol Bivins, Stefan Faderl, et al. 2012. "Phase 2 Study of the JAK Kinase Inhibitor Ruxolitinib in Patients with Refractory Leukemias, Including Postmyeloproliferative Neoplasm Acute Myeloid Leukemia." *Blood* 119 (20): 4614–18. doi: 10.1182/blood-2011-12-400051.

Englund, Annika, Ingrid Glimelius, Karin E. Klaus Rostgaard, Sandra Eloranta Smedby, Daniel Molin, Thomas Kuusk, et al. 2018. "Hodgkin Lymphoma in Children, Adolescents and Young Adults – A Comparative Study of Clinical Presentation and Treatment Outcome." *Acta Oncologica (Stockholm, Sweden)* 57 (2): 276–82. doi: 10.1080/0284186X.2017.1355563.

Davis, R. Eric, Keith D. Brown, Ulrich Siebenlist, and Louis M. Staudt. 2001. "Constitutive Nuclear Factor KB Activity Is Required for Survival of Activated B Cell–Like Diffuse Large B Cell Lymphoma Cells." *Journal of Experimental Medicine* 194 (12): 1861–74. doi: 10.1084/jem.194.12.1861.

Farruggia, Piero, Paolo D'Angelo, Maria La Rosa, Nunzia Scibetta, Giuseppe Santangelo, Antonio Lo Bello, Elena Duner, Maria Luigia Randi, Maria Caterina Putti, and Alessandra Santoro. 2013. "MPL W515L Mutation in Pediatric Essential Thrombocythemia." *Pediatric Blood & Cancer* 60 (8): E52–4. doi: 10.1002/pbc.24500.

Feenstra, Jelena D., Harini Milosevic, Heinz Gisslinger Nivarthi, Emilie Leroy, Elisa Rumi, Ilyas Chachoua, Klaudia Bagienski, et al. 2016. "Whole-Exome Sequencing Identifies Novel MPL and JAK2 Mutations in Triple-Negative Myeloproliferative Neoplasms." *Blood* 127 (3): 325–32. doi: 10.1182/blood-2015-07-661835.

Fenaux, P., M. Simon, M. T. Caulier, J. L. Lai, J. Goudemand, and F. Bauters. 1990. "Clinical Course of Essential Thrombocythemia in 147 Cases." *Cancer* 66 (3): 549–56.

Flex, Elisabetta, Valentina Petrangeli, Lorenzo Stella, Sabina Chiaretti, Tekla Hornakova, Laurent Knoops, Cristina Ariola, et al. 2008. "Somatically Acquired JAK1 Mutations in Adult Acute Lymphoblastic Leukemia." *Journal of Experimental Medicine* 205 (4): 751–8. doi: 10.1084/jem.20072182.

Forestier, Erik, Shai Izraeli, Berna Beverloo, Oskar Haas, Andrea Pession, Kyra Michalová, Batia Stark, Christine J. Harrison, Andrea Teigler-Schlegel, and Bertil Johansson. 2008. "Cytogenetic Features of Acute Lymphoblastic and Myeloid Leukemias in Pediatric Patients with Down Syndrome: An IBFM-SG Study." *Blood* 111 (3): 1575–83. doi: 10.1182/blood-2007-09-114231.

Fu, R., D. Liu, Z. Cao, S. Zhu, H. Li, H. Su, L. Zhang, et al. 2016. "Distinct Molecular Abnormalities Underlie Unique Clinical Features of Essential Thrombocythemia in Children." *Leukemia* 30 (3): 746–9. doi: 10.1038/leu.2015.167.

Fu, Rongfeng, Lei Zhang, and Renchi Yang. 2013. "Paediatric Essential Thrombocythaemia: Clinical and Molecular Features, Diagnosis and Treatment." *British Journal of Haematology* 163 (3): 295–302. doi: 10.1111/bjh.12530.

Gaikwad, Amos, Cassia L. Rye, Meenakshi Devidas, Nyla A. Heerema, Andrew J. Carroll, Shai Izraeli, Sharon E. Plon, Giuseppe Basso, Andrea Pession, and Karen R. Rabin. 2009. "Prevalence and Clinical Correlates of JAK2 Mutations in Down Syndrome Acute Lymphoblastic Leukaemia." *British Journal of Haematology* 144 (6): 930–2. doi: 10.1111/j.1365-2141.2008.07552.x.

Gamis, Alan S., Todd A. Alonzo, Soheil Meshinchi, Lillian Sung, Robert B. Gerbing, Susana C. Raimondi, Betsy A. Hirsch, et al. 2014. "Gemtuzumab Ozogamicin in Children and Adolescents with De Novo Acute Myeloid Leukemia Improves Event-Free Survival by Reducing Relapse Risk: Results from the Randomized Phase III Children's Oncology Group Trial AAML0531." *Journal of Clinical Oncology: Official Journal of the American Society of Clinical Oncology* 32 (27): 3021–32. doi: 10.1200/JCO.2014.55.3628.

Gamis, Alan S., Todd A. Alonzo, John P. Perentesis, and Soheil Meshinchi. 2013. "Children's Oncology Group's 2013 Blueprint for Research: Acute Myeloid Leukemia." *Pediatric Blood & Cancer* 60 (6): 964–71. doi: 10.1002/pbc.24432.

Garçon, Loïc, Christine Rivat, Chloé James, Catherine Lacout, Valérie Camara-Clayette, Valérie Ugo, Yann Lecluse, Annelise Bennaceur-Griscelli, and William Vainchenker. 2006. "Constitutive Activation of STAT5 and Bcl-XL Overexpression Can Induce Endogenous Erythroid Colony Formation in Human Primary Cells." *Blood* 108 (5): 1551–4. doi: 10.1182/blood-2005-10-009514.

Giona, Fiorina, Luciana Teofili, Sara Capodimonti, Marica Laurino, Maurizio Martini, Deborah Marzella, Giovanna Palumbo, Daniela Diverio, Robin Foà, and Luigi Maria Larocca. 2014. "CALR Mutations in Patients with Essential Thrombocythemia Diagnosed in Childhood and Adolescence." *Blood* 123 (23): 3677–9. doi: 10.1182/blood-2014-04-572040.

Giona, Fiorina, Luciana Teofili, Maria Luisa Moleti, Maurizio Martini, Giovanna Palumbo, Angela Amendola, Maria Gabriella Mazzucconi, et al. 2012. "Thrombocythemia and Polycythemia in Patients Younger than 20 Years at Diagnosis: Clinical and Biologic Features, Treatment, and Long-Term Outcome." *Blood* 119 (10): 2219–27. doi: 10.1182/blood-2011-08-371328.

Goto, Hiroaki, Takeshi Inukai, Hiroyasu Inoue, Chitose Ogawa, Takashi Fukushima, Miharu Yabe, Akira Kikuchi, et al. 2011. "Acute Lymphoblastic Leukemia and Down Syndrome: The Collaborative Study of the Tokyo Children's Cancer Study Group and the Kyushu Yamaguchi Children's Cancer Study Group." *International Journal of Hematology* 93 (2): 192–8. doi: 10.1007/s12185-011-0765-3.

Green, Michael R., Stefano Monti, Scott J. Rodig, Przemyslaw Juszczynski, Treeve Currie, Evan O'Donnell, Bjoern Chapuy, et al. 2010. "Integrative Analysis Reveals Selective 9p24.1 Amplification, Increased PD-1 Ligand Expression, and Further Induction via JAK2 in Nodular Sclerosing Hodgkin Lymphoma and Primary Mediastinal Large B-Cell Lymphoma." *Blood* 116 (17): 3268–77. doi: 10.1182/blood-2010-05-282780.

Groner, Bernd, and Viktoria von Manstein. 2017. "Jak Stat Signaling and Cancer: Opportunities, Benefits and Side Effects of Targeted Inhibition." *Molecular and Cellular Endocrinology, JAK–Stat Signaling and Cancer* 451 (August): 1–14. doi: 10.1016/j.mce.2017.05.033.

Guglielmelli, P., T. L. Lasho, G. Rotunno, J. Score, C. Mannarelli, A. Pancrazzi, F. Biamonte, et al. 2014. "The Number of Prognostically Detrimental Mutations and Prognosis in Primary Myelofibrosis: An International Study of 797 Patients." *Leukemia* 28 (9): 1804–10. doi: 10.1038/leu.2014.76.

Gunawan, Arief, Patrick Harrington, Natalia Garcia-Curto, Donal McLornan, Deepti Radia, and Claire Harrison. 2018. "Ruxolitinib for the Treatment of Essential Thrombocythemia." *HemaSphere* 2 (4): e56. doi: 10.1097/HS9.0000000000000056.

Gunawardana, Jay, Fong Chun Chan, Adèle Telenius, Bruce Woolcock, Robert Kridel, King L. Tan, Susana Ben-Neriah, et al. 2014. "Recurrent Somatic Mutations of *PTPN1* in PR and Hodgkin Lymphoma." *Nature Genetics* 46 (4): 329–35. doi: 10.1038/ng.2900.

Gupta, Mamta, Jing Jing Han, Mary Stenson, Matthew Maurer, Linda Wellik, Guangzhen Hu, Steve Ziesmer, Ahmet Dogan, and Thomas E. Witzig. 2012. "Elevated Serum IL-10 Levels in Diffuse Large B-Cell Lymphoma: A Mechanism of Aberrant JAK2 Activation." *Blood* 119 (12): 2844–53. doi: 10.1182/blood-2011-10-388538.

Hama, Asahito, Hiroshi Yagasaki, Yoshiyuki Takahashi, Kimikazu Matsumoto, Hitoshi Kiyoi, and Seiji Kojima. 2008. "Response: Mutations of JAK2, JAK3 and GATA1 in Acute Megakaryoblastic Leukemia of down Syndrome." *Blood* 111 (4): 2493–4. doi: 10.1182/blood-2007-10-117614.

Hamlin, Paul, Bruce Cheson, Charles Farber, Tatyana Feldman, Timothy Fenske, Greg Coffey, and John Curnutte. 2018. "The Dual SYK/JAK Inhibitor Cerdulatinib Demonstrates Rapid Tumor Responses in a Phase 2 Study in Patients with Relapsed/Refractory B- and T-Cell Non-Hodgkin Lymphoma (NHL)." *Journal of Clinical Oncology* ASCO 2018 Abstract.

Hao, Yansheng, Bjoern Chapuy, Stefano Monti, Heather H. Sun, Scott J. Rodig, and Margaret A. Ship. 2014. "Selective JAK2 Inhibition Specifically Decreases Hodgkin Lymphoma and Mediastinal Large B-Cell Lymphoma Growth in Vitro and in Vivo." *Clinical Cancer Research* 20 (10): 2674–83. doi: 10.1158/1078-0432.CCR-13-3007.

Harrison, Claire N., Adam J. Mead, Anesh Panchal, Sonia Fox, Christina Yap, Emmanouela Gbandi, Aimee Houlton, et al. 2017a. "Ruxolitinib vs Best Available Therapy for ET Intolerant or Resistant to Hydroxycarbamide." *Blood* 130 (17): 1889–97. doi: 10.1182/blood-2017-05-785790.

Harrison, Claire N, Nicolaas Schaap, Alessandro M Vannucchi, Jean-Jacques Kiladjian, Ramon V Tiu, Pierre Zachee, Eric Jourdan, et al. 2017b. "Janus Kinase-2 Inhibitor Fedratinib in Patients with Myelofibrosis Previously Treated with Ruxolitinib (JAKARTA-2): A Single-Arm, Open-Label, Non-Randomised, Phase 2, Multicentre Study." *The Lancet Haematology* 4 (7): e317–24. doi: 10.1016/S2352-3026(17)30088-1.

Hart, S., K. C. Goh, V. Novotny-Diermayr, C. Y. Hu, H. Hentze, Y. C. Tan, B. Madan, et al. 2011. "SB1518, a Novel Macrocyclic Pyrimidine-Based JAK2 Inhibitor for the Treatment of Myeloid and Lymphoid Malignancies." *Leukemia* 25 (11): 1751–9. doi: 10.1038/leu.2011.148.

Harvey, Richard C., Charles G. Mullighan, I.-Ming Chen, Walker Wharton, Fady M. Mikhail, Andrew J. Carroll, Huining Kang, et al. 2010. "Rearrangement of CRLF2 Is Associated with Mutation of JAK Kinases, Alteration of IKZF1, Hispanic/Latino Ethnicity, and a Poor Outcome in Pediatric B-Progenitor Acute Lymphoblastic Leukemia." *Blood* 115 (26): 5312–21. doi: 10.1182/blood-2009-09-245944.

Hasle, H., G. Kerndrup, and B. B. Jacobsen. 1995. "Childhood Myelodysplastic Syndrome in Denmark: Incidence and Predisposing Conditions." *Leukemia* 9 (9): 1569–72.

Hasle, H., L. D. Wadsworth, B. G. Massing, M. McBride, and K. R. Schultz. 1999. "A Population-Based Study of Childhood Myelodysplastic Syndrome in British Columbia, Canada." *British Journal of Haematology* 106 (4): 1027–32.

Hasle, Henrik. 2000. "Incidence of Essential Thrombocythaemia in Children." *British Journal of Haematology* 110 (3): 751. doi: 10.1046/j.1365-2141.2000.02239-7.x.

Hasle, Henrik, Inge Haunstrup Clemmensen, and Margareta Mikkelsen. 2000. "Risks of Leukaemia and Solid Tumours in Individuals with Down's Syndrome." *The Lancet* 355 (9199): 165–9. doi: 10.1016/S0140-6736(99)05264-2.

Hasselbalch, Hans Carl. 2012. "Perspectives on Chronic Inflammation in Essential Thrombocythemia, Polycythemia Vera, and Myelofibrosis: Is Chronic Inflammation a Trigger and Driver of Clonal Evolution and Development of Accelerated Atherosclerosis and Second Cancer?" *Blood* 119 (14): 3219–25. doi: 10.1182/blood-2011-11-394775.

Hasselbalch, Hans Carl. 2013. "Chronic Inflammation as a Promotor of Mutagenesis in Essential Thrombocythemia, Polycythemia Vera and Myelofibrosis. A Human Inflammation Model for Cancer Development?" *Leukemia Research* 37 (2): 214–20. doi: 10.1016/j.leukres.2012.10.020.

Hennighausen, Lothar, and Gertraud W. Robinson. 2008. "Interpretation of Cytokine Signaling through the Transcription Factors STAT5A and STAT5B." *Genes & Development* 22 (6): 711–21. doi: 10.1101/gad.1643908.

Hertzberg, Libi, Elena Vendramini, Ithamar Ganmore, Gianni Cazzaniga, Maike Schmitz, Jane Chalker, Ruth Shiloh, et al. 2010. "Down Syndrome Acute Lymphoblastic Leukemia, A Highly Heterogeneous Disease in Which Aberrant Expression of CRLF2 Is Associated with Mutated JAK2: A Report from the International BFM Study Group." *Blood* 115 (5): 1006–17. doi: 10.1182/blood-2009-08-235408.

Hidalgo-López, Juliana E., L. Rashmi Kanagal-Shamanna, Jeffrey Medeiros, C. Zeev Estrov, Cameron Yin, Srdan Verstovsek, Sergej Konoplev, et al. 2017. "Morphologic and Molecular Characteristics of De Novo AML with JAK2 V617F Mutation." *Journal of the National Comprehensive Cancer Network: JNCCN* 15 (6): 790–6. doi: 10.6004/jnccn.2017.0106.

Hornakova, Tekla, Judith Staerk, Yohan Royer, Elisabetta Flex, Stefan N. Marco Tartaglia, Laurent Knoops Constantinescu, and Jean-Christophe Renauld. 2009. "Acute Lymphoblastic Leukemia-Associated JAK1 Mutants Activate the Janus Kinase/STAT Pathway via Interleukin-9 Receptor α Homodimers." *Journal of Biological Chemistry* 284 (11): 6773–81. doi: 10.1074/jbc.M807531200.

Hu, Guangzhen, Thomas E. Witzig, and Mamta Gupta. 2013. "A Novel Missense (M206K) STAT3 Mutation in Diffuse Large B Cell Lymphoma Deregulates STAT3 Signaling." *PLoS One* 8 (7): e67851. doi: 10.1371/journal.pone.0067851.

Hunger, Stephen P., and Charles G. Mullighan. 2015a. "Acute Lymphoblastic Leukemia in Children." *New England Journal of Medicine* 373 (16): 1541–52. doi: 10.1056/NEJMra1400972.

Iacobucci, Ilaria, Li Yongjin, Kathryn G. Roberts, Stephanie M. Dobson, Jaeseung C. Kim, Debbie Payne-Turner, Richard C. Harvey, et al. 2016. "Truncating Erythropoietin Receptor Rearrangements in Acute Lymphoblastic Leukemia." *Cancer Cell* 29 (2): 186–200. doi: 10.1016/j.ccell.2015.12.013.

Isaksen, Deborah E., Heinz Baumann, Patty A. Trobridge, Andrew G. Farr, Steven D. Levin, and Steven F. Ziegler. 1999. "Requirement for Stat5 in Thymic Stromal Lymphopoietin-Mediated Signal Transduction." *The Journal of Immunology* 163 (11): 5971–7. www.jimmunol.org/content/163/11/5971.

Isaksen, Deborah E., Heinz Baumann, Baohua Zhou, Sebastien Nivollet, Andrew G. Farr, Steven D. Levin, and Steven F. Ziegler. 2002. "Uncoupling of Proliferation and Stat5 Activation in Thymic Stromal Lymphopoietin-Mediated Signal Transduction." *The Journal of Immunology* 168 (7): 3288–94. doi: 10.4049/jimmunol.168.7.3288.

Ishikawa, Chie, Masachika Senba, and Naoki Mori. 2018. "Anti-Adult T-Cell Leukemia/Lymphoma Activity of Cerdulatinib, a Dual SYK/JAK Kinase Inhibitor." *International Journal of Oncology* 53 (4): 1681–90. doi: 10.3892/ijo.2018.4513.

Ismael, Olfat, Akira Shimada, Asahito Hama, Hiroshi Sakaguchi, Sayoko Doisaki, Hideki Muramatsu, Nao Yoshida, et al. 2012. "Mutations Profile of Polycythemia Vera and Essential Thrombocythemia among Japanese Children." *Pediatric Blood & Cancer* 59 (3): 530–5. doi: 10.1002/pbc.23409.

Jain, Nitin, Kathryn G. Roberts, Elias Jabbour, Keyur Patel, Agda Karina Eterovic, Ken Chen, Patrick Zweidler-McKay, et al. 2017. "Ph-like Acute Lymphoblastic Leukemia: A High-Risk Subtype in Adults." *Blood* 129 (5): 572–81. doi: 10.1182/blood-2016-07-726588.

James, Chloé, Valérie Ugo, Jean-Pierre Le Couédic, Judith Staerk, François Delhommeau, Catherine Lacout, Loïc Garçon, et al. 2005. "A Unique Clonal *JAK2* Mutation Leading to Constitutive Signalling Causes Polycythaemia Vera." *Nature* 434 (7037): 1144–8. doi: 10.1038/nature03546.

Jamieson, Catriona H. M., Jeffrey A. Jason Gotlib, Mark P. Durocher, Chao M. Rajan Mariappan, Marla Lay, Carol Jones, James L. Zehnder, Stan L. Lilleberg, and Irving L. Weissman. 2006. "The JAK2 V617F Mutation Occurs in Hematopoietic Stem Cells in Polycythemia Vera and Predisposes toward Erythroid Differentiation." *Proceedings of the National Academy of Sciences* 103 (16): 6224–9. doi: 10.1073/pnas.0601462103.

Jelinek, Jaroslav, Yasuhiro Oki, Vazganush Gharibyan, Josef T. Carlos Bueso-Ramos, Srdan Verstovsek Prchal, Miloslav Beran, Elihu Estey, Hagop M. Kantarjian, and Jean-Pierre J. Issa. 2005. "JAK2 Mutation 1849G>T Is Rare in Acute Leukemias but Can Be Found in CMML, Philadelphia Chromosome–Negative CML, and Megakaryocytic Leukemia." *Blood* 106 (10): 3370–3. doi: 10.1182/blood-2005-05-1800.

Jensen, M. K., P. de Nully Brown, O. J. Nielsen, and H. C. Hasselbalch. 2000. "Incidence, Clinical Features and Outcome of Essential Thrombocythaemia in a Well Defined Geographical Area." *European Journal of Haematology* 65 (2): 132–9.

Jones, Amy V., Peter J. Campbell, Philip A. Beer, Alessandro M. Susanne Schnittger, Katerina Zoi Vannucchi, MelanieJ. Percy, et al. 2010. "The JAK2 46/1 Haplotype Predisposes to MPL-Mutated Myeloproliferative Neoplasms." *Blood* 115 (22): 4517–23. doi: 10.1182/blood-2009-08-236448.

Ju, Wei, Kelli M. Meili Zhang, Michael N. Wilson, Richard N. Petrus, Xiaohu Zhang Bamford, Rajarshi Guha, Marc Ferrer, Craig J. Thomas, and Thomas A. Waldmann. 2016. "Augmented Efficacy of Brentuximab Vedotin Combined with Ruxolitinib and/or Navitoclax in a Murine Model of Human Hodgkin's Lymphoma." *Proceedings of the National Academy of Sciences* 113 (6): 1624–9. doi: 10.1073/pnas.1524668113.

Karjalainen, Riikka, Tea Pemovska, Mihaela Popa, Komal K. Minxia Liu, Muntasir M. Javarappa, Bhagwan Yadav Majumder, et al. 2017. "JAK1/2 and BCL2 Inhibitors Synergize to Counteract Bone Marrow Stromal Cell–Induced Protection of AML." *Blood* 130 (6): 789–802. doi: 10.1182/blood-2016-02-699363.

Kawamura, Machiko, Tomohiko Taki, Hidefumi Kaku, Kentaro Ohki, and Yasuhide Hayashi. 2015. "Identification of SPAG9 as a Novel JAK2 Fusion Partner Gene in Pediatric Acute Lymphoblastic Leukemia with t(9;17)(P24;Q21)." *Genes, Chromosomes and Cancer* 54 (7): 401–8. doi: 10.1002/gcc.22251.

Kearney, Lyndal, David Gonzalez De Castro, Jenny Yeung, Julia Procter, Sharon W. Horsley, Minenori Eguchi-Ishimae, Caroline M. Bateman, et al. 2009. "Specific JAK2 Mutation (JAK2R683) and Multiple Gene Deletions in Down Syndrome Acute Lymphoblastic Leukemia." *Blood* 113 (3): 646–8. doi: 10.1182/blood-2008-08-170928.

Kelly, Kevin, Corrina McMahon, Stephen Langabeer, Hassan Eliwan, Aengus O'Marcaigh, and Owen P. Smith. 2008. "Congenital JAK2V617F Polycythemia Vera: Where Does the Genotype-Phenotype Diversity End?" *Blood* 112 (10): 4356–7. doi: 10.1182/blood-2008-08-175620.

Keohane, Clodagh, Shahram Kordasti, Thomas Seidl, Pilar Perez, Nicholas Abellan, S. B. Thomas, Claire N. Harrison, Donal P. McLornan, and Ghulam J. Mufti. 2015. "JAK Inhibition Induces Silencing of T Helper Cytokine Secretion and a Profound Reduction in T Regulatory Cells." *British Journal of Haematology* 171 (1): 60–73. doi: 10.1111/bjh.13519.

Khan, I., Z. Huang, Q. Wen, M. J. Stankiewicz, L. Gilles, B. Goldenson, R. Schultz, et al. 2013. "AKT Is a Therapeutic Target in Myeloproliferative Neoplasms." *Leukemia* 27 (9): 1882–90. doi: 10.1038/leu.2013.167.

Khwaja, Asim. 2006. "The Role of Janus Kinases in Haemopoiesis and Haematological Malignancy." *British Journal of Haematology* 134 (4): 366–84. doi: 10.1111/j.1365-2141.2006.06206.x.

Kim, Sang-Kyu, Deborah A. Knight, Lisa R. Jones, Ashley P. Stephin Vervoort, John F. Ng, James E. Seymour, Michaela Waibel Bradner, Lev Kats, and Ricky W. Johnstone. 2018. "JAK2 Is Dispensable for Maintenance of JAK2 Mutant B-Cell Acute Lymphoblastic Leukemias." *Genes & Development* 32 (11–12): 849–64. doi: 10.1101/gad.307504.117.

Kim, Seok Jin, Hye Jin Kang, Dong-Yeop Shin, Dok Hyun Yoon, Kana Sakamoto, Jee Hyun Kong, Young Hyeh Ko, Kengo Takeuchi, Cheolwon Suh, and Won Seog Kim. 2014. "Pilot Study of Ruxolitinib in Relapsed or Refractory Hodgkin Lymphoma and Primary Mediastinal Large B-Cell Lymphoma." *Blood* 124 (21): 4443 ASH 2014 Abstract. www.bloodjournal.org/content/124/21/4443.

Kiyoi, H., S. Yamaji, S. Kojima, and T. Naoe. 2007. "JAK3 Mutations Occur in Acute Megakaryoblastic Leukemia Both in Down Syndrome Children and Non-Down Syndrome Adults." *Leukemia* 21 (3): 574–6. doi: 10.1038/sj.leu.2404527.

Klampfl, Thorsten, Ashot Harutyunyan, Tiina Berg, Bettina Gisslinger, Martin Schalling, Klaudia Bagienski, Damla Olcaydu, et al. 2011. "Genome Integrity of Myeloproliferative Neoplasms in Chronic Phase and during Disease Progression." *Blood* 118 (1): 167–76. doi: 10.1182/blood-2011-01-331678.

Klampfl, Thorsten, Ashot S. Heinz Gisslinger, Harini Nivarthi Harutyunyan, Jelena D. Elisa Rumi, Nicole C. C. Them Milosevic, et al. 2013. "Somatic Mutations of Calreticulin in Myeloproliferative Neoplasms." *The New England Journal of Medicine* 369 (25): 2379–90. doi: 10.1056/NEJMoa1311347.

Kleppe, Maria, Thomas Tousseyn, Eva Geissinger, Zeynep Kalender Atak, Stein Aerts, Andreas Rosenwald, Iwona Wlodarska, and Jan Cools. 2011. "Mutation Analysis of the Tyrosine Phosphatase PTPN2 in Hodgkin's Lymphoma and T-Cell Non-Hodgkin's Lymphoma." *Haematologica* 96 (11): 1723–7. doi: 10.3324/haematol.2011.041921.

Koo, Ghee Chong, Soo Yong Tan, Tiffany Tang, Song Ling Poon, George E. Allen, Leonard Tan, Soo Ching Chong, et al. 2012. "Janus Kinase 3–Activating Mutations Identified in Natural Killer/T-Cell Lymphoma." *Cancer Discovery* 2 (7): 591–7. doi: 10.1158/2159-8290.CD-12-0028.

Koppikar, Priya, Neha Bhagwat, Outi Kilpivaara, Taghi Manshouri, Mazhar Adli, Todd Hricik, Fan Liu, et al. 2012. "Heterodimeric JAK–STAT Activation as a Mechanism of Persistence to JAK2 Inhibitor Therapy." *Nature* 489 (7414): 155–9. doi: 10.1038/nature11303.

Koskela, Hanna L.M., Samuli Eldfors, Arjan J. Pekka Ellonen, Heikki van Adrichem, Emma I. Kuusanmäki, Sonja Lagström Andersson, et al. 2012. "Somatic STAT3 Mutations in Large Granular Lymphocytic Leukemia." *New England Journal of Medicine* 366 (20): 1905–13. doi: 10.1056/NEJMoa1114885.

Kralovics, Robert, Yongli Guan, and Josef T Prchal. 2002. "Acquired Uniparental Disomy of Chromosome 9p Is a Frequent Stem Cell Defect in Polycythemia Vera." *Experimental Hematology* 30 (3): 229–36. doi: 10.1016/S0301-472X(01)00789-5.

Kralovics, Robert, Francesco Passamonti, S. Buser Andreas, Soon-Siong Teo, Ralph Tiedt, Jakob R. Passweg, Andre Tichelli, Mario Cazzola, and Radek C. Skoda. 2005. "A Gain-of-Function Mutation of JAK2 in Myeloproliferative Disorders." *The New England Journal of Medicine* 352 (17): 1779–90. doi: 10.1056/NEJMoa051113.

Kralovics, Robert, David W. Stockton, and Josef T. Prchal. 2003. "Clonal Hematopoiesis in Familial Polycythemia Vera Suggests the Involvement of Multiple Mutational Events in the Early Pathogenesis of the Disease." *Blood* 102 (10): 3793–6. doi: 10.1182/blood-2003-03-0885.

Lacout, Catherine, Didier F. Pisani, Micheline Tulliez, Françoise Moreau Gachelin, William Vainchenker, and Jean-Luc Villeval. 2006. "JAK2V617F Expression in Murine Hematopoietic Cells Leads to MPD Mimicking Human PV with Secondary Myelofibrosis." *Blood* 108 (5): 1652–60. doi: 10.1182/blood-2006-02-002030.

Lacronique, Virginie, Anthony Boureux, Véronique Della Valle, Hélène Poirel, Christine Tran Quang, Martine Mauchauffé, Christian Berthou, et al. 1997. "A TEL-JAK2 Fusion Protein with Constitutive Kinase Activity in Human Leukemia." *Science* 278 (5341): 1309–12. doi: 10.1126/science.278.5341.1309.

Lai, Catherine, and Kieron Dunleavy. 2013. "NK/T-Cell Lymphomas in Children." *Best Practice & Research. Clinical Haematology* 26 (1): 33–41. doi: 10.1016/j.beha.2013.04.004.

Lam, Lloyd T., R. George Wright, Eric Davis, Georg Lenz, Pedro Farinha, John W. Lenny Dang, Andreas Rosenwald Chan, Randy D. Gascoyne, and Louis M. Staudt. 2008. "Cooperative Signaling through the Signal Transducer and Activator of Transcription 3 and Nuclear Factor-KB Pathways in Subtypes of Diffuse Large B-Cell Lymphoma." *Blood* 111 (7): 3701–13. doi: 10.1182/blood-2007-09-111948.

Lasho, Terra L., Rebecca F. Animesh Pardanani, Ruben A. McClure, Ross L. Mesa, Levine D. Gary Gilliland, and Ayalew Tefferi. 2006. "Concurrent MPL515 and JAK2V617F Mutations in Myelofibrosis: Chronology of Clonal Emergence and Changes in Mutant Allele Burden over Time." *British Journal of Haematology* 135 (5): 683–7. doi: 10.1111/j.1365-2141.2006.06348.x.

Lee, Sanghoon, Tishi Shah, Changhong Yin, Jessica Hochberg, Janet Ayello, Erin Morris, Carmella van de Ven, and Mitchell S. Cairo. 2018. "Ruxolitinib Significantly Enhances in Vitro Apoptosis in Hodgkin Lymphoma and Primary Mediastinal B-Cell Lymphoma and Survival in a Lymphoma Xenograft Murine Model." *Oncotarget* 9 (11): 9776–88. doi: 10.18632/oncotarget.24267.

Lenglet, Marion, Florence Robriquet, Klaus Schwarz, Carme Camps, Anne Couturier, David Hoogewijs, Alexandre Buffet, et al. 2018. "Identification of a New VHL Exon and Complex Splicing

Alterations in Familial Erythrocytosis or Von Hippel-Lindau Disease." *Blood* 132 (5): 469–83. doi: 10.1182/blood-2018-03-838235.

Lenz, Georg, George W. Wright, N. C. Tolga Emre, Sandeep S. Holger Kohlhammer, Dave R. Eric Davis, Shannon Carty, et al. 2008. "Molecular Subtypes of Diffuse Large B-Cell Lymphoma Arise by Distinct Genetic Pathways." *Proceedings of the National Academy of Sciences* 105 (36): 13520–5. doi: 10.1073/pnas.0804295105.

Levine, Ross L., Martha Wadleigh, Benjamin L. Jan Cools, Gerlinde Wernig Ebert, Brian J. P. Huntly, Titus J. Boggon, et al. 2005. "Activating Mutation in the Tyrosine Kinase JAK2 in Polycythemia Vera, Essential Thrombocythemia, and Myeloid Metaplasia with Myelofibrosis." *Cancer Cell* 7 (4): 387–97. doi: 10.1016/j.ccr.2005.03.023.

Lippert, Eric, Marjorie Boissinot, Robert Kralovics, François Girodon, Irène Dobo, Vincent Praloran, Nathalie Boiret-Dupré, Radek C. Skoda, and Sylvie Hermouet. 2006. "The JAK2-V617F Mutation Is Frequently Present at Diagnosis in Patients with Essential Thrombocythemia and Polycythemia Vera." *Blood* 108 (6): 1865–7. doi: 10.1182/blood-2006-01-013540.

Liu, Yu, John Easton, Ying Shao, Jamie Maciaszek, Mark R. Zhaoming Wang, Kelly McCastlain Wilkinson, et al. 2017. "The Genomic Landscape of Pediatric and Young Adult T-Lineage Acute Lymphoblastic Leukemia." *Nature Genetics* 49 (8): 1211–18. doi: 10.1038/ng.3909.

Loh, Mignon L., Sarah K. Tasian, Karen R. Rabin, Patrick Brown, Joel M. Daniel Magoon, Xuejun Chen Reid, Charlotte H. Ahern, Brenda J. Weigel, and Susan M. Blaney. 2015. "A Phase 1 Dosing Study of Ruxolitinib in Children with Relapsed or Refractory Solid Tumors, Leukemias, or Myeloproliferative Neoplasms: A Children's Oncology Group Phase 1 Consortium Study (ADVL1011)." *Pediatric Blood & Cancer* 62 (10): 1717–24. doi: 10.1002/pbc.25575.

Lohr, Jens G., Michael S. Petar Stojanov, Daniel Auclair Lawrence, Bjoern Chapuy, Carrie Sougnez, Peter Cruz-Gordillo, et al. 2012. "Discovery and Prioritization of Somatic Mutations in Diffuse Large B-Cell Lymphoma (DLBCL) by Whole-Exome Sequencing." *Proceedings of the National Academy of Sciences* 109 (10): 3879–84. doi: 10.1073/pnas.1121343109.

Lorsbach, Robert B., Dominic Shay-Seymore, Jennifer Moore, Peter M. Banks, Robert P. Hasserjian, John T. Sandlund, and Frederick G. Behm. 2002. "Clinicopathologic Analysis of Follicular Lymphoma Occurring in Children." *Blood* 99 (6): 1959–64. doi: 10.1182/blood.V99.6.1959.

Louissaint, Abner, Kristian T. Schafernak, Julia T. Geyer, Alexandra E. Kovach, Mahmoud Ghandi, Dita Gratzinger, Christine G. Roth, et al. 2016. "Pediatric-Type Nodal Follicular Lymphoma: A Biologically Distinct Lymphoma with Frequent MAPK Pathway Mutations." *Blood* 128 (8): 1093–100. doi: 10.1182/blood-2015-12-682591.

Lundberg, Pontus, Axel Karow, Ronny Nienhold, Renate Looser, Hui Hao-Shen, Ina Nissen, Sabine Girsberger, et al. 2014. "Clonal Evolution and Clinical Correlates of Somatic Mutations in Myeloproliferative Neoplasms." *Blood* 123 (14): 2220–8. doi: 10.1182/blood-2013-11-537167.

Ma, Jiao, Wei Xing, Greg Coffey, Karen Dresser, Lu Kellie, Ailin Guo, Gordana Raca, et al. 2015. "Cerdulatinib, a Novel Dual SYK/JAK Kinase Inhibitor, Has Broad Anti-Tumor Activity in Both ABC and GCB Types of Diffuse Large B Cell Lymphoma." *Oncotarget* 6 (41): 43881–96. doi: 10.18632/oncotarget.6316.

Malinge, Sebastien, Raouf Ben-Abdelali, Catherine Settegrana, Isabelle Radford-Weiss, Marianne Debre, Kheira Beldjord, Elizabeth A. Macintyre, et al. 2007. "Novel Activating JAK2 Mutation in a Patient with Down Syndrome and B-Cell Precursor Acute Lymphoblastic Leukemia." *Blood* 109 (5): 2202–4. doi: 10.1182/blood-2006-09-045963.

Malinge, Sébastien, Christine Ragu, Veronique Della-Valle, Stefan N. Didier Pisani, Christelle Constantinescu, Jean-Luc Villeval Perez, et al. 2008. "Activating Mutations in Human Acute Megakaryoblastic Leukemia." *Blood* 112 (10): 4220–6. doi: 10.1182/blood-2008-01-136366.

Marchioli, Roberto, Guido Finazzi, Raffaele Landolfi, Jack Kutti, Heinz Gisslinger, Carlo Patrono, Raphael Marilus, Ana Villegas, Gianni Tognoni, and Tiziano Barbui. 2005. "Vascular and Neoplastic Risk in a Large Cohort of Patients with Polycythemia Vera." *Journal of Clinical Oncology* 23 (10): 2224–32. doi: 10.1200/JCO.2005.07.062.

Marshall, John L., Hedy Kindler, John Deeken, Nicholas J. Pankaj Bhargava, Naiyer Rizvi Vogelzang, Taina Luhtala, et al. 2005. "Phase I Trial of Orally Administered CEP-701, a Novel Neurotrophin Receptor-Linked Tyrosine Kinase Inhibitor." *Investigational New Drugs* 23 (1): 31–7. doi: 10.1023/B:DRUG.0000047103.64335.b0.

Martinaud, Christophe, Patrick Brisou, and Marie-Joelle Mozziconacci. 2010. "Is the JAK2(V617F) Mutation Detectable in Healthy Volunteers?" *American Journal of Hematology* 85 (4): 287–8. doi: 10.1002/ajh.21627.

Martínez-Avilés, Luz, Carles Besses, Alberto Álvarez-Larrán, Francisco Cervantes, Juan Carlos Hernández-Boluda, and Beatriz Bellosillo. 2007. "JAK2 Exon 12 Mutations in Polycythemia Vera or Idiopathic Erythrocytosis." *Haematologica* 92 (12): 1717–18. doi: 10.3324/haematol.12011.

Marty, Caroline, Christian Pecquet, Harini Nivarthi, Mira El-Khoury, Ilyas Chachoua, Micheline Tulliez, Jean-Luc Villeval, et al. 2016. "Calreticulin Mutants in Mice Induce an MPL-Dependent Thrombocytosis with Frequent Progression to Myelofibrosis." *Blood* 127 (10): 1317–24. doi: 10.1182/blood-2015-11-679571.

Mascarenhas, John, Ronald Hoffman, Aaron T. Moshe Talpaz, Brady Stein Gerds, Vikas Gupta, Anita Szoke, et al. 2018. "Pacritinib Vs Best Available Therapy, Including Ruxolitinib, in Patients with Myelofibrosis: A Randomized Clinical Trial." *JAMA Oncology* 4 (5): 652–9. doi: 10.1001/jamaoncol.2017.5818.

Maslah, N., B. Cassinat, E. Verger, J.-J. Kiladjian, and L. Velazquez. 2017. "The Role of LNK/SH2B3 Genetic Alterations in Myeloproliferative Neoplasms and Other Hematological Disorders." *Leukemia* 31 (8): 1661–70. doi: 10.1038/leu.2017.139.

Mathas, Stephan, Sylvia Hartmann, and Ralf Küppers. 2016. "Hodgkin Lymphoma: Pathology and Biology." *Seminars in Hematology, Hodgkin Lymphoma* 53 (3): 139–47. doi: 10.1053/j.seminhematol.2016.05.007.

Maude, Shannon L., Sarah K. Tasian, Junior W. Tiffaney Vincent, Cecilia Sheen Hall, Kathryn G. Roberts, AlixE Seif, et al. 2012. "Targeting JAK1/2 and MTOR in Murine Xenograft Models of Ph-Like Acute Lymphoblastic Leukemia." *Blood* 120 (17): 3510–18. doi: 10.1182/blood-2012-03-415448.

McGirt, Laura Y., Peilin Jia, Devin A. Baerenwald, Robert J. Duszynski, Kimberly B. Dahlman, John A. Zic, Jeffrey P. Zwerner, et al. 2015. "Whole-Genome Sequencing Reveals Oncogenic Mutations in Mycosis Fungoides." *Blood* 126 (4): 508–19. doi: 10.1182/blood-2014-11-611194.

McMullin, Mary Frances, and Holger Cario. 2016. "LNK Mutations and Myeloproliferative Disorders." *American Journal of Hematology* 91 (2): 248–51. doi: 10.1002/ajh.24259.

McNally, R. J., D. Rowland, E. Roman, and R. A. Cartwright. 1997. "Age and Sex Distributions of Hematological Malignancies in the U.K." *Hematological Oncology* 15 (4): 173–89.

Mead, Adam J., Michelle J. Rugless, W. Jacobsen Sten Eirik, and Anna Schuh. 2012. "Germline JAK2 Mutation in a Family with Hereditary Thrombocytosis." *The New England Journal of Medicine* 366 (10): 967–9. doi: 10.1056/NEJMc1200349.

Meier, Cecile, Sylvia Hoeller, Caroline Bourgau, Petra Hirschmann, Juerg Schwaller, Stefano A. Philip Went, Andreas Reiter Pileri, Stephan Dirnhofer, and Alexandar Tzankov. 2009. "Recurrent Numerical Aberrations of *JAK2* and Deregulation of the JAK2-STAT Cascade in Lymphomas." *Modern Pathology* 22 (3): 476–87. doi: 10.1038/modpathol.2008.207.

Melzner, I., M. A. Weniger, C. K. Menz, and P. Möller. 2006. "Absence of the JAK2 V617F Activating Mutation in Classical Hodgkin Lymphoma and Primary Mediastinal B-Cell Lymphoma." *Leukemia* 20 (1): 157–8. doi: 10.1038/sj.leu.2404036.

Melzner, Ingo, Alexandra Juliana Bucur, Silke Brüderlein, Karola Dorsch, Thomas F. Cornelia Hasel, E. Barth, Frank Leithäuser, and Möller. Peter. 2005. "Biallelic Mutation of SOCS-1 Impairs JAK2 Degradation and Sustains Phospho-JAK2 Action in the MedB-1 Mediastinal Lymphoma Line." *Blood* 105 (6): 2535–42. doi: 10.1182/blood-2004-09-3701.

Mesa, Ruben A., John V. Jean-Jacques Kiladjian, Timothy Devos Catalano, Miklos Egyed, Andrzei Hellmann, Donal McLornan, et al. 2017. "SIMPLIFY-1: A Phase III Randomized Trial of Momelotinib versus Ruxolitinib in Janus Kinase Inhibitor–Naïve Patients with Myelofibrosis." *Journal of Clinical Oncology* 35 (34): 3844–50. doi: 10.1200/JCO.2017.73.4418.

Morak, Maria, Andishe Attarbaschi, Susanna Fischer, Christine Nassimbeni, Reinhard Grausenburger, Stephan Bastelberger, Stefanie Krentz, et al. 2012. "Small Sizes and Indolent Evolutionary Dynamics Challenge the Potential Role of P2RY8-CRLF2–Harboring Clones as Main Relapse-Driving Force in Childhood ALL." *Blood* 120 (26): 5134–42. doi: 10.1182/blood-2012-07-443218.

Morin, Ryan D., Andrew J. Maria Mendez-Lago, Rodrigo Goya Mungall, Karen L. Mungall, Richard D. Corbett, Nathalie A. Johnson, et al. 2011. "Frequent Mutation of Histone-Modifying Genes in Non-Hodgkin Lymphoma." *Nature* 476 (7360): 298–303. doi: 10.1038/nature10351.

Mottok, Anja, Christoph Renné, Marc Seifert, Elsie Oppermann, Wolf Bechstein, Martin-Leo Hansmann, Ralf Küppers, and Bräuninger Andreas. 2009. "Inactivating SOCS1 Mutations are Caused by Aberrant Somatic Hypermutation and Restricted to a Subset of B-Cell Lymphoma Entities." *Blood* 114 (20): 4503–6. doi: 10.1182/blood-2009-06-225839.

Mottok, Anja, Christoph Renné, Klaus Willenbrock, Martin-Leo Hansmann, and Bräuninger Andreas. 2007. "Somatic Hypermutation of SOCS1 in Lymphocyte-Predominant Hodgkin Lymphoma Is Accompanied by High JAK2 Expression and Activation of STAT6." *Blood* 110 (9): 3387–90. doi: 10.1182/blood-2007-03-082511.

Mullighan, Charles G., Richard C. Jinghui Zhang, Harvey J. Racquel Collins-Underwood, Brenda A. Schulman, Letha A. Phillips, Sarah K. Tasian, et al. 2009a. "JAK Mutations in High-Risk Childhood Acute Lymphoblastic Leukemia." *Proceedings of the National Academy of Sciences* 106 (23): 9414–18. doi: 10.1073/pnas.0811761106.

Mullighan, Charles G., J. Racquel Collins-Underwood, A. A. Letha, Michael G. Phillips, Wei Liu Loudin, Jinghui Zhang, Jing Ma, et al. 2009b. "Rearrangement of *CRLF2* in B-Progenitor– And down Syndrome–Associated Acute Lymphoblastic Leukemia." *Nature Genetics* 41 (11): 1243–6. doi: 10.1038/ng.469.

Mullighan, Charles G., Su Xiaoping, Jinghui Zhang, Letha A. Ina Radtke, A. Phillips, Christopher B. Miller, Ma Jing, et al. 2009c. "Deletion of IKZF1 and Prognosis in Acute Lymphoblastic Leukemia." *The New England Journal of Medicine* 360 (5): 470–80. doi: 10.1056/NEJMoa0808253.

Nairismägi, M.-L., M. E. Gerritsen, Z. M. Li, G. C. Wijaya, B. K. H. Chia, Y. Laurensia, J. Q. Lim, et al. 2018. "Oncogenic Activation of JAK3-STAT Signaling Confers Clinical Sensitivity to PRN371, a Novel Selective and Potent JAK3 Inhibitor, in Natural Killer/T-Cell Lymphoma." *Leukemia* 32 (5): 1147–56. doi: 10.1038/s41375-017-0004-x.

Nangalia, J., C. E. Massie, E. J. Baxter, F. L. Nice, G. Gundem, D. C. Wedge, E. Avezov, et al. 2013. "Somatic CALR Mutations in Myeloproliferative Neoplasms with Nonmutated JAK2." *The New England Journal of Medicine* 369 (25): 2391–405. doi: 10.1056/NEJMoa1312542.

Neste, Eric, Van den, Marc André, Thomas Gastinne, Aspasia Stamatoullas, Corinne Haioun, Amine Belhabri, Oumedaly Reman, et al. 2018. "Phase II Study of Oral JAK1/JAK2 Inhibitor Ruxolitinib in Advanced Relapsed/Refractory Hodgkin Lymphoma." *Haematologica*, January, haematol.2017.180554. doi: 10.3324/haematol.2017.180554.

Ngo, Vu N., Ryan M. Young, Roland Schmitz, Sameer Jhavar, Wenming Xiao, Kian-Huat Lim, Holger Kohlhammer, et al. 2011. "Oncogenically Active *MYD88* Mutations in Human Lymphoma." *Nature* 470 (7332): 115–19. doi: 10.1038/nature09671.

Nguyen, Melody H.-H., M.-Y. Ho Jenny, Bryan K. Beattie, and Dwayne L. Barber. 2001. "TEL-JAK2 Mediates Constitutive Activation of the Phosphatidylinositol 3′-Kinase/protein Kinase B Signaling Pathway." *Journal of Biological Chemistry* 276 (35): 32704–13. doi: 10.1074/jbc.M103100200.

Nielsen, Camilla, Henrik S. Birgens, Børge G. Nordestgaard, Lasse Kjær, and Stig E. Bojesen. 2011. "The JAK2 V617F Somatic Mutation, Mortality and Cancer Risk in the General Population." *Haematologica* 96 (3): 450–3. doi: 10.3324/haematol.2010.033191.

Noone, A. M., N. Howlader, M. Krapcho, D. Miller, Brest Brest, M. Yu, J. Ruhl, et al. 2018. "SEER Cancer Statistics Review." *National Cancer Institute, Bethesda, MD* May, 1975–2015.

Norton, Alice, Chris Fisher, Hui Liu, Qiang Wen, Gina Mundschau, Jose Luis Fuster, Henrik Hasle, et al. 2007. "Analysis of JAK3, JAK2, and C-MPL Mutations in Transient Myeloproliferative Disorder and Myeloid Leukemia of down Syndrome Blasts in Children with down Syndrome." *Blood* 110 (3): 1077–9. doi: 10.1182/blood-2007-03-080374.

Nussenzveig, Roberto H., Sabina I. Swierczek, Jaroslav Jelinek, Amos Gaikwad, Enli Liu, Jaroslav F. Prchal Srdan Verstovsek, and Josef T. Prchal. 2007. "Polycythemia Vera Is Not Initiated by JAK2V617F Mutation." *Experimental Hematology* 35 (1): 32.e1–32.e9. doi: 10.1016/j.exphem.2006.11.012.

Oh, Stephen T., Erin F. Simonds, Matthew B. Carol Jones, Yury Goltsev Hale, Kenneth D. Gibbs, Jason D. Merker, James L. Zehnder, Garry P. Nolan, and Jason Gotlib. 2010. "Novel Mutations in the Inhibitory Adaptor Protein LNK Drive JAK–STAT Signaling in Patients with Myeloproliferative Neoplasms." *Blood* 116 (6): 988–92. doi: 10.1182/blood-2010-02-270108.

Oschlies, Ilske, Wolfram Klapper, Martin Zimmermann, Matthias Krams, Hans-Heinrich Wacker, Birgit Burkhardt, Lana Harder, Reiner Siebert, Alfred Reiter, and Reza Parwaresch. 2006. "Diffuse

Large B-Cell Lymphoma in Pediatric Patients Belongs Predominantly to the Germinal-Center Type B-Cell Lymphomas: A Clinicopathologic Analysis of Cases Included in the German BFM (berlin-frankfurt-münster) Multicenter Trial." *Blood* 107 (10): 4047–52. doi: 10.1182/blood-2005-10-4213.

Parampalli, Yajnanarayana Sowmya, Thomas Stübig, Isabelle Cornez, Haefaa Alchalby, Kathrin Schönberg, Janna Rudolph, Ioanna Triviai, et al. 2015. "JAK1/2 Inhibition Impairs T Cell Function in Vitro and in Patients with Myeloproliferative Neoplasms." *British Journal of Haematology* 169 (6): 824–33. doi: 10.1111/bjh.13373.

Pardanani, A., T. Lasho, C. Finke, S. T. Oh, J. Gotlib, and A. Tefferi. 2010. "LNK Mutation Studies in Blast-Phase Myeloproliferative Neoplasms, and in Chronic-Phase Disease with *TET2, IDH, JAK2* or *MPL* Mutations." *Leukemia* 24 (10): 1713–18. doi: 10.1038/leu.2010.163.

Pardanani, A., T. L. Lasho, C. Finke, C. A. Hanson, and A. Tefferi. 2007. "Prevalence and Clinicopathologic Correlates of *JAK2* Exon 12 Mutations in *JAK2*V617F-Negative Polycythemia Vera." *Leukemia* 21 (9): 1960–3. doi: 10.1038/sj.leu.2404810.

Pardanani, Animesh, Jorge E. Claire Harrison, Francisco Cervantes Cortes, Ruben A. Mesa, Donald Milligan, Tamás Masszi, et al. 2015. "Safety and Efficacy of Fedratinib in Patients with Primary or Secondary Myelofibrosis: A Randomized Clinical Trial." *JAMA Oncology* 1 (5): 643–51. doi: 10.1001/jamaoncol.2015.1590.

Pardanani, Animesh D., Ross L. Levine, Terra Lasho, Ruben A. Yana Pikman, Martha Wadleigh Mesa, David P. Steensma, et al. 2006. "MPL515 Mutations in Myeloproliferative and Other Myeloid Disorders: A Study of 1182 Patients." *Blood* 108 (10): 3472–6. doi: 10.1182/blood-2006-04-018879.

Passamonti, Francesco, Chiara Elena, Radek C. Susanne Schnittger, Anthony R. Skoda, François Girodon Green, Jean-Jacques Kiladjian, et al. 2011. "Molecular and Clinical Features of the Myeloproliferative Neoplasm Associated with JAK2 Exon 12 Mutations." *Blood* 117 (10): 2813–16. doi: 10.1182/blood-2010-11-316810.

Passamonti, Francesco, Martin Griesshammer, Francesca Palandri, Miklos Egyed, Giulia Benevolo, Timothy Devos, Jeannie Callum, et al. 2017. "Ruxolitinib for the Treatment of Inadequately Controlled Polycythaemia Vera without Splenomegaly (RESPONSE-2): A Randomised, Open-Label, Phase 3b Study." *The Lancet Oncology* 18 (1): 88–99. doi: 10.1016/S1470-2045(16)30558-7.

Peeters, Pieter, Sophie D. Raynaud, Jan Cools, Iwona Wlodarska, Josiane Grosgeorge, Patrick Philip, Fabrice Monpoux, et al. 1997. "Fusion of TEL, the ETS-Variant Gene 6 (ETV6), to the Receptor-Associated Kinase JAK2 as a Result of T(9; 12) in a Lymphoid and T(9; 15; 12) in a Myeloid Leukemia." *Blood* 90 (7): 2535–40. www.bloodjournal.org/content/90/7/2535.

Pemmaraju, Naveen, Hagop Kantarjian, Tapan Kadia, Jorge Cortes, Gautam Borthakur, Kate Newberry, Guillermo Garcia-Manero, et al. 2015. "A Phase I/II Study of the Janus Kinase (JAK)1 and 2 Inhibitor Ruxolitinib in Patients with Relapsed or Refractory Acute Myeloid Leukemia." *Clinical Lymphoma Myeloma and Leukemia* 15 (3): 171–6. doi: 10.1016/j.clml.2014.08.003.

Perez-Andreu, Virginia, Kathryn G. Roberts, Richard C. Harvey, Wenjian Yang, Cheng Cheng, Deqing Pei, Xu Heng, et al. 2013. "Inherited GATA3 Variants are Associated with Ph-like Childhood Acute Lymphoblastic Leukemia and Risk of Relapse." *Nature Genetics* 45 (12): 1494–8. doi: 10.1038/ng.2803.

Perez-Andreu, Virginia, Kathryn G. Roberts, Xu Heng, Colton Smith, Hui Zhang, Wenjian Yang, Richard C. Harvey, et al. 2015. "A Genome-Wide Association Study of Susceptibility to Acute Lymphoblastic Leukemia in Adolescents and Young Adults." *Blood* 125 (4): 680–6. doi: 10.1182/blood-2014-09-595744.

Pietra, Daniela, Matteo C. Ilaria Casetti, Chiara Da Vià, Chiara Milanesi Elena, and Elisa Rumi. 2012. "JAK2 GGCC Haplotype in MPL Mutated Myeloproliferative Neoplasms." *American Journal of Hematology* 87 (7): 746–7. doi: 10.1002/ajh.23229.

Pietra, Daniela, Li Sai, Angela Brisci, Francesco Passamonti, Elisa Rumi, Alexandre Theocharides, Maurizio Ferrari, et al. 2008. "Somatic Mutations of JAK2 Exon 12 in Patients with JAK2 (v617f)-negative Myeloproliferative Disorders." *Blood* 111 (3): 1686–9. doi: 10.1182/blood-2007-07-101576.

Pikman, Yana, Benjamin H. Lee, Thomas Mercher, Benjamin L. Elizabeth McDowell, Maricel Gozo Ebert, Adam Cuker, et al. 2006. "MPLW515L Is a Novel Somatic Activating Mutation in Myelofibrosis with Myeloid Metaplasia." *PLoS Medicine* 3 (7): e270. doi: 10.1371/journal.pmed.0030270.

Pillai, Vinodh, Michael Tallarico, Michael R. Bishop, and Megan S. Lim. 2016. "Mature T- and NK-Cell Non-Hodgkin Lymphoma in Children and Young Adolescents." *British Journal of Haematology* 173 (4): 573–81. doi: 10.1111/bjh.14044.

Porcu, Michaël, Olga Gielen, Jan Cools, and De Keersmaecker Kim. 2009. "JAK1 Mutation Analysis in T-Cell Acute Lymphoblastic Leukemia Cell Lines." *Haematologica* 94 (3): 435–7. doi: 10.3324/haematol.13587.

Prutsch, Nicole, Elisabeth Gurnhofer, Tobias Suske, Huan Chang Liang, Michaela Schlederer, Simone Roos, Lawren C. Wu, et al. 2018. "Dependency on the TYK2/STAT1/MCL1 Axis in Anaplastic Large Cell Lymphoma." *Leukemia*, August, 1. doi: 10.1038/s41375-018-0239-1.

Pui, Ching-Hon, Kathryn G. Roberts, Jun J. Yang, and Charles G. Mullighan. 2017. "Philadelphia Chromosome-Like Acute Lymphoblastic Leukemia." *Clinical Lymphoma, Myeloma & Leukemia* 17 (8): 464–70. doi: 10.1016/j.clml.2017.03.299.

Putti, Maria Caterina, Marco Pizzi, Irene Bertozzi, Elena Sabattini, Concetta Micalizzi, Piero Farruggia, Ugo Ramenghi, et al. 2017. "Bone Marrow Histology for the Diagnosis of Essential Thrombocythemia in Children: A Multicenter Italian Study." *Blood* 129 (22): 3040–42. doi: 10.1182/blood-2017-01-761767.

Qin, Haiying, Monica Cho, Waleed Haso, Sarah K. Ling Zhang, Htoo Tasian, Oo Zarni, Gian Luca Negri, et al. 2015. "Eradication of B-ALL Using Chimeric Antigen Receptor–Expressing T Cells Targeting the TSLPR Oncoprotein." *Blood* 126 (5): 629–39. doi: 10.1182/blood-2014-11-612903.

Rampal, Raajit, Fatima Al-Shahrour, Omar Abdel-Wahab, Jay P. Patel, Jean-Philippe Brunel, Craig H. Mermel, AdamJ. Bass, et al. 2014. "Integrated Genomic Analysis Illustrates the Central Role of JAK–STAT Pathway Activation in Myeloproliferative Neoplasm Pathogenesis." *Blood* 123 (22): e123–33. doi: 10.1182/blood-2014-02-554634.

Randi, Maria L., Giulia Geranio, Irene Bertozzi, Concetta Micalizzi, Ugo Ramenghi, Fabio Tucci, Lucia D. Notarangelo, et al. 2015. "Are All Cases of Paediatric Essential Thrombocythaemia Really Myeloproliferative Neoplasms? Analysis of a Large Cohort." *British Journal of Haematology* 169 (4): 584–9. doi: 10.1111/bjh.13329.

Redell, Michele S., Marcos J. Ruiz, Todd A. Alonzo, Robert B. Gerbing, and David J. Tweardy. 2011. "Stat3 Signaling in Acute Myeloid Leukemia: Ligand-Dependent and -independent Activation and Induction of Apoptosis by a Novel Small-Molecule Stat3 Inhibitor." *Blood* 117 (21): 5701–9. doi: 10.1182/blood-2010-04-280123.

Redell, Michele S., Marcos J. Ruiz, Robert B. Gerbing, Todd A. Alonzo, Beverly J. Lange, David J. Tweardy, and Soheil Meshinchi. 2013. "FACS Analysis of Stat3/5 Signaling Reveals Sensitivity to G-CSF and IL-6 as A Significant Prognostic Factor in Pediatric AML: A Children's Oncology Group Report." *Blood* 121 (7): 1083–93. doi: 10.1182/blood-2012-04-421925.

Reshmi, Shalini C., Richard C. Harvey, Kathryn G. Roberts, Eileen Stonerock, Amy Smith, Heather Jenkins, I.-Ming Chen, et al. 2017. "Targetable Kinase Gene Fusions in High-Risk B-ALL: A Study from the Children's Oncology Group." *Blood* 129 (25): 3352–61. doi: 10.1182/blood-2016-12-758979.

Ritz, Olga, Chrystelle Guiter, Flavia Castellano, Karola Dorsch, Julia Melzner, Jean-Philippe Jais, Gwendoline Dubois, Philippe Gaulard, Peter Möller, and Karen Leroy. 2009. "Recurrent Mutations of the STAT6 DNA Binding Domain in Primary Mediastinal B-Cell Lymphoma." *Blood* 114 (6): 1236–42. doi: 10.1182/blood-2009-03-209759.

Roberts, Kathryn G., Ryan D. Morin, Jinghui Zhang, Martin Hirst, Yongjun Zhao, Su Xiaoping, Shann-Ching Chen, et al. 2012. "Genetic Alterations Activating Kinase and Cytokine Receptor Signaling in High-Risk Acute Lymphoblastic Leukemia." *Cancer Cell* 22 (2): 153–66. doi: 10.1016/j.ccr.2012.06.005.

Roberts, Kathryn G., Shalini C. Reshmi, Richard C. Harvey, I.-Ming Chen, Kinnari Patel, Eileen Stonerock, Heather Jenkins, et al. 2018. "Genomic and Outcome Analyses of Ph-like ALL in NCI Standard-Risk Patients: A Report from the Children's Oncology Group." *Blood* 132 (8): 815–24. doi: 10.1182/blood-2018-04-841676.

Roberts, Kathryn G., Yung-Li Yang, Debbie Payne-Turner, Jacob K. Wenwei Lin, Kirsten Dickerson Files, Zhaohui Gu, et al. 2017. "Oncogenic Role and Therapeutic Targeting of ABL-Class and JAK–STAT Activating Kinase Alterations in Ph-like ALL." *Blood Advances* 1 (20): 1657–71. doi: 10.1182/bloodadvances.2017011296.

Roberts, Kathryn G., Li Yongjin, Debbie Payne-Turner, Richard C. Harvey, Yung-Li Yang, Deqing Pei, Kelly McCastlain, et al. 2014. "Targetable Kinase-Activating Lesions in Ph-like Acute Lymphoblastic Leukemia." *The New England Journal of Medicine* 371 (11): 1005–15. doi: 10.1056/NEJMoa1403088.

Roberts, Kathryn G., Gu Zhaohui, Debbie Payne-Turner, Richard C. Kelly McCastlain, I-Ming Chen Harvey, Deqing Pei, et al. 2016. "High Frequency and Poor Outcome of Philadelphia Chromosome–Like Acute Lymphoblastic Leukemia in Adults." *Journal of Clinical Oncology* 35 (4): 394–401. doi: 10.1200/JCO.2016.69.0073.

Roemer, Margaretha G.M., Ranjana H. Advani, Azra H. Ligon, Robert A. Yasodha Natkunam, Heather Homer Redd, Courtney F. Connelly, et al. 2016. "PD-L1 and PD-L2 Genetic Alterations Define Classical Hodgkin Lymphoma and Predict Outcome." *Journal of Clinical Oncology* 34 (23): 2690–7. doi: 10.1200/JCO.2016.66.4482.

Rooij, Jasmijn D. E. de, C. Michel Zwaan, and Marry van den Heuvel-Eibrink. 2015. "Pediatric AML: From Biology to Clinical Management." *Journal of Clinical Medicine* 4 (1): 127–49. doi: 10.3390/jcm4010127.

Roosbroeck, Katrien Van, Luk Cox, Thomas Tousseyn, Idoya Lahortiga, Olga Gielen, Barbara Cauwelier, Pascale De Paepe, et al. 2011. "JAK2 Rearrangements, Including the Novel SEC31A-JAK2 Fusion, are Recurrent in Classical Hodgkin Lymphoma." *Blood* 117 (15): 4056–64. doi: 10.1182/blood-2010-06-291310.

Rosenwald, Andreas, George Wright, Karen Leroy, Yu Xin, Randy D. Philippe Gaulard, Wing C. Chan Gascoyne, et al. 2003. "Molecular Diagnosis of Primary Mediastinal B Cell Lymphoma Identifies a Clinically Favorable Subgroup of Diffuse Large B Cell Lymphoma Related to Hodgkin Lymphoma." *Journal of Experimental Medicine* 198 (6): 851–62. doi: 10.1084/jem.20031074.

Rui, Lixin, N. C. Tolga Emre, Michael J. Kruhlak, Hye-Jung Chung, Christian Steidl, Graham Slack, George W. Wright, et al. 2010. "Cooperative Epigenetic Modulation by Cancer Amplicon Genes." *Cancer Cell* 18 (6): 590–605. doi: 10.1016/j.ccr.2010.11.013.

Rumi, Elisa, Daniela Pietra, Virginia Ferretti, Thorsten Klampfl, Ashot S. Harutyunyan, Jelena D. Milosevic, Nicole C. C. Them, et al. 2014a. "JAK2 or CALR Mutation Status Defines Subtypes of Essential Thrombocythemia with Substantially Different Clinical Course and Outcomes." *Blood* 123 (10): 1544–51. doi: 10.1182/blood-2013-11-539098.

Rumi, Elisa, Daniela Pietra, Cristiana Pascutto, Paola Guglielmelli, Alejandra Martínez-Trillos, Ilaria Casetti, Dolors Colomer, et al. 2014b. "Clinical Effect of Driver Mutations of JAK2, CALR, or MPL in Primary Myelofibrosis." *Blood* 124 (7): 1062–9. doi: 10.1182/blood-2014-05-578435.

Russell, Lisa J., Melania Capasso, Inga Vater, Olivier A. Takashi Akasaka, Maria Bernard, Jose Calasanz, Thiruppavaii Chandrasekaran, et al. 2009. "Deregulated Expression of Cytokine Receptor Gene, CRLF2, Is Involved in Lymphoid Transformation in B-Cell Precursor Acute Lymphoblastic Leukemia." *Blood* 114 (13): 2688–98. doi: 10.1182/blood-2009-03-208397.

Saksena, Annapurna, Prerna Arora, G. R. Sethi Nita Khurana, and Tejinder Singh. 2014. "Paediatric Idiopathic Myelofibrosis." *Indian Journal of Hematology & Blood Transfusion: an Official Journal of Indian Society of Hematology and Blood Transfusion* 30 (Suppl 1): 363–5. doi: 10.1007/s12288-014-0412-2.

Sangkhae, Veena S., Leah Etheridge, Kenneth Kaushansky, and Ian S. Hitchcock. 2014. "The Thrombopoietin Receptor, MPL, Is Critical for Development of a JAK2V617F-Induced Myeloproliferative Neoplasm." *Blood* 124 (26): 3956–63. doi: 10.1182/blood-2014-07-587238.

Santos, Nuno R. dos, David S. Rickman, Aurélien de Reynies, Françoise Cormier, Maryvonne Williame, Camille Blanchard, Marc-Henri Stern, and Jacques Ghysdael. 2007. "Pre-TCR Expression Cooperates with TEL-JAK2 to Transform Immature Thymocytes and Induce T-Cell Leukemia." *Blood* 109 (9): 3972–81. doi: 10.1182/blood-2006-09-048801.

Sato, Tomohiko, Tsutomu Toki, Rika Kanezaki, Xu Gang, Kiminori Terui, Hirokazu Kanegane, Masayoshi Miura, et al. 2008. "Functional Analysis of JAK3 Mutations in Transient Myeloproliferative Disorder and Acute Megakaryoblastic Leukaemia Accompanying down Syndrome." *British Journal of Haematology* 141 (5): 681–8. doi: 10.1111/j.1365-2141.2008.07081.x.

Savage, Kerry J., Jeffery L. Stefano Monti, Giorgio Cattoretti Kutok, Donna Neuberg, Laurence de Leval, Paul Kurtin, et al. 2003. "The Molecular Signature of Mediastinal Large B-Cell Lymphoma

Differs from that of Other Diffuse Large B-Cell Lymphomas and Shares Features with Classical Hodgkin Lymphoma." *Blood* 102 (12): 3871–9. doi: 10.1182/blood-2003-06-1841.

Savino, A. M., J. Sarno, L. Trentin, M. Vieri, G. Fazio, M. Bardini, C. Bugarin, et al. 2017. "The Histone Deacetylase Inhibitor Givinostat (ITF2357) Exhibits Potent Anti-Tumor Activity against *CRLF2*-Rearranged BCP-ALL." *Leukemia* 31 (11): 2365–75. doi: 10.1038/leu.2017.93.

Scheeren, Ferenc A., Anja U. van Lent, Maho Nagasawa, Kees Weijer, Hergen Spits, Nicolas Legrand, and Bianca Blom. 2010. "Thymic Stromal Lymphopoietin Induces Early Human B-Cell Proliferation and Differentiation." *European Journal of Immunology* 40 (4): 955–65. doi: 10.1002/eji.200939419.

Schinnerl, Dagmar, Klaus Fortschegger, João R. Maximilian Kauer, M. Marchante, Monique L. Reinhard Kofler, Den Boer, and Sabine Strehl. 2015. "The Role of the Janus-Faced Transcription Factor PAX5-JAK2 in Acute Lymphoblastic Leukemia." *Blood* 125 (8): 1282–91. doi: 10.1182/blood-2014-04-570960.

Schmitz, Roland, George W. Wright, Calvin A. Da Wei Huang, James D. Johnson, James Q. Phelan, Sandrine Roulland Wang, et al. 2018. "Genetics and Pathogenesis of Diffuse Large B-Cell Lymphoma." *The New England Journal of Medicine* 378 (15): 1396–407. doi: 10.1056/NEJMoa1801445.

Schultz, K. R., A. Carroll, N. A. Heerema, W. P. Bowman, A. Aledo, W. B. Slayton, H. Sather, et al. 2014. "Long-Term Follow-up of Imatinib in Pediatric Philadelphia Chromosome-Positive Acute Lymphoblastic Leukemia: Children's Oncology Group Study AALL0031." *Leukemia* 28 (7): 1467–71. doi: 10.1038/leu.2014.30.

Schwaller, J., J. Frantsve, J. Aster, I. R. Williams, M. H. Tomasson, T. S. Ross, P. Peeters, et al. 1998. "Transformation of Hematopoietic Cell Lines to Growth-Factor Independence and Induction of a Fatal Myelo- and Lymphoproliferative Disease in Mice by Retrovirally Transduced TEL/JAK2 Fusion Genes." *The EMBO Journal* 17 (18): 5321–33. doi: 10.1093/emboj/17.18.5321.

Schwartzman, Omer, Angela Maria Savino, Michael Gombert, Chiara Palmi, Gunnar Cario, Martin Schrappe, Cornelia Eckert, et al. 2017. "Suppressors and Activators of JAK–STAT Signaling at Diagnosis and Relapse of Acute Lymphoblastic Leukemia in down Syndrome." *Proceedings of the National Academy of Sciences of the United States of America* 114 (20): E4030–39. doi: 10.1073/pnas.1702489114.

Scott, Linda M., Peter J. Campbell, E. Joanna Baxter, Tony Todd, Philip Stephens, Sarah Edkins, Michael R. Richard Wooster, Stratton P. Andrew Futreal, and Anthony R. Green. 2005. "The V617F JAK2 Mutation Is Uncommon in Cancers and in Myeloid Malignancies Other than the Classic Myeloproliferative Disorders." *Blood* 106 (8): 2920–1. doi: 10.1182/blood-2005-05-2087.

Scott, Linda M., Mike A. Scott, Peter J. Campbell, and Anthony R. Green. 2006. "Progenitors Homozygous for the V617F Mutation Occur in Most Patients with Polycythemia Vera, but Not Essential Thrombocythemia." *Blood* 108 (7): 2435–7. doi: 10.1182/blood-2006-04-018259.

Scott, Linda M., Ross L. Wei Tong, Mike A. Levine, Philip A. Scott, Michael R. Beer, Stratton P. Andrew Futreal, et al. 2007. "*JAK2* Exon 12 Mutations in Polycythemia Vera and Idiopathic Erythrocytosis." *New England Journal of Medicine* 356 (5): 459–68. doi: 10.1056/NEJMoa065202.

Sekiya, Yuko, Yusuke Okuno, Hideki Muramatsu, Olfat Ismael, Nozomu Kawashima, Atsushi Narita, Xinan Wang, et al. 2016. "JAK2, MPL and CALR Mutations in Children with Essential Thrombocythemia." *International Journal of Hematology* 104 (2): 266–7. doi: 10.1007/s12185-016-2022-2.

Semenza, Gregg L. 2009. "Involvement of Oxygen-Sensing Pathways in Physiologic and Pathologic Erythropoiesis." *Blood* 114 (10): 2015–19. doi: 10.1182/blood-2009-05-189985.

Senkevitch, Emilee, Li Wenqing, Julie A. Hixon, Sarah D. Caroline Andrews, Gary T. Cramer, Timothy Back Pauly, Kelli Czarra, and Scott K. Durum. 2018. "Inhibiting Janus Kinase 1 and BCL-2 to Treat T Cell Acute Lymphoblastic Leukemia with IL7-Rα Mutations." *Oncotarget* 9 (32): 22605–17. doi: 10.18632/oncotarget.25194.

Shah, Niketa, Ali Al-Ahmari, Arwa Al-Yamani, Lee Dupuis, Derek Stephens, and Johann Hitzler. 2009. "Outcome and Toxicity of Chemotherapy for Acute Lymphoblastic Leukemia in Children with Down Syndrome." *Pediatric Blood & Cancer* 52 (1): 14–19. doi: 10.1002/pbc.21737.

Shochat, Chen, Obul R. Noa Tal, Chiara Palmi Bandapalli, Ithamar Ganmore, Geertruy Te Kronnie, Gunnar Cario, et al. 2011. "Gain-of-Function Mutations in Interleukin-7 Receptor-α (IL7R) in

Childhood Acute Lymphoblastic Leukemias." *Journal of Experimental Medicine* 208 (5): 901–8. doi: 10.1084/jem.20110580.

Sidon, P., H. El Housni, B. Dessars, and P. Heimann. 2006. "The JAK2V617F Mutation Is Detectable at Very Low Level in Peripheral Blood of Healthy Donors." *Leukemia* 20 (9): 1622. doi: 10.1038/sj.leu.2404292.

Silva, M., C. Richard, A. Benito, C. Sanz, I. Olalla, and J. L. Fernández-Luna. 1998. "Expression of Bcl-x in Erythroid Precursors from Patients with Polycythemia Vera." *The New England Journal of Medicine* 338 (9): 564–71. doi: 10.1056/NEJM199802263380902.

Silvennoinen, Olli, and Stevan R. Hubbard. 2015. "Molecular Insights into Regulation of JAK2 in Myeloproliferative Neoplasms." *Blood* 125 (22): 3388–92. doi: 10.1182/blood-2015-01-621110.

Skoda, Radek C. 2010. "Hereditary Myeloproliferative Disorders." *Haematologica* 95 (1): 6–8. doi: 10.3324/haematol.2009.015941.

Slayton, William B., Kirk R. Schultz, John A. Kairalla, Meenakshi Devidas, Mi Xinlei, Michael A. Pulsipher, Bill H. Chang, et al. 2018. "Dasatinib Plus Intensive Chemotherapy in Children, Adolescents, and Young Adults with Philadelphia Chromosome-Positive Acute Lymphoblastic Leukemia: Results of Children's Oncology Group Trial AALL0622." *Journal of Clinical Oncology: Official Journal of the American Society of Clinical Oncology* 36 (22): 2306–14. doi: 10.1200/JCO.2017.76.7228.

Smith, Malcolm A., Nita L. Seibel, Sean F. Altekruse, A. G. Lynn, Danielle L. Ries, Maura O'Leary Melbert, Franklin O. Smith, and Gregory H. Reaman. 2010. "Outcomes for Children and Adolescents with Cancer: Challenges for the Twenty-First Century." *Journal of Clinical Oncology* 28 (15): 2625–34. doi: 10.1200/JCO.2009.27.0421.

Song, Tammy Linlin, Maarja-Liisa Nairismägi, Yurike Laurensia, Jing-Quan Lim, Jing Tan, Zhi-Mei Li, Wan-Lu Pang, et al. 2018. "Oncogenic Activation of STAT3 Pathway Drives PD-L1 Expression in Natural Killer/T Cell Lymphoma." *Blood*, January, blood-2018-01-829424. doi: 10.1182/blood-2018-01-829424.

Springuel, Lorraine, Tekla Hornakova, Elisabeth Losdyck, Fanny Lambert, Stefan N. Emilie Leroy, Elisabetta Flex Constantinescu, Marco Tartaglia, Laurent Knoops, and Jean-Christophe Renauld. 2014. "Cooperating JAK1 and JAK3 Mutants Increase Resistance to JAK Inhibitors." *Blood* 124 (26): 3924–31. doi: 10.1182/blood-2014-05-576652.

Staerk, Judith, Anders Kallin, Jean-Baptiste Demoulin, William Vainchenker, and Stefan N. Constantinescu. 2005. "JAK1 and Tyk2 Activation by the Homologous Polycythemia Vera JAK2 V617F Mutation CROSS-TALK WITH IGF1 RECEPTOR." *Journal of Biological Chemistry* 280 (51): 41893–9. doi: 10.1074/jbc.C500358200.

Staerk, Judith, Catherine Lacout, Steven O. Takeshi Sato, William Vainchenker Smith, and Stefan N. Constantinescu. 2006. "An Amphipathic Motif at the Transmembrane-Cytoplasmic Junction Prevents Autonomous Activation of the Thrombopoietin Receptor." *Blood* 107 (5): 1864–71. doi: 10.1182/blood-2005-06-2600.

Steensma, David P., Gordon W. Dewald, Terra L. Lasho, Heather L. Powell, Rebecca F. McClure, Ross L. Levine, D. Gary Gilliland, and Ayalew Tefferi. 2005. "The JAK2 V617F Activating Tyrosine Kinase Mutation Is an Infrequent Event in Both 'Atypical' Myeloproliferative Disorders and Myelodysplastic Syndromes." *Blood* 106 (4): 1207–9. doi: 10.1182/blood-2005-03-1183.

Steensma, D. P., R. F. McClure, J. E. Karp, A. Tefferi, T. L. Lasho, H. L. Powell, G. W. DeWald, and S. H. Kaufmann. 2006. "JAK2 V617F Is a Rare Finding in *De Novo* Acute Myeloid Leukemia, but STAT3 Activation Is Common and Remains Unexplained." *Leukemia* 20 (6): 971–8. doi: 10.1038/sj.leu.2404206.

Stevens, Alexandra M., Marcos J. Ruiz, Robert B. Gerbing, Todd A. Alonzo, Alan S. Gamis, and Michele S. Redell. 2015. "Ligand-Induced STAT3 Signaling Increases at Relapse and Is Associated with Outcome in Pediatric Acute Myeloid Leukemia: A Report from the Children's Oncology Group." *Haematologica* 100 (12): e496–500. doi: 10.3324/haematol.2015.131508.

Suryani, Santi, S. Lauryn, Richard C. Bracken, Keith C. Harvey, S. Sia Hernan Carol, I.-Ming Chen, Kathryn Evans, et al. 2015. "Evaluation of the In Vitro and In Vivo Efficacy of the JAK Inhibitor AZD1480 against JAK–Mutated Acute Lymphoblastic Leukemia." *Molecular Cancer Therapeutics* 14 (2): 364–74. doi: 10.1158/1535-7163.MCT-14-0647.

Swerdlow, S. H., E. Campo, N. L. Harris, E. S. Jaffe, S. A. Pileri, H. Stein, J. Thiele, and J. W. Vardiman. 2008. "WHO Classification of Tumours of Haematopoietic and Lymphoid Tissues." *IARC Publications.* http://publications.iarc.fr/Book-And-Report-Series/Who-Iarc-Classification-Of-Tumours/WHO-Classification-Of-Tumours-Of-Haematopoietic-And-Lymphoid-Tissues-2008.

Swerdlow, Steven H., Stefano A. Elias Campo, Nancy Pileri, Lee Harris, Harald Stein, Reiner Siebert, Ranjana Advani, et al. 2016. "The 2016 Revision of the World Health Organization Classification of Lymphoid Neoplasms." *Blood* 127 (20): 2375–90. doi: 10.1182/blood-2016-01-643569.

Szpurka, H., L. P. Gondek, S. R. Mohan, E. D. Hsi, K. S. Theil, and J. P. Maciejewski. 2009. "UPD1p Indicates the Presence of MPL W515L Mutation in RARS-T, a Mechanism Analogous to UPD9p and JAK2 V617F Mutation." *Leukemia* 23 (3): 610–14. doi: 10.1038/leu.2008.249.

Tasian, Sarah K., Deborah S. Albert Assad, Yining Du Hunter, and Mignon L. Loh. 2018. "A Phase 2 Study of Ruxolitinib with Chemotherapy in Children with Philadelphia Chromosome-Like Acute Lymphoblastic Leukemia (INCB18424-269/AALL1521): Dose-Finding Results from the Part 1 Safety Phase." *Blood* 132 (Suppl 1): 555. doi: 10.1182/blood-2018-99-110221.

Tasian, Sarah K., Michelle Y. Doral, Michael J. Borowitz, Brent L. Wood, I.-Ming Chen, Richard C. Harvey, Julie M. Gastier-Foster, et al. 2012. "Aberrant STAT5 and PI3K/MTOR Pathway Signaling Occurs in Human CRLF2-Rearranged B-Precursor Acute Lymphoblastic Leukemia." *Blood* 120 (4): 833–42. doi: 10.1182/blood-2011-12-389932.

Tasian, Sarah K., David T. Teachey, Li Yong, Feng Shen, Richard C. Harvey, I.-Ming Chen, Theresa Ryan, et al. 2017. "Potent Efficacy of Combined PI3K/MTOR and JAK or ABL Inhibition in Murine Xenograft Models of Ph-like Acute Lymphoblastic Leukemia." *Blood* 129 (2): 177–87. doi: 10.1182/blood-2016-05-707653.

Tefferi, A. 2010. "Novel Mutations and Their Functional and Clinical Relevance in Myeloproliferative Neoplasms: *JAK2, MPL, TET2, ASXL1, CBL, IDH* and *IKZF1*." *Leukemia* 24 (6): 1128–38. doi: 10.1038/leu.2010.69.

Tefferi, A., T. L. Lasho, C. M. Finke, R. A. Knudson, R. Ketterling, C. H. Hanson, M. Maffioli, D. Caramazza, F. Passamonti, and A. Pardanani. 2014a. "*CALR* Vs *JAK2* Vs *MPL*-Mutated or Triple-Negative Myelofibrosis: Clinical, Cytogenetic and Molecular Comparisons." *Leukemia* 28 (7): 1472–7. doi: 10.1038/leu.2014.3.

Tefferi, A., E. Rumi, G. Finazzi, H. Gisslinger, A. M. Vannucchi, F. Rodeghiero, M. L. Randi, et al. 2013. "Survival and Prognosis among 1545 Patients with Contemporary Polycythemia Vera: An International Study." *Leukemia* 27 (9): 1874–81. doi: 10.1038/leu.2013.163.

Tefferi, Ayalew. 2006. "Classification, Diagnosis and Management of Myeloproliferative Disorders in the JAK2V617F Era." *ASH Education Program Book* 2006 (1): 240–5. doi: 10.1182/asheducation-2006.1.240.

Tefferi, Ayalew, Juergen Thiele, Attilio Orazi, Hans Michael Kvasnicka, Curtis A. Tiziano Barbui, Giovanni Barosi Hanson, et al. 2007. "Proposals and Rationale for Revision of the World Health Organization Diagnostic Criteria for Polycythemia Vera, Essential Thrombocythemia, and Primary Myelofibrosis: Recommendations from an Ad Hoc International Expert Panel." *Blood* 110 (4): 1092–97. doi: 10.1182/blood-2007-04-083501.

Tefferi, Ayalew, Rakhee Vaidya, Domenica Caramazza, Christy Finke, Terra Lasho, and Animesh Pardanani. 2011. "Circulating Interleukin (IL)-8, IL-2R, IL-12, and IL-15 Levels Are Independently Prognostic in Primary Myelofibrosis: A Comprehensive Cytokine Profiling Study." *Journal of Clinical Oncology* 29 (10): 1356–63. doi: 10.1200/JCO.2010.32.9490.

Tefferi, Ayalew, and William Vainchenker. 2011. "Myeloproliferative Neoplasms: Molecular Pathophysiology, Essential Clinical Understanding, and Treatment Strategies." *Journal of Clinical Oncology* 29 (5): 573–82. doi: 10.1200/JCO.2010.29.8711.

Tefferi, Ayalew, Emnet A. Wassie, Paola Guglielmelli, Alem A. Naseema Gangat, Terra L. Belachew, Christy Finke Lasho, et al. 2014b. "Type 1 versus Type 2 Calreticulin Mutations in Essential Thrombocythemia: A Collaborative Study of 1027 Patients." *American Journal of Hematology* 89 (8): E121–4. doi: 10.1002/ajh.23743.

Teofili, Luciana, Tonia Cenci, Maurizio Martini, Sara Capodimonti, Lorenza Torti, Fiorina Giona, Angela Amendola, et al. 2009. "The Mutant JAK2V617F Allele Burden in Children with Essential Thrombocythemia." *British Journal of Haematology* 145 (3): 430–2. doi: 10.1111/j.1365-2141.2009.07591.x.

Teofili, Luciana, Robin Foà, Fiorina Giona, and Luigi Maria Larocca. 2008. "Childhood Polycythemia Vera and Essential Thrombocythemia: Does Their Pathogenesis Overlap with that of Adult Patients?" *Haematologica* 93 (2): 169–72. doi: 10.3324/haematol.12002.

Teofili, Luciana, Fiorina Giona, Maurizio Martini, Tonia Cenci, Francesco Guidi, Lorenza Torti, Giovanna Palumbo, Angela Amendola, Robin Foà, and Luigi M. Larocca. 2007a. "Markers of Myeloproliferative Diseases in Childhood Polycythemia Vera and Essential Thrombocythemia." *Journal of Clinical Oncology* 25 (9): 1048–53. doi: 10.1200/JCO.2006.08.6884.

Teofili, Luciana, Fiorina Giona, Maurizio Martini, Tonia Cenci, Francesco Guidi, Lorenza Torti, Giovanna Palumbo, Angela Amendola, Giuseppe Leone, et al. 2007b. "The Revised WHO Diagnostic Criteria for Ph-Negative Myeloproliferative Diseases are Not Appropriate for the Diagnostic Screening of Childhood Polycythemia Vera and Essential Thrombocythemia." *Blood* 110 (9): 3384–6. doi: 10.1182/blood-2007-06-094276.

Teofili, Luciana, Fiorina Giona, Lorenza Torti, Tonia Cenci, Bianca Maria Ricerca, Carlo Rumi, Vittorio Nunes, et al. 2010. "Hereditary Thrombocytosis Caused by MPLSer505Asn Is Associated with a High Thrombotic Risk, Splenomegaly and Progression to Bone Marrow Fibrosis." *Haematologica* 95 (1): 65–70. doi: 10.3324/haematol.2009.007542.

Teofili, Luciana, Maurizio Martini, Tonia Cenci, Giovanna Petrucci, Lorenza Torti, Sergio Storti, Francesco Guidi, Giuseppe Leone, and Luigi Maria Larocca. 2007c. "Different STAT-3 and STAT-5 Phosphorylation Discriminates among Ph-Negative Chronic Myeloproliferative Diseases and Is Independent of the V617F JAK-2 Mutation." *Blood* 110 (1): 354–9. doi: 10.1182/blood-2007-01-069237.

Thomson, A. B., and W. H. B. Wallace. 2002. "Treatment of Paediatric Hodgkin's Disease. A Balance of Risks." *European Journal of Cancer (Oxford, England: 1990)* 38 (4): 468–77.

Tiacci, Enrico, Erik Ladewig, Gianluca Schiavoni, Alex Penson, Elisabetta Fortini, Valentina Pettirossi, Yuchun Wang, et al. 2018. "Pervasive Mutations of JAK–STAT Pathway Genes in Classical Hodgkin Lymphoma." *Blood* 131 (22): 2454–65. doi: 10.1182/blood-2017-11-814913.

Tiedt, Ralph, Marta A. Hui Hao-Shen, Renate Looser Sobas, Stephan Dirnhofer, Jürg Schwaller, and Radek C. Skoda. 2008. "Ratio of Mutant JAK2-V617F to Wild-Type Jak2 Determines the MPD Phenotypes in Transgenic Mice." *Blood* 111 (8): 3931–40. doi: 10.1182/blood-2007-08-107748.

Tirado, Carlos A., Weina Chen, Lily Jun-shen Huang, Matthew C. Carrie Laborde, Federico Valdez Hiemenz, Ho Kevin, Naomi Winick, Zhenjun Lou, and Prasad Koduru. 2010. "Novel JAK2 Rearrangement Resulting from a t(9;22)(P24;Q11.2) In B-Acute Lymphoblastic Leukemia." *Leukemia Research* 34 (12): 1674–6. doi: 10.1016/j.leukres.2010.05.031.

Tomasson, Michael H., Zhifu Xiang, Richard Walgren, Yu Zhao, Yumi Kasai, Tracie Miner, Rhonda E. Ries, et al. 2008. "Somatic Mutations and Germline Sequence Variants in the Expressed Tyrosine Kinase Genes of Patients with De Novo Acute Myeloid Leukemia." *Blood* 111 (9): 4797–808. doi: 10.1182/blood-2007-09-113027.

Tortolani, P. J., J. A. Johnston, C. M. Bacon, D. W. McVicar, A. Shimosaka, D. Linnekin, D. L. Longo, and J. J. O'Shea. 1995. "Thrombopoietin Induces Tyrosine Phosphorylation and Activation of the Janus Kinase, JAK2." *Blood* 85 (12): 3444–51. www.bloodjournal.org/content/85/12/3444.

Tran, Thai Hoa, Marian H. Harris, Jonathan V. Nguyen, Traci M. Blonquist, Kristen E. Stevenson, Eileen Stonerock, Barbara L. Asselin, et al. 2018. "Prognostic Impact of Kinase-Activating Fusions and IKZF1 Deletions in Pediatric High-Risk B-Lineage Acute Lymphoblastic Leukemia." *Blood Advances* 2 (5): 529–33. doi: 10.1182/bloodadvances.2017014704.

Treon, Steven P., Xu Lian, Guang Yang, Yangsheng Zhou, Xia Liu, Yang Cao, Patricia Sheehy, et al. 2012. "MYD88 L265P Somatic Mutation in Waldenström's Macroglobulinemia." *New England Journal of Medicine* 367 (9): 826–33. doi: 10.1056/NEJMoa1200710.

Ugo, Valérie, Christophe Marzac, Irène Teyssandier, Frédéric Larbret, Yann Lécluse, Najet Debili, William Vainchenker, and Nicole Casadevall. 2004. "Multiple Signaling Pathways Are Involved in Erythropoietin-Independent Differentiation of Erythroid Progenitors in Polycythemia Vera." *Experimental Hematology* 32 (2): 179–87. doi: 10.1016/j.exphem.2003.11.003.

Ungureanu, Daniela, Wu Jinhua, Tuija Pekkala, Yashavanthi Niranjan, Ole N. Clifford Young, Chong-Feng Xu Jensen, et al. 2011. "The Pseudokinase Domain of JAK2 Is a Dual-Specificity Protein Kinase that Negatively Regulates Cytokine Signaling." *Nature Structural & Molecular Biology* 18 (9): 971–6. doi: 10.1038/nsmb.2099.

Ungureanu, Daniela, Pipsa Saharinen, Ilkka Junttila, Douglas J. Hilton, and Olli Silvennoinen. 2002. "Regulation of Jak2 through the Ubiquitin-Proteasome Pathway Involves Phosphorylation of Jak2 on Y1007 and Interaction with SOCS-1." *Molecular and Cellular Biology* 22 (10): 3316–26.

Vainchenker, William, Emilie Leroy, Laure Gilles, Caroline Marty, Isabelle Plo, and Stefan N. Constantinescu. 2018. "JAK Inhibitors for the Treatment of Myeloproliferative Neoplasms and Other Disorders." *F1000Research* 7 (January): 82. doi: 10.12688/f1000research.13167.1.

Vannucchi, Alessandro M., Elisabetta Antonioli, Paola Guglielmelli, Alessandro Pancrazzi, Vittoria Guerini, Giovanni Barosi, Marco Ruggeri, et al. 2008. "Characteristics and Clinical Correlates of MPL 515W>L/K Mutation in Essential Thrombocythemia." *Blood* 112 (3): 844–7. doi: 10.1182/blood-2008-01-135897.

Vannucchi, Alessandro M., Jean Jacques Kiladjian, Martin Griesshammer, Tamas Masszi, Simon Durrant, Francesco Passamonti, Claire N. Harrison, et al. 2015. "Ruxolitinib versus Standard Therapy for the Treatment of Polycythemia Vera." *The New England Journal of Medicine* 372 (5): 426–35. doi: 10.1056/NEJMoa1409002.

Vardiman, James W., Nancy Lee Harris, and Richard D. Brunning. 2002. "The World Health Organization (WHO) Classification of the Myeloid Neoplasms." *Blood* 100 (7): 2292–302. doi: 10.1182/blood-2002-04-1199.

Velusamy, Thirunavukkarasu, Mark J. Kiel, Anagh A. Sahasrabuddhe, Delphine Rolland, Catherine A. Dixon, Nathanael G. Bailey, Bryan L. Betz, et al. 2014. "A Novel Recurrent NPM1-TYK2 Gene Fusion in Cutaneous CD30-Positive Lymphoproliferative Disorders." *Blood* 124 (25): 3768–71. doi: 10.1182/blood-2014-07-588434.

Verstovsek, Srdan, Ruben A. Hagop Kantarjian, Animesh D. Mesa, Jorge Pardanani, Deborah A. Cortes-Franco, Zeev Estrov Thomas, et al. 2010. "Safety and Efficacy of INCB018424, a JAK1 and JAK2 Inhibitor, in Myelofibrosis." *The New England Journal of Medicine* 363 (12): 1117–27. doi: 10.1056/NEJMoa1002028.

Verstovsek, Srdan, Francesco Passamonti, Alessandro Rambaldi, Giovanni Barosi, Elisa Rumi, Elisabetta Gattoni, Lisa Pieri, et al. 2017. "Ruxolitinib for Essential Thrombocythemia Refractory to or Intolerant of Hydroxyurea: Long-Term Phase 2 Study Results." *Blood* 130 (15): 1768–71. doi: 10.1182/blood-2017-02-765032.

Vesely, C., C. Frech, C. Eckert, G. Cario, A. Mecklenbräuker, U. Zur Stadt, K. Nebral, et al. 2017. "Genomic and Transcriptional Landscape of *P2RY8-CRLF2*-Positive Childhood Acute Lymphoblastic Leukemia." *Leukemia* 31 (7): 1491–501. doi: 10.1038/leu.2016.365.

Vita, Serena De, Claire Mulligan, Suzanne McElwaine, Franca Dagna-Bricarelli, Monica Spinelli, Giuseppe Basso, Dean Nizetic, and Jürgen Groet. 2007. "Loss-of-Function JAK3 Mutations in TMD and AMKL of down Syndrome." *British Journal of Haematology* 137 (4): 337–41. doi: 10.1111/j.1365-2141.2007.06574.x.

Walz, Christoph, Wesam Ahmed, Katherine Lazarides, Monica Betancur, Nihal Patel, Lothar Hennighausen, Virginia M. Zaleskas, and Richard A. Van Etten. 2012. "Essential Role for Stat5a/b in Myeloproliferative Neoplasms Induced by BCR-ABL1 and JAK2V617F in Mice." *Blood* 119 (15): 3550–60. doi: 10.1182/blood-2011-12-397554.

Weigert, Oliver, Andrew A. Lane, Liat Bird, Nadja Kopp, Bjoern Chapuy, Diederik van Bodegom, Angela V. Toms, et al. 2012. "Genetic Resistance to JAK2 Enzymatic Inhibitors Is Overcome by HSP90 Inhibition." *Journal of Experimental Medicine* 209 (2): 259–73. doi: 10.1084/jem.20111694.

Weniger, M. A., I. Melzner, C. K. Menz, S. Wegener, A. J. Bucur, K. Dorsch, T. Mattfeldt, T. F. E. Barth, and P. Möller. 2006. "Mutations of the Tumor Suppressor Gene *SOCS-1* in Classical Hodgkin Lymphoma Are Frequent and Associated with Nuclear Phospho-STAT5 Accumulation." *Oncogene* 25 (18): 2679–84. doi: 10.1038/sj.onc.1209151.

Wernig, Gerlinde, Thomas Mercher, Ross L. Rachel Okabe, Benjamin H. Lee Levine, and D. Gary Gilliland. 2006. "Expression of Jak2V617F Causes a Polycythemia Vera–Like Disease with Associated Myelofibrosis in a Murine Bone Marrow Transplant Model." *Blood* 107 (11): 4274–81. doi: 10.1182/blood-2005-12-4824.

Whitlock, James A., Harland N. Sather, Leslie L. Paul Gaynon, Robert J. Robison, Michael Trigg Wells, Nyla A. Heerema, and Smita Bhatia. 2005. "Clinical Characteristics and Outcome of Children with

Down Syndrome and Acute Lymphoblastic Leukemia: A Children's Cancer Group Study." *Blood* 106 (13): 4043–9. doi: 10.1182/blood-2003-10-3446.

Winter, Stuart S., Kimberly P. Dunsmore, Brent L. Meenakshi Devidas, Natia Esiashvili Wood, Zhiguo Chen, Nancy Eisenberg, et al. 2018. "Improved Survival for Children and Young Adults with T-Lineage Acute Lymphoblastic Leukemia: Results from the Children's Oncology Group AALL0434 Methotrexate Randomization." *Journal of Clinical Oncology: Official Journal of the American Society of Clinical Oncology* 36 (29): 2926–34. doi: 10.1200/JCO.2018.77.7250.

Wu, David, Bethany Dutra, Neal Lindeman, Hidenobu Takahashi, Nancy L. Kunihiko Takeyama, Geraldine S. Harris, Janina Longtine Pinkus, Margaret Shipp, and Jeffery L. Kutok. 2009. "No Evidence for the JAK2 (V617F) or JAK2 Exon 12 Mutations in Primary Mediastinal Large B-Cell Lymphoma." *Diagnostic Molecular Pathology* 18 (3): 144–9. doi: 10.1097/PDM.0b013e3181855c7f.

Wu, Shuo-Chieh, Loretta S. Li, Nadja Kopp, Joan Montero, Bjoern Chapuy, Akinori Yoda, Amanda L. Christie, et al. 2015. "Activity of the Type II JAK2 Inhibitor CHZ868 in B Cell Acute Lymphoblastic Leukemia." *Cancer Cell* 28 (1): 29–41. doi: 10.1016/j.ccell.2015.06.005.

Xia, Z., M.R. Baer, A.W. Block, H. Baumann, and M. Wetzler. 1998. "Expression of Signal Transducers and Activators of Transcription Proteins in Acute Myeloid Leukemia Blasts." *Cancer Research* 58 (14): 3173–80.

Xiang, Zhifu, Yu Zhao, Daved H. Vesselin Mitaksov, Yumi Kasai Fremont, AnnaLynn Molitoris, Rhonda E. Ries, et al. 2008. "Identification of Somatic JAK1 Mutations in Patients with Acute Myeloid Leukemia." *Blood* 111 (9): 4809–12. doi: 10.1182/blood-2007-05-090308.

Xu, Xuesong, Qi Zhang, Jian Luo, Shu Xing, Li Qingshan, Sanford B. Krantz, Fu Xueqi, and Zhizhuang Joe Zhao. 2007. "JAK2V617F: Prevalence in a Large Chinese Hospital Population." *Blood* 109 (1): 339–42. doi: 10.1182/blood-2006-03-009472.

Yamamoto, Yukiya, Hitoshi Kiyoi, Yasuyuki Nakano, Ritsuro Suzuki, Yoshihisa Kodera, Shuichi Miyawaki, Norio Asou, et al. 2001. "Activating Mutation of D835 within the Activation Loop of FLT3 in Human Hematologic Malignancies." *Blood* 97 (8): 2434–9. doi: 10.1182/blood.V97.8.2434.

Yan, Dongqing, Robert E. Hutchison, and Golam Mohi. 2012. "Critical Requirement for Stat5 in a Mouse Model of Polycythemia Vera." *Blood* 119 (15): 3539–49. doi: 10.1182/blood-2011-03-345215.

Yano, Mio, Toshihiko Imamura, Daisuke Asai, Nobutaka Kiyokawa, Kazuhiko Nakabayashi, Kenji Matsumoto, Takao Deguchi, et al. 2015. "Identification of Novel Kinase Fusion Transcripts in Paediatric B Cell Precursor Acute Lymphoblastic Leukaemia with IKZF1 Deletion." *British Journal of Haematology* 171 (5): 813–17. doi: 10.1111/bjh.13757.

Yildiz, Mehmet, Li Hongxiu, Nisar A. Denzil Bernard, Peter Ouillette Amin, Siân Jones, Kamlai Saiya-Cork, et al. 2015. "Activating STAT6 Mutations in Follicular Lymphoma." *Blood* 125 (4): 668–79. doi: 10.1182/blood-2014-06-582650.

Yoda, Akinori, Yuka Yoda, Sabina Chiaretti, Michal Bar-Natan, Kartik Mani, Scott J. Rodig, Nathan West, et al. 2010. "Functional Screening Identifies CRLF2 in Precursor B-Cell Acute Lymphoblastic Leukemia." *Proceedings of the National Academy of Sciences* 107 (1): 252–7. doi: 10.1073/pnas.0911726107.

Younes, Anas, Jorge Romaguera, Michelle Fanale, Peter McLaughlin, Frederick Hagemeister, Amanda Copeland, Sattva Neelapu, et al. 2012. "Phase I Study of a Novel Oral Janus Kinase 2 Inhibitor, SB1518, in Patients with Relapsed Lymphoma: Evidence of Clinical and Biologic Activity in Multiple Lymphoma Subtypes." *Journal of Clinical Oncology* 30 (33): 4161–7. doi: 10.1200/JCO.2012.42.5223.

Zahn, Malena, Ralf Marienfeld, Ingo Melzner, Janine Heinrich, Benjamin Renner, Silke Wegener, Anna Mießner, et al. 2017. "A Novel PTPN1 Splice Variant Upregulates JAK/STAT Activity in Classical Hodgkin Lymphoma Cells." *Blood* 129 (11): 1480–90. doi: 10.1182/blood-2016-06-720516.

Zeller, Bernward, Göran Gustafsson, Erik Forestier, Jonas Abrahamsson, Niels Clausen, Jesper Heldrup, Liisa Hovi, et al. 2005. "Acute Leukaemia in Children with down Syndrome: A Population-Based Nordic Study." *British Journal of Haematology* 128 (6): 797–804. doi: 10.1111/j.1365-2141.2005.05398.x.

Zenatti, Priscila P., Daniel Ribeiro, Li Wenqing, Milene C. Linda Zuurbier, Maddalena Paganin Silva, Julia Tritapoe, et al. 2011. "Oncogenic *IL7R* Gain-of-Function Mutations in Childhood T-Cell Acute Lymphoblastic Leukemia." *Nature Genetics* 43 (10): 932–9. doi: 10.1038/ng.924.
Zhao, Runxiang, Shu Xing, Li Zhe, Fu Xueqi, Li Qingshan, Sanford B. Krantz, and Zhizhuang Joe Zhao. 2005. "Identification of an Acquired JAK2 Mutation in Polycythemia Vera." *Journal of Biological Chemistry* 280 (24): 22788–92. doi: 10.1074/jbc.C500138200.

12

Aberrant JAK-STAT Signaling in Hematopoietic Malignancies

Parvis Sadjadian
University Clinic for Hematology, Oncology, Hemostaseology and Palliative Care
Johannes Wesling Medical Center Minden, University of Bochum
Minden, Germany

12.1 Introduction

Many functions of the cells of the lympho-hematopoetic system like proliferation, differentiation, and survival are regulated via cytokines and interferons (Commins et al. 2010). As most cytokines lack intrinsic kinase activity, signaling is enabled using Janus kinases (JAK), a family of four cytoplasmic non-receptor kinases (JAK1, JAK2, JAK3, TYK2). Downstream, JAK phosphorylates cytoplasmic proteins called STAT (signal transducers and activators of transcription), a group of seven (STAT1, STAT2, STAT3, STAT4, STAT5A, STAT5B and STAT6) transcription factors, which after activation dimerize and translocate to the nucleus to bind DNA for regulation of gene expression and initiation of transcription (Figure 12.1). The JAK-STAT pathway is an extremely fast membrane-to-nucleus signaling system, whereby extracellular factors control gene expression. Interestingly, STAT activity in the nucleus can be detected only minutes after cytokine binding (Leonard and O'Shea 1998). JAK-mediated signaling is essential for both short-chain cytokines (like IL-2, IL-3, IL-4, IL-5, GM-CSF, IL-7, IL-9, IL-13, IL-15) and long-chain cytokines (like IL-6, IL-11, OSM, CNTF, CT-1, growth hormone, prolactin, erythropoietin, thrombopoietin) (Leonard and O'Shea 1998). Distinct cytokines and their associated receptors use selective JAKs and STATs for signaling, explaining their different in vivo roles (O'Shea et al. 2015).

More than 25 years ago, the JAK-STAT pathway was first described investigating interferon signaling (Velazquez et al. 1992, Darnell et al. 1994). Enabling cytokine-regulated cell proliferation, differentiation and survival, it soon became clear that this pathway has a critical role in tumorigenesis (Bromberg et al. 1999, Levy and Darnell 2002), both in solid tumors and hematologic malignancies.

The JAK-STAT pathway can be pathologically activated by several mechanisms like autocrine/paracrine cytokine production, mutations in upstream oncogenes (like *BCR-ABL* in CML or *FLT3* in AML), or rare activating mutations of STATs. This overview will primarily focus on mutations of JAKs leading to various hematopoetic malignancies. In addition, therapeutic options to target the JAK-STAT pathway will be elucidated, with both experimental approaches and therapies already implemented in clinical routine.

12.2 Dysregulation of the JAK-STAT Pathway in Hematological Malignancies

Alterations of the JAK-STAT pathway are of relevance in hematopoietic malignancies (Benekli et al. 2003). For example, oncogenic JAK2 rearrangements with multiple fusion gene partners have been reported in a variety of hematologic malignancies (Hammaren et al. 2019). However, these

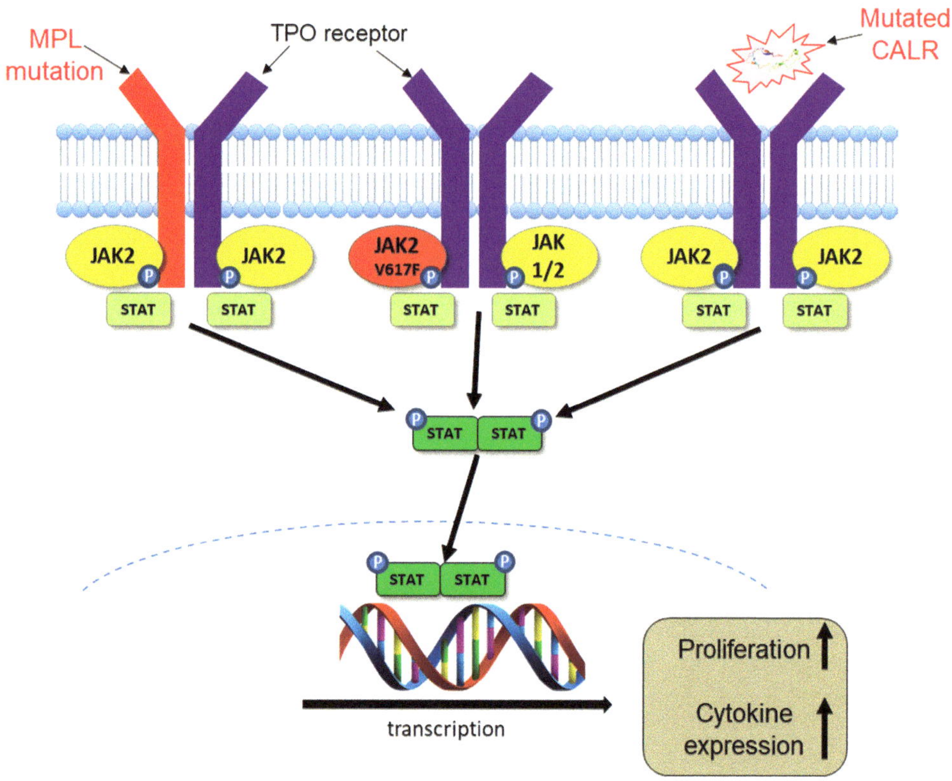

FIGURE 12.1 Illustration of the role of driver mutations in the pathogenesis of MPN. Left: Activation mutation in the thrombopoetin receptor. Middle: JAK2 (V617F or exon12) gain of function mutation. Right: Calreticulin mutation (TPO: thrombopoietin; JAK2: Janus kinase 2; CALR: calreticulin; MPL: myeloproliferative leukemia virus oncogene; STAT: signal transducer and activator of transcription).

translocations are rare: the Mayo Clinic cytogenetics database spanning the years 1989–2008 with 24,262 patients showed JAK2 translocations in only 0.06% of hematopoietic neoplasms (Patnaik et al. 2010). In contrast, *JAK2 V617F* mutations in exon 14 of JAK2 is prevalent in 50–60% of patients with MPNs (in polycythemia vera (PV) even 95%), as compared to 0.1% in the general population (Xu et al. 2007, Nielsen et al. 2014).

In the following sections, the impact of JAK-STAT alterations on the pathogenesis, course, and prognosis of major myeloid and lymphoid malignancies are discussed.

12.2.1 Myeloid Malignancies

12.2.1.1 Myeloproliferative Neoplasms

The classical *BCR-ABL*-negative MPN essential thrombocythemia (ET), PV, and primary myelofibrosis (PMF) are closely related stem-cell-derived disorders characterized by increased proliferation of erythroid, megakaryocytic, or granulocytic cells in the absence of dysplasia (Dameshek 1951). The genetic basis of relationship was discovered in 2005 with the identification of the acquired mutation in exon 14 of *JAK2* (*V617F*, a substitution of valine to phenylalanine at codon 617), which can be found in almost all patients with PV and more than 50% of patients with ET or PMF (Baxter et al. 2005, James et al. 2005, Kralovics et al. 2005, Levine et al. 2005b). *JAK2 V617F* is a gain-of-function mutation in the JH2 (JAK homology 2) pseudokinase domain, a region that inhibits JAK2 kinase activity (Saharinen and Silvennoinen 2002, Saharinen et al. 2003).

Subsequently, other driver mutations in calreticulin (CALR, an endoplasmic reticulin chaperone) or myeloproliferative leukemia virus oncogene (MPL, the thrombopoietin receptor) were identified (Chaligne et al. 2007, Klampfl et al. 2013, Nangalia et al. 2013, Pardanani et al. 2006). All lead to a deregulated JAK-STAT pathway via increased JAK2 signaling (Figure 12.1), resulting in abnormally high circulating levels of pro-inflammatory cytokines (Rampal et al. 2014, Skoda et al. 2015).

Within the three classical MPNs, the frequency of driver mutations is varying (Figure 12.2) (Griesshammer and Sadjadian 2017). Interestingly, even in rare MPN cases not carrying one of the three driver mutations (approximately 10% of PMF and 15% of ET cases), there is an overactive JAK-STAT pathway, too (Skoda et al. 2015). Thus, JAK-inhibiting strategies emerged as a promising concept of treating MPN soon after the discovery of the *JAK2 V617F* point mutation in 2005 (Griesshammer and Sadjadian 2017).

12.2.1.2 Acute Myeloid Leukemia

Acute leukemias—both acute myeloid leukemia (AML) and acute lymphoblastic leukemia (ALL)—are characterized by a differentiation block causing the accumulation of immature blast cells in the bone marrow and/or peripheral blood.

Disturbances of the JAK-STAT pathway are uncommon in AML. In a larger series of genotypic analyses of patients with hematopoietic malignancies only four of 222 patients (1.8%) with AML were positive for *JAK2 V617F* mutations, three of whom had been previously diagnosed with MPN (Levine et al. 2005b). A group from Cambridge reported *JAK2 V617F* mutations in five of 90 (4%) primary hematopoietic cells (Scott et al. 2005). Steesma et al. found *JAK2 V617F* mutations in 13 of 162 AML samples (8%), but 10 of 13 were secondary post MPN AML patients and only three were de novo AML patients (Steensma et al. 2006).

Several analogous JAK2 gene fusions have been described in AML (Murati et al. 2005, Reiter et al. 2005). The mechanism of JAK2 activation is assumed to be similar in all cases, with the JAK2 fusion partner promoting dimerization and constitutive activation of the JAK2 tyrosine kinase (Chen et al. 2012).

Interestingly, in a study of 77 AML patients, the detection of high levels of activated JAK2 (by immunohistochemistry using a phosphor specific antibody against JAK2) predicted a poor outcome (Ikezoe et al. 2011).

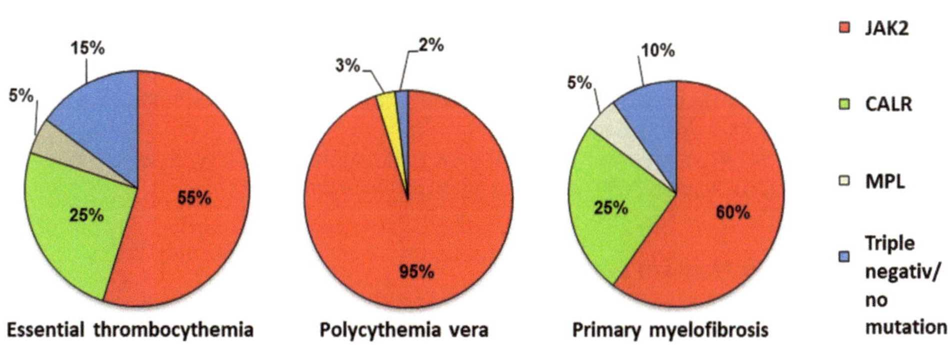

FIGURE 12.2 Frequencies of driver mutations JAK2 exon12 and exon14, CALR and MPL in ET, PV and PMF. In PMF and ET about 10–15% do not have any driver mutation (so-called triple negative), in PV around 2% do not carry a JAK2 (neither exon 12 nor exon 14) mutation.

12.2.1.3 Myelodysplastic Syndrome

Myelodysplastic syndromes (MDS) are a group of clonal disorders of the hematopoietic stem cell characterized by dysplasia of bone marrow and peripheral blood cells, hematopoietic insufficiency (mostly anemia), and an elevated risk of transformation to AML.

JAK2 mutations also rare in MDS, only two of 48 patients (4.2%) with MDS were tested positive for *JAK2 V617F* mutation (Levine et al. 2005b). Interestingly, in MDS/MPN with ring sideroblasts and thrombocytosis (formerly: refractory anemia with ring sideroblasts and thrombocytosis, RARS-T), a myeloid neoplasm showing clinical and morphological features that overlap between MDS and MPN [study from Arber et al. 2016], *JAK2 V617F* mutations could be found in 54 of 93 patients (58%) (Wardrop and Steensma 2009, Arber et al. 2016).

12.2.1.4 Chronic Myelomonocytic Leukemia

Chronic myelomonocytic leukemia (CMML), one of the MDS/MPN overlap syndromes, is characterized by peripheral blood monocytosis, bone marrow dysplasia with transfusion-dependent cytopenias, and an elevated risk of progression to AML. Typically, CMML is dependent on aberrant signaling through the granulocyte-macrophage colony-stimulating factor pathway, which requires JAK-STAT (Padron et al. 2013, Bose and Verstovsek 2017). Acknowledging the myeloproliferative component of the disease makes JAK-inhibition a reasonable approach (Solary 2016). A phase I trial of ruxolitinib in CMML showed encouraging results with hematological or spleen responses in seven of 20 patients (35%) (Padron et al. 2016).

12.2.2 Lymphoid Malignancies

Activation of the JAK-STAT pathway by cytokines such as IL-2, IL-4, IL-7, IL-9, IL-15, and IL-21 is essential in the formation and function of lymphoid cells like T-cells, B-cells, and NK-cells (Liao et al. 2011). Receptor signaling (JAK1 via the common cytokine-binding chains, JAK3 via the common γ-chain) activates typically STAT3 and STAT5 (Vainchenker and Constantinescu 2013).

12.2.2.1 Acute Lymphoblastic Leukemia

Acute lymphoblastic leukemia was the first malignancy that was associated with JAK-signaling pathway disturbances. In 1997, Lacronique et al. discovered a TEL-JAK2 fusion gene as a product of a t(9;12)(p24;p13) translocation in a patient with a T-ALL (Lacronique et al. 1997). Subsequently, other JAK2 fusion proteins with different fusion partners like PCM1-JAK2, PAX5-JAK2 or SSBP2-JAK2 were discovered in ALL, all with the same mechanism of JAK2-activation via facilitating dimerization and constitutive activation of the JAK2 tyrosine kinase component (Reiter et al. 2005, Poitras et al. 2008, Nebral et al. 2009).

Sequencing of exon14 of the *JAK2* gene in patients with Down syndrome-associated B-progenitor ALL revealed acquired somatic *JAK2* mutations (R683G being the most common) in 18–28%, also in 7% of high-risk pediatric B-ALL patients irrespective of having Down syndrome (Bercovich et al. 2008, Kearney et al. 2009, Mullighan et al. 2009).

JAK3 mutations in ALL are rare; however, JAK1 mutations have been reported in 4–18% of patients with T-ALL (Flex et al. 2008, Asnafi et al. 2010).

12.2.2.2 B-Cell Non-Hodgkin Lymphoma

Non-Hodgkin lymphomas (NHL) such as diffuse large B-cell lymphoma (DLBCL), follicular lymphoma, marginal zone lymphoma, and mantle cell lymphoma are a heterogeneous group of malignant lymphoproliferative disorders predominantly originating in B-lymphocytes.

In activated B-cell-like diffuse large B-cell lymphoma (ABC-DLBCL), one of the two phenotypic subtypes, high-level STAT3 expression and activation can be detected (Ding et al. 2008).

Activation of STAT3 and its kinase, JAK1, is caused by autocrine production of IL-6 and IL-10. STAT3 regulates multiple oncogenic signaling pathways, including NF-κB, a cell-cycle checkpoint, PI3K/AKT/mTORC1, and STAT3 itself. In addition, STAT3 negatively regulates the lethal type I IFN signaling pathway (Lu et al. 2018). In DLBCL, STAT3 activity is associated with a poor prognosis (Huang et al. 2013, Ok et al. 2014).

12.2.2.3 Chronic Lymphocytic Leukemia

Chronic lymphocytic leukemia (CLL) is an indolent B-cell non-Hodgkin lymphoma characterized by a leukemic course of disease, representing the most common form of adult leukemia in western countries.

The *JAK2 V617F*-mutation is not present in CLL, but constitutive phosphorylation and activation of STATs has been described (Frank et al. 1997, Levine et al. 2005a).

12.2.2.4 Hodgkin Lymphoma

Hodgkin lymphoma is a rare malignant disease involving lymph nodes and the lymphatic system. Biopsies typically show a limited number of the malignant Reed-Sternberg cells surrounded by numerous reactive bystander cells.

The JAK/STAT pathway is an essential regulator of Hodgkin lymphoma cell viability, and JAK-mediated signaling has been demonstrated in a significant number of patients (Baus and Pfitzner 2006, Navarro et al. 2009). In Hodgkin's lymphoma, JAK-STAT is primarily activated by an aberrant deregulation of a network of cytokines and chemokines in the microenvironment (Derenzini and Younes 2013).

JAK2 amplifications and gain-of-function mutations have been observed in up to a third of patients with Hodgkin's lymphoma. The same alterations were found in 30–50% of patients with primary mediastinal B-cell lymphoma (Joos et al. 2000, Rosenwald et al. 2003, Green et al. 2010).

JAK2 rearrangements occur in a minority of classical Hodgkin lymphoma cases; screening of 131 patients identified one case with a SEC31A-JAK2 rearrangement and two additional cases with rearrangements involving JAK2 (Van Roosbroeck et al. 2011).

12.2.2.5 Multiple Myeloma

Multiple myeloma is a heterogeneous hematologic neoplasia characterized by monoclonal proliferation and accumulation of malignant plasma cells in the bone marrow leading to bone marrow insufficiency, bone destruction, and increased production of complete and incomplete immunoglobulins.

The therapeutic armamentarium in this disease, which accounts for approximately 10% of all hematologic malignancies, has dramatically improved over the past years. However, most of these patients eventually relapse. One of the pathways involved in the pathogenesis is the JAK-STAT3 axis via STAT3 activation by several cytokines including IL-6, which can be secreted by bone marrow stromal cells or myeloma cells themselves (Mitsiades et al. 2006, Podar et al. 2009, Sansone and Bromberg 2012, Hu and Hu 2018). Preclinical experiments demonstrated decreased survival of primary myeloma cells or myeloma cell lines using a STAT3 inhibitor (Nelson et al. 2008). Combining the JAK1/2 inhibitor ruxolitinib with the proteasome inhibitor bortezomib, one of the standard drugs in myeloma treatment, showed a decrease of many JAK-STAT pathway genes (de Oliveira et al. 2017).

12.2.3 Non-Malignant Hematologic Disorders: Immune Thrombocytopenic Purpura (Idiopathic Thrombocytopenia, ITP)

ITP is an acquired form of thrombocytopenia induced by an autoimmune reaction targeting thrombocytes and megakaryocytes. The underlying cause is mostly unknown and secondary forms (e.g. drug-related) account for approximately 20% (Moulis et al. 2014).

A recent publication from China elucidated the role of the JAK-STAT pathway in the pathogenesis of immune thrombocytopenia (ITP) (Zhang et al. 2018). In total, 26 patients with confirmed ITP were compared with 24 healthy controls concerning the expression of JAK3, pJAK3 mRNA, STAT3, and pSTAT3. In addition, changes in levels of *IL21* mRNA, IL-21 cell secretion after dexamethasone intervention and AG490 (a specific and potent inhibitor of JAK2) blockade were measured. Compared with the healthy controls, patients with ITP had a significantly higher expression of JAK3, pJAK3 mRNA, STAT3, and pSTAT3 protein, which were significantly lower after blocking of AG490. The expression of IL-21 mRNA and the secretion of IL-21 decreased after Dexamethasone intervention, but significantly increased after AG490 blocking. These results suggest that the IL-21 activated JAK-STAT signaling pathway may play a role both in the pathogenesis and in the regulation of ITP (Zhang et al. 2018).

Therefore, activation of the JAK-STAT pathway could be used to treat ITP (Evangelista et al. 2007, Garnock-Jones and Keam 2009, Zhang and Kolesar 2011, Kuter 2013): eltrombopag, that activates the JAK-STAT pathway as a thrombopoietin (TPO) non-peptide mimetic, has been evaluated in clinical use and was approved by the FDA (2008) and EMA (2010) to treat patients with ITP who have had an insufficient response to corticosteroids or immunoglobulin therapy.

12.3 Therapeutic Implications

12.3.1 JAK Inhibitors in the Treatment of Hematopoietic Malignancies

12.3.1.1 Ruxolitinib

Ruxolitinib (INCB018424/INC424) is a potent and selective oral inhibitor of JAK 1 and JAK2. Currently, Ruxolitinib is licensed for the treatment of myelofibrosis by the US Food and Drug Administration (FDA) since 2011 and by the European Medicines Agency (EMA) since 2012, followed by the approval for the treatment of hydroxyurea-resistant or -intolerant PV in 2014.

Approval of ruxolitinib for myelofibrosis was based on the two pivotal phase III trials COMFORT I (ruxolitinib vs. placebo) and COMFORT-II (ruxolitinib vs. best available therapy) (Harrison et al. 2012, Verstovsek et al. 2012). In both COMFORT trials, crossover was allowed. Ruxolitinib was superior in reducing spleen size, improving myelofibrosis-associated symptoms and quality of life. The most frequent hematological side effects of ruxolitinib treatment were dose-dependent anemia and thrombocytopenia. Long-term follow-up demonstrated an on-going preserved treatment effect of ruxolitinib (Harrison et al. 2016). In pooled analyses of COMFORT-1 and COMFORT-2, a prolonged overall survival of ruxolitinib treated patients was suggested compared with controls (Verstovsek et al. 2017).

In PV, ruxolitinib was licensed based on the RESPONSE-1 and RESPONSE-2 phase III trials (Vannucchi 2015, Verstovsek et al. 2016, Passamonti et al. 2017, Griesshammer et al. 2018). In PV patients who had an inadequate response to or unacceptable side effects from previous hydroxyurea treatment, ruxolitinib was superior at controlling hematocrit levels and the number of phlebotomy-free patients was doubled. Additionally, ruxolitinib led to an improvement both in symptom burden and quality of life (Passamonti et al. 2017).

Ruxolitinib has been evaluated in various hematologic indications, albeit clinical data are limited. In a phase II trial, 38 patients with refractory leukemia (12 with *JAK2 V617F* mutation) received ruxolitinib (Eghtedar et al. 2012). The anti-leukemic potential of ruxolitinib was modest, and three patients showed a significant response (all of which with a secondary post-MPN AML).

As there is a strong rationale for targeting JAK-STAT signaling, ruxolitinib was evaluated in relapsed/refractory Hodgkin lymphoma (Baus and Pfitzner 2006, Lee et al. 2018b). In a recent phase II trial, ruxolitinib demonstrated a rapid B-symptom relief and a hematological response in six of 32 patients (18.8%), although lasting only little more than 7 months (Van Den Neste et al. 2018). Now, ruxolitinib is evaluated in various non-MPN indications in several clinical trials (see Table 12.1).

TABLE 12.1

Interventional Clinical Trials Evaluating Ruxolitinib in the Setting of Leukemia or Lymphoma (https://clinicaltrials.gov, accessed November 1, 2018)

Identifier	Indication	Phase	Intervention	Status
NCT03722407	CMML	2	Ruxolitinib	Not yet recruiting
NCT03681561	r/r classical Hodgkin lymphoma	1/2	Ruxolitinib/nivolumab	Recruiting
NCT03571321	ph-like ALL (young adults)	1	Ruxolitinib	Not yet recruiting
NCT03110822	r/r multiple myeloma	1	Ruxolitinib/lenalidomide	Recruiting
NCT02974647	T or NK cell lymphoma	2	Ruxolitinib	Recruiting
NCT02912754	CLL	1/2	Ruxolitinib/ibrutinib	Not yet recruiting
NCT02723994	pediatric ALL	2	Ruxolitinib	Recruiting
NCT02613598	r/r lymphoma	1	Ruxolitinib/bortezomid	Recruiting
NCT02257138	r/r or post-MPN AML	1/2	Ruxolitinib/decitabine	Recruiting
NCT02164500	classical Hodgkin lymphoma	2	Ruxolitinib	Recruiting
NCT02131584	CLL	2	Ruxolitinib	Active, not recruiting
NCT02015208	CLL	1/2	Ruxolitinib	Completed
NCT01776723	CMML	1/2	Ruxolitinib	Active, not recruiting
NCT01712659	T-cell leukemia (adults)	2	Ruxolitinib	Recruiting
NCT01431209	r/r DLBCL or T-cell-NHL	2	Ruxolitinib	Active, not recruiting
NCT00674479	AML, ALL, CML, MDS	2	Ruxolitinib	Completed

Ruxolitinib like most other JAK inhibitors in clinical evaluation is a type I kinase inhibitor that binds in the ATP binding pocket of the kinase domain in its active configuration (Liu and Gray 2006). It ameliorates splenomegaly and improves constitutional symptom burden, but has only a modest effect on the mutant allele burden and has a limited disease modifying potential (Harrison et al. 2012, Verstovsek et al. 2012, Vannucchi et al. 2017). However, a recently published trial in myelofibrosis patients receiving long-term ruxolitinib demonstrated a stabilization or improvement of fibrosis versus a control cohort receiving best available therapy (Kvasnicka et al. 2018). Type II JAK inhibitors like CHZ868, which preferentially inhibit *V617F* mutant versus wildtype JAK2 thus stabilizing JAK2 in the inactive conformation, showed increased efficacy in preclinical MPN models and reduced mutant allele burden in JAK2-/MPL mutant MPN models (Meyer et al. 2015, Silvennoinen and Hubbard 2015).

12.3.1.2 Other JAK Inhibitors

Dose escalation of ruxolitinib is limited due to myelotoxicity, excluding many patients with baseline cytopenias from treatment. Novel JAK inhibitors with less potential to induce thrombocytopenia and even properties to improve anemia could offer additional treatment options for a number of patients, especially those with baseline cytopenias (Griesshammer and Sadjadian 2017).

Pacritinib (SB-1518) is a small molecular weight macrocyclic oral JAK2-selective inhibitor (JAK2 wild type and *V617F* mutant) that also targets FLT3 (FMS-like tyrosine kinase 3, an important target in the treatment of AML), CSF1R and IRAK1, and no meaningful inhibition of JAK1 (Poulsen et al. 2012, Hatzimichael et al. 2014). Pacritinib was evaluated in two recent phase III trials (PERSIST-1 and PERSIST-2) in patients with myelofibrosis regardless of thrombocyte count and in patients with thrombocytopenic myelofibrosis (Mesa et al. 2017b, Mascarenhas et al. 2018). Compared to best available therapy, pacritinib met the primary end point of superior spleen volume reduction, but failed to show a significant improvement in total symptom score which was the co-primary endpoint.

Pacritinib has encouraging activity in lymphoma. In a phase I trial of 34 patients with relapsed/refractory Hodgkin or non-Hodgkin lymphoma, three patients had a partial remission and 15 patients had a stable disease (Younes et al. 2012).

Pacritinib could also be an interesting option in the therapy of AML as up to a third of AML patients harbor FLT3 mutations. An activation of the JAK-STAT pathway has been observed after initiation of FLT3 tyrosine kinase inhibitor therapy (Hart et al. 2011). Thus, pacritinib could reduce the development of secondary resistance or serve as a second-line therapy to re-sensitize resistant cells to FLT3 inhibition (Weisberg et al. 2012). Clinical results with the potential to change treatment algorithms are eagerly awaited.

Another important drug in advanced clinical development is momelotinib (CYT387), a dual JAK1/2 inhibitor (Pardanani et al. 2009, Tyner et al. 2010). Momelotinib demonstrated a unique property of improving myelofibrosis-associated anemia (Pardanani et al. 2013). However, in two phase III trials in myelofibrosis (SIMPLIFY-1 and SIMPLIFY-2), momelotinib was not superior to best available therapy concerning spleen size reduction, but it significantly improved disease-related symptoms and reduced transfusion dependency (Mesa et al. 2017a, Harrison et al. 2018).

12.3.2 STAT Inhibitors

Blocking the JAK-STAT pathway with STAT inhibitors is an intriguing approach for anticancer therapy. Limitations are the selectivity, bioavailability, similar homology (e.g. STAT3 and STAT1), and poor in vivo efficacy (O'Shea et al. 2015).

Several STAT inhibitors are in clinical development; however, due to the problems mentioned above none has been successfully introduced into clinical routine (Mandal et al. 2011). Different strategies may be used to design inhibitors targeting STAT: e.g. blocking of phosphorylation, preventing of dimerization by blocking the SH2 domain, or interfering with DNA-binding (Seif et al. 2017, Lee et al. 2018a).

Many publications and studies have demonstrated the important role of STAT (especially STAT3) in numerous aspects of tumorigenesis like differentiation, proliferation, angiogenesis, apoptosis, and metastasis (Jove 2000, Levy and Inghirami 2006, Yu et al. 2009). Recently, a study from Taiwan showed that JAK-STAT3 signaling is the major pathway in DLBCL (diffuse large B-cell lymphoma) amoeboid movement, which is critical for early dissemination of the disease (Pan et al. 2018). STAT3 is activated in response to cytokines (like IL-6, IL-10, IL-21, IL-27, leptin or G-CSF) or receptor and non-receptor tyrosine-kinases (Sansone and Bromberg 2012).

AZD9150 (ISIS481464) is an antisense oligonucleotide inhibitor of STAT3. At ASCO 2013, Hong and colleagues presented a phase I study in metastatic cancer that showed promising results using AZD9150 as a single agent. In addition, two out of three patients with chemotherapy-refractory diffuse large B-cell lymphoma had prolonged partial responses (Seth et al. 2010, Hong et al. 2015). As STAT3 blocking rather slows tumor progression than having antineoplastic effects, AZD9150 is now evaluated in several clinical trials in combination with checkpoint inhibitors (Kroemer et al. 2016, Pianko et al. 2017, Syed 2017, Cullberg et al. 2018).

Moreover, in vitro tests with existing drugs inhibiting STAT5 like the psychotropic drug pimozide have shown that STAT5 inhibitors may be effective in improving treatment outcome in acute or chronic myelogenous leukemia (Nelson et al. 2011, 2012).

12.4 Summary

The JAK-STAT signaling pathway is playing a central role in the formation of blood cells via cytokine activation of transmembrane receptors with transmission of anti-apoptotic, proliferative, or differentiation signals.

Several hematopoietic malignancies have been associated with mutations of JAK, cytokine receptors, or other components of the JAK-STAT signaling pathway. All these alterations are able to drive the disease. Today, most data are available in MPNs, where the presence of mutations in JAK2, calreticulin, or MPL has also been implemented as major diagnostic criteria (Arber et al. 2016). Moreover, ruxolitinib is currently the only licensed drug successfully targeting the JAK-STAT pathway in MPN and is now a milestone in the armamentarium of MPN therapy.

STAT inhibitors, particularly targeting STAT3, are now evaluated in many hematopoietic malignancies and could potentially emerge as valuable treatment options in the future.

REFERENCES

Arber, D. A., A. Orazi, R. Hasserjian, J. Thiele, M. J. Borowitz, M. M. Le Beau, C. D. Bloomfield, M. Cazzola, and J. W. Vardiman. 2016. "The 2016 revision to the World Health Organization classification of myeloid neoplasms and acute leukemia." *Blood* 127 (20): 2391–2405. doi: 10.1182/blood-2016-03-643544.

Asnafi, V., S. Le Noir, L. Lhermitte, C. Gardin, F. Legrand, X. Vallantin, J. V. Malfuson, N. Ifrah, H. Dombret, and E. Macintyre. 2010. "JAK1 mutations are not frequent events in adult T-ALL: a GRAALL study." *Br J Haematol* 148 (1): 178–179. doi: 10.1111/j.1365-2141.2009.07912.x.

Baus, D., and E. Pfitzner. 2006. "Specific function of STAT3, SOCS1, and SOCS3 in the regulation of proliferation and survival of classical Hodgkin lymphoma cells." *Int J Cancer* 118 (6):1404–1413. doi: 10.1002/ijc.21539.

Baxter, E. J., L. M. Scott, P. J. Campbell, C. East, N. Fourouclas, S. Swanton, G. S. Vassiliou, A. J. Bench, E. M. Boyd, N. Curtin, M. A. Scott, W. N. Erber, A. R. Green, and Project Cancer Genome. 2005. "Acquired mutation of the tyrosine kinase JAK2 in human myeloproliferative disorders." *Lancet* 365 (9464):1054–1061. doi: 10.1016/S0140-6736(05)71142-9.

Benekli, M., M. R. Baer, H. Baumann, and M. Wetzler. 2003. "Signal transducer and activator of transcription proteins in leukemias." *Blood* 101 (8):2940–2954. doi: 10.1182/blood-2002-04-1204.

Bercovich, D., I. Ganmore, L. M. Scott, G. Wainreb, Y. Birger, A. Elimelech, C. Shochat, G. Cazzaniga, A. Biondi, G. Basso, G. Cario, M. Schrappe, M. Stanulla, S. Strehl, O. A. Haas, G. Mann, V. Binder, A. Borkhardt, H. Kempski, J. Trka, B. Bielorei, S. Avigad, B. Stark, O. Smith, N. Dastugue, J. P. Bourquin, N. B. Tal, A. R. Green, and S. Izraeli. 2008. "Mutations of JAK2 in acute lymphoblastic leukaemias associated with Down's syndrome." *Lancet* 372 (9648): 1484–1492. doi: 10.1016/S0140-6736(08)61341-0.

Bose, P., and S. Verstovsek. 2017. "JAK2 inhibitors for myeloproliferative neoplasms: what is next?" *Blood* 130 (2): 115–125. doi: 10.1182/blood-2017-04-742288.

Bromberg, J. F., M. H. Wrzeszczynska, G. Devgan, Y. Zhao, R. G. Pestell, C. Albanese, and J. E. Darnell, Jr. 1999. "Stat3 as an oncogene." *Cell* 98 (3):295–303. doi: 10.1016/s0092-8674(00)81959-5.

Chaligne, R., C. James, C. Tonetti, R. Besancenot, J. P. Le Couedic, F. Fava, F. Mazurier, I. Godin, K. Maloum, F. Larbret, Y. Lecluse, W. Vainchenker, and S. Giraudier. 2007. "Evidence for MPL W515L/K mutations in hematopoietic stem cells in primitive myelofibrosis." *Blood* 110 (10):3735–3743. doi: 10.1182/blood-2007-05-089003.

Chen, E., L. M. Staudt, and A. R. Green. 2012. "Janus kinase deregulation in leukemia and lymphoma." *Immunity* 36 (4): 529–541. doi: 10.1016/j.immuni.2012.03.017.

Commins, S. P., L. Borish, and J. W. Steinke. 2010. "Immunologic messenger molecules: cytokines, interferons, and chemokines." *J Allergy Clin Immunol* 125 (2 Suppl 2): S53–S72. doi: 10.1016/j.jaci.2009.07.008.

Cullberg, M., C. Arfvidsson, B. Larsson, A. Malmgren, P. Mitchell, U. Wahlby Hamren, and H. Wray. 2018. "Pharmacokinetics of the Oral Selective CXCR2 Antagonist AZD5069: a summary of eight phase I studies in healthy volunteers." *Drugs R D* 18 (2): 149–159. doi: 10.1007/s40268-018-0236-x.

Dameshek, W. 1951. "Some speculations on the myeloproliferative syndromes." *Blood* 6 (4): 372–375.

Darnell, J. E., Jr., I. M. Kerr, and G. R. Stark. 1994. "Jak-STAT pathways and transcriptional activation in response to IFNs and other extracellular signaling proteins." *Science* 264 (5164): 1415–1421. doi: 10.1126/science.8197455.

de Oliveira, M. B., V. L. Fook-Alves, A. I. P. Eugenio, R. C. Fernando, L. F. G. Sanson, M. F. de Carvalho, W. M. T. Braga, F. E. Davies, and G. W. B. Colleoni. 2017. "Anti-myeloma effects of ruxolitinib combined with bortezomib and lenalidomide: a rationale for JAK/STAT pathway inhibition in myeloma patients." *Cancer Lett* 403: 206–215. doi: 10.1016/j.canlet.2017.06.016.

Derenzini, E., and A. Younes. 2013. "Targeting the JAK-STAT pathway in lymphoma: a focus on pacritinib." *Expert Opin Investig Drugs* 22 (6):775–785. doi: 10.1517/13543784.2013.775244.

Ding, B. B., J. J. Yu, R. Y. Yu, L. M. Mendez, R. Shaknovich, Y. Zhang, G. Cattoretti, and B. H. Ye. 2008. "Constitutively activated STAT3 promotes cell proliferation and survival in the activated B-cell subtype of diffuse large B-cell lymphomas." *Blood* 111 (3): 1515–1523. doi: 10.1182/blood-2007-04-087734.

Eghtedar, A., S. Verstovsek, Z. Estrov, J. Burger, J. Cortes, C. Bivins, S. Faderl, A. Ferrajoli, G. Borthakur, S. George, P. A. Scherle, R. C. Newton, H. M. Kantarjian, and F. Ravandi. 2012. "Phase 2 study of the JAK kinase inhibitor ruxolitinib in patients with refractory leukemias, including postmyeloproliferative neoplasm acute myeloid leukemia." *Blood* 119 (20): 4614–4618. doi: 10.1182/blood-2011-12-400051.

Evangelista, M. L., S. Amadori, and R. Stasi. 2007. "Biologic aspects of thrombopoietin and the development of novel thrombopoietic agents for clinical use." *Curr Drug Discov Technol* 4 (3): 162–173.

Flex, E., V. Petrangeli, L. Stella, S. Chiaretti, T. Hornakova, L. Knoops, C. Ariola, V. Fodale, E. Clappier, F. Paoloni, S. Martinelli, A. Fragale, M. Sanchez, S. Tavolaro, M. Messina, G. Cazzaniga, A. Camera, G. Pizzolo, A. Tornesello, M. Vignetti, A. Battistini, H. Cave, B. D. Gelb, J. C. Renauld, A. Biondi, S. N. Constantinescu, R. Foa, and M. Tartaglia. 2008. "Somatically acquired JAK1 mutations in adult acute lymphoblastic leukemia." *J Exp Med* 205 (4): 751–758. doi: 10.1084/jem.20072182.

Frank, D. A., S. Mahajan, and J. Ritz. 1997. "B lymphocytes from patients with chronic lymphocytic leukemia contain signal transducer and activator of transcription (STAT) 1 and STAT3 constitutively phosphorylated on serine residues." *J Clin Invest* 100 (12): 3140–3148. doi: 10.1172/JCI119869.

Garnock-Jones, K. P., and S. J. Keam. 2009. "Eltrombopag." *Drugs* 69 (5): 567–576. doi: 10.2165/00003495-200969050-00005.

Green, M. R., S. Monti, S. J. Rodig, P. Juszczynski, T. Currie, E. O'Donnell, B. Chapuy, K. Takeyama, D. Neuberg, T. R. Golub, J. L. Kutok, and M. A. Ship. 2010. "Integrative analysis reveals selective 9p24.1 amplification, increased PD-1 ligand expression, and further induction via JAK2 in nodular sclerosing Hodgkin lymphoma and primary mediastinal large B-cell lymphoma." *Blood* 116 (17): 3268–3277. doi: 10.1182/blood-2010-05-282780.

Griesshammer, M., and P. Sadjadian. 2017. "The BCR-ABL1-negative myeloproliferative neoplasms: a review of JAK inhibitors in the therapeutic armamentarium." *Expert Opin Pharmacother* 18 (18): 1929–1938. doi: 10.1080/14656566.2017.1404574.

Griesshammer, M., G. Saydam, F. Palandri, G. Benevolo, M. Egyed, J. Callum, T. Devos, S. Sivgin, P. Guglielmelli, C. Bensasson, M. Khan, J. P. Ronco, and F. Passamonti. 2018. "Ruxolitinib for the treatment of inadequately controlled polycythemia vera without splenomegaly: 80-week follow-up from the RESPONSE-2 trial." *Ann Hematol* 97 (9): 1591–1600. doi: 10.1007/s00277-018-3365-y.

Hammaren, H. M., A. T. Virtanen, J. Raivola, and O. Silvennoinen. 2019. "The regulation of JAKs in cytokine signaling and its breakdown in disease." *Cytokine* 118: 48–63. doi: 10.1016/j.cyto.2018.03.041.

Harrison, C., J. J. Kiladjian, H. K. Al-Ali, H. Gisslinger, R. Waltzman, V. Stalbovskaya, M. McQuitty, D. S. Hunter, R. Levy, L. Knoops, F. Cervantes, A. M. Vannucchi, T. Barbui, and G. Barosi. 2012. "JAK inhibition with ruxolitinib versus best available therapy for myelofibrosis." *N Engl J Med* 366 (9): 787–798. doi: 10.1056/NEJMoa1110556.

Harrison, C. N., A. M. Vannucchi, J. J. Kiladjian, H. K. Al-Ali, H. Gisslinger, L. Knoops, F. Cervantes, M. M. Jones, K. Sun, M. McQuitty, V. Stalbovskaya, P. Gopalakrishna, and T. Barbui. 2016. "Long-term findings from COMFORT-II, a phase 3 study of ruxolitinib vs best available therapy for myelofibrosis." *Leukemia* 30 (8): 1701–1707. doi: 10.1038/leu.2016.148.

Harrison, C. N., A. M. Vannucchi, U. Platzbecker, F. Cervantes, V. Gupta, D. Lavie, F. Passamonti, E. F. Winton, H. Dong, J. Kawashima, J. D. Maltzman, J. J. Kiladjian, and S. Verstovsek. 2018. "Momelotinib versus best available therapy in patients with myelofibrosis previously treated with ruxolitinib (SIMPLIFY 2): a randomised, open-label, phase 3 trial." *Lancet Haematol* 5 (2): e73–e81. doi: 10.1016/S2352-3026(17)30237-5.

Hart, S., K. C. Goh, V. Novotny-Diermayr, Y. C. Tan, B. Madan, C. Amalini, L. C. Ong, B. Kheng, A. Cheong, J. Zhou, W. J. Chng, and J. M. Wood. 2011. "Pacritinib (SB1518), a JAK2/FLT3 inhibitor for the treatment of acute myeloid leukemia." *Blood Cancer J* 1 (11): e44. doi: 10.1038/bcj.2011.43.

Hatzimichael, E., E. Tsolas, and E. Briasoulis. 2014. "Profile of pacritinib and its potential in the treatment of hematologic disorders." *J Blood Med* 5:143–152. doi: 10.2147/JBM.S51253.

Hong, D., R. Kurzrock, Y. Kim, R. Woessner, A. Younes, J. Nemunaitis, N. Fowler, T. Zhou, J. Schmidt, M. Jo, S. J. Lee, M. Yamashita, S. G. Hughes, L. Fayad, S. Piha-Paul, M. V. Nadella, M. Mohseni, D. Lawson, C. Reimer, D. C. Blakey, X. Xiao, J. Hsu, A. Revenko, B. P. Monia, and A. R. MacLeod. 2015. "AZD9150, a next-generation antisense oligonucleotide inhibitor of STAT3 with early evidence of clinical activity in lymphoma and lung cancer." *Sci Transl Med* 7 (314):314ra185. doi: 10.1126/scitranslmed.aac5272.

Hu, J., and W. X. Hu. 2018. "Targeting signaling pathways in multiple myeloma: pathogenesis and implication for treatments." *Cancer Lett* 414:214–221. doi: 10.1016/j.canlet.2017.11.020.

Huang, X., B. Meng, J. Iqbal, B. B. Ding, A. M. Perry, W. Cao, L. M. Smith, C. Bi, C. Jiang, T. C. Greiner, D. D. Weisenburger, L. Rimsza, A. Rosenwald, G. Ott, J. Delabie, E. Campo, R. M. Braziel, R. D. Gascoyne, J. R. Cook, R. R. Tubbs, E. S. Jaffe, J. O. Armitage, J. M. Vose, L. M. Staudt, T. W. McKeithan, W. C. Chan, B. H. Ye, and K. Fu. 2013. "Activation of the STAT3 signaling pathway is associated with poor survival in diffuse large B-cell lymphoma treated with R-CHOP." *J Clin Oncol* 31 (36):4520–4528. doi: 10.1200/JCO.2012.45.6004.

Ikezoe, T., S. Kojima, M. Furihata, J. Yang, C. Nishioka, A. Takeuchi, M. Isaka, H. P. Koeffler, and A. Yokoyama. 2011. "Expression of p-JAK2 predicts clinical outcome and is a potential molecular target of acute myelogenous leukemia." *Int J Cancer* 129 (10):2512–2521. doi: 10.1002/ijc.25910.

James, C., V. Ugo, J. P. Le Couedic, J. Staerk, F. Delhommeau, C. Lacout, L. Garcon, H. Raslova, R. Berger, A. Bennaceur-Griscelli, J. L. Villeval, S. N. Constantinescu, N. Casadevall, and W. Vainchenker. 2005. "A unique clonal JAK2 mutation leading to constitutive signalling causes polycythaemia vera." *Nature* 434 (7037):1144–1148. doi: 10.1038/nature03546.

Joos, S., M. Kupper, S. Ohl, F. von Bonin, G. Mechtersheimer, M. Bentz, P. Marynen, P. Moller, M. Pfreundschuh, L. Trumper, and P. Lichter. 2000. "Genomic imbalances including amplification of the tyrosine kinase gene JAK2 in CD30+ Hodgkin cells." *Cancer Res* 60 (3):549–552.

Jove, R. 2000. "Preface: STAT signaling." *Oncogene* 19 (21):2466–2467. doi: 10.1038/sj.onc.1203549.

Kearney, L., D. Gonzalez De Castro, J. Yeung, J. Procter, S. W. Horsley, M. Eguchi-Ishimae, C. M. Bateman, K. Anderson, T. Chaplin, B. D. Young, C. J. Harrison, H. Kempski, C. W. So, A. M. Ford, and M. Greaves. 2009. "Specific JAK2 mutation (JAK2R683) and multiple gene deletions in Down syndrome acute lymphoblastic leukemia." *Blood* 113 (3):646–648. doi: 10.1182/blood-2008-08-170928.

Klampfl, T., H. Gisslinger, A. S. Harutyunyan, H. Nivarthi, E. Rumi, J. D. Milosevic, N. C. Them, T. Berg, B. Gisslinger, D. Pietra, D. Chen, G. I. Vladimer, K. Bagienski, C. Milanesi, I. C. Casetti, E. Sant'Antonio, V. Ferretti, C. Elena, F. Schischlik, C. Cleary, M. Six, M. Schalling, A. Schonegger, C. Bock, L. Malcovati, C. Pascutto, G. Superti-Furga, M. Cazzola, and R. Kralovics. 2013. "Somatic mutations of calreticulin in myeloproliferative neoplasms." *N Engl J Med* 369 (25):2379–2390. doi: 10.1056/NEJMoa1311347.

Kralovics, R., F. Passamonti, A. S. Buser, S. S. Teo, R. Tiedt, J. R. Passweg, A. Tichelli, M. Cazzola, and R. C. Skoda. 2005. "A gain-of-function mutation of JAK2 in myeloproliferative disorders." *N Engl J Med* 352 (17):1779–1790. doi: 10.1056/NEJMoa051113.

Kroemer, G., L. Galluzzi, and L. Zitvogel. 2016. "STAT3 inhibition for cancer therapy: cell-autonomous effects only?." *Oncoimmunology* 5 (5):e1126063. doi: 10.1080/2162402X.2015.1126063.

Kuter, D. J. 2013. "The biology of thrombopoietin and thrombopoietin receptor agonists." *Int J Hematol* 98 (1):10–23. doi: 10.1007/s12185-013-1382-0.

Kvasnicka, H. M., J. Thiele, C. E. Bueso-Ramos, W. Sun, J. Cortes, H. M. Kantarjian, and S. Verstovsek. 2018. "Long-term effects of ruxolitinib versus best available therapy on bone marrow fibrosis in patients with myelofibrosis." *J Hematol Oncol* 11 (1):42. doi: 10.1186/s13045-018-0585-5.

Lacronique, V., A. Boureux, V. D. Valle, H. Poirel, C. T. Quang, M. Mauchauffe, C. Berthou, M. Lessard, R. Berger, J. Ghysdael, and O. A. Bernard. 1997. "A TEL-JAK2 fusion protein with constitutive kinase activity in human leukemia." *Science* 278 (5341):1309–1312. doi: 10.1126/science.278.5341.1309.

Lee, D. S., R. A. O'Keefe, P. K. Ha, J. R. Grandis, and D. E. Johnson. 2018a. "Biochemical Properties of a Decoy Oligodeoxynucleotide Inhibitor of STAT3 Transcription Factor." *Int J Mol Sci* 19:6. doi: 10.3390/ijms19061608.

Lee, S., T. Shah, C. Yin, J. Hochberg, J. Ayello, E. Morris, C. van de Ven, and M. S. Cairo. 2018b. "Ruxolitinib significantly enhances in vitro apoptosis in Hodgkin lymphoma and primary mediastinal B-cell lymphoma and survival in a lymphoma xenograft murine model." *Oncotarget* 9 (11):9776–9788. doi: 10.18632/oncotarget.24267.

Leonard, W. J., and J. J. O'Shea. 1998. "Jaks and STATs: biological implications." *Annu Rev Immunol* 16:293–322. doi: 10.1146/annurev.immunol.16.1.293.

Levine, R. L., M. Loriaux, B. J. Huntly, M. L. Loh, M. Beran, E. Stoffregen, R. Berger, J. J. Clark, S. G. Willis, K. T. Nguyen, N. J. Flores, E. Estey, N. Gattermann, S. Armstrong, A. T. Look, J. D. Griffin, O. A. Bernard, M. C. Heinrich, D. G. Gilliland, B. Druker, and M. W. Deininger. 2005a. "The JAK2V617F activating mutation occurs in chronic myelomonocytic leukemia and acute myeloid leukemia, but not in acute lymphoblastic leukemia or chronic lymphocytic leukemia." *Blood* 106 (10):3377–3379. doi: 10.1182/blood-2005-05-1898.

Levine, R. L., M. Wadleigh, J. Cools, B. L. Ebert, G. Wernig, B. J. Huntly, T. J. Boggon, I. Wlodarska, J. J. Clark, S. Moore, J. Adelsperger, S. Koo, J. C. Lee, S. Gabriel, T. Mercher, A. D'Andrea, S. Frohling, K. Dohner, P. Marynen, P. Vandenberghe, R. A. Mesa, A. Tefferi, J. D. Griffin, M. J. Eck, W. R. Sellers, M. Meyerson, T. R. Golub, S. J. Lee, and D. G. Gilliland. 2005b. "Activating mutation in the tyrosine kinase JAK2 in polycythemia vera, essential thrombocythemia, and myeloid metaplasia with myelofibrosis." *Cancer Cell* 7 (4):387–397. doi: 10.1016/j.ccr.2005.03.023.

Levy, D. E., and J. E. Darnell, Jr. 2002. "Stats: transcriptional control and biological impact." *Nat Rev Mol Cell Biol* 3 (9):651–662. doi: 10.1038/nrm909.

Levy, D. E., and G. Inghirami. 2006. "STAT3: a multifaceted oncogene." *Proc Natl Acad Sci U S A* 103 (27):10151–10152. doi: 10.1073/pnas.0604042103.

Liao, W., J. X. Lin, and W. J. Leonard. 2011. "IL-2 family cytokines: new insights into the complex roles of IL-2 as a broad regulator of T helper cell differentiation." *Curr Opin Immunol* 23 (5):598–604. doi: 10.1016/j.coi.2011.08.003.

Liu, Y., and N. S. Gray. 2006. "Rational design of inhibitors that bind to inactive kinase conformations." *Nat Chem Biol* 2 (7):358–364. doi: 10.1038/nchembio799.

Lu, L., F. Zhu, M. Zhang, Y. Li, A. C. Drennan, S. Kimpara, I. Rumball, C. Selzer, H. Cameron, A. Kellicut, A. Kelm, F. Wang, T. A. Waldmann, and L. Rui. 2018. "Gene regulation and suppression of type I interferon signaling by STAT3 in diffuse large B cell lymphoma." *Proc Natl Acad Sci U S A* 115 (3):E498–E505. doi: 10.1073/pnas.1715118115.

Mandal, P. K., F. Gao, Z. Lu, Z. Ren, R. Ramesh, J. S. Birtwistle, K. K. Kaluarachchi, X. Chen, R. C. Bast, Jr., W. S. Liao, and J. S. McMurray. 2011. "Potent and selective phosphopeptide mimetic prodrugs targeted to the Src homology 2 (SH2) domain of signal transducer and activator of transcription 3." *J Med Chem* 54 (10):3549–3563. doi: 10.1021/jm2000882.

Mascarenhas, J., R. Hoffman, M. Talpaz, A. T. Gerds, B. Stein, V. Gupta, A. Szoke, M. Drummond, A. Pristupa, T. Granston, R. Daly, S. Al-Fayoumi, J. A. Callahan, J. W. Singer, J. Gotlib, C. Jamieson, C. Harrison, R. Mesa, and S. Verstovsek. 2018. "Pacritinib vs best available therapy, including ruxolitinib, in patients with myelofibrosis: a randomized clinical trial." *JAMA Oncol* 4 (5):652–659. doi: 10.1001/jamaoncol.2017.5818.

Mesa, R. A., J. J. Kiladjian, J. V. Catalano, T. Devos, M. Egyed, A. Hellmann, D. McLornan, K. Shimoda, E. F. Winton, W. Deng, R. L. Dubowy, J. D. Maltzman, F. Cervantes, and J. Gotlib. 2017a. "SIMPLIFY-1: a phase iii randomized trial of momelotinib versus ruxolitinib in janus kinase inhibitor-naive patients with myelofibrosis." *J Clin Oncol* 35 (34):3844–3850. doi: 10.1200/JCO.2017.73.4418.

Mesa, R. A., A. M. Vannucchi, A. Mead, M. Egyed, A. Szoke, A. Suvorov, J. Jakucs, A. Perkins, R. Prasad, J. Mayer, J. Demeter, P. Ganly, J. W. Singer, H. Zhou, J. P. Dean, P. A. Te Boekhorst, J. Nangalia, J. J. Kiladjian, and C. N. Harrison. 2017b. "Pacritinib versus best available therapy for the treatment of myelofibrosis irrespective of baseline cytopenias (PERSIST-1): an international, randomised, phase 3 trial." *Lancet Haematol* 4 (5):e225–e236. doi: 10.1016/S2352-3026(17)30027-3.

Meyer, S. C., M. D. Keller, S. Chiu, P. Koppikar, O. A. Guryanova, F. Rapaport, K. Xu, K. Manova, D. Pankov, R. J. O'Reilly, M. Kleppe, A. S. McKenney, A. H. Shih, K. Shank, J. Ahn, E. Papalexi,

B. Spitzer, N. Socci, A. Viale, E. Mandon, N. Ebel, R. Andraos, J. Rubert, E. Dammassa, V. Romanet, A. Dolemeyer, M. Zender, M. Heinlein, R. Rampal, R. S. Weinberg, R. Hoffman, W. R. Sellers, F. Hofmann, M. Murakami, F. Baffert, C. Gaul, T. Radimerski, and R. L. Levine. 2015. "CHZ868, a Type II JAK2 inhibitor, reverses Type I JAK inhibitor persistence and demonstrates efficacy in myeloproliferative neoplasms." *Cancer Cell* 28 (1):15–28. doi: 10.1016/j.ccell.2015.06.006.

Mitsiades, C. S., N. S. Mitsiades, N. C. Munshi, P. G. Richardson, and K. C. Anderson. 2006. "The role of the bone microenvironment in the pathophysiology and therapeutic management of multiple myeloma: interplay of growth factors, their receptors and stromal interactions." *Eur J Cancer* 42 (11):1564–1573. doi: 10.1016/j.ejca.2005.12.025.

Moulis, G., A. Palmaro, J. L. Montastruc, B. Godeau, M. Lapeyre-Mestre, and L. Sailler. 2014. "Epidemiology of incident immune thrombocytopenia: a nationwide population-based study in France." *Blood* 124 (22):3308–3315. doi: 10.1182/blood-2014-05-578336.

Mullighan, C. G., J. Zhang, R. C. Harvey, J. R. Collins-Underwood, B. A. Schulman, L. A. Phillips, S. K. Tasian, M. L. Loh, X. Su, W. Liu, M. Devidas, S. R. Atlas, I. M. Chen, R. J. Clifford, D. S. Gerhard, W. L. Carroll, G. H. Reaman, M. Smith, J. R. Downing, S. P. Hunger, and C. L. Willman. 2009. "JAK mutations in high-risk childhood acute lymphoblastic leukemia." *Proc Natl Acad Sci U S A* 106 (23):9414–9418. doi: 10.1073/pnas.0811761106.

Murati, A., V. Gelsi-Boyer, J. Adelaide, C. Perot, P. Talmant, S. Giraudier, L. Lode, A. Letessier, B. Delaval, V. Brunel, M. Imbert, R. Garand, L. Xerri, D. Birnbaum, M. J. Mozziconacci, and M. Chaffanet. 2005. "PCM1-JAK2 fusion in myeloproliferative disorders and acute erythroid leukemia with t(8;9) translocation." *Leukemia* 19 (9):1692–1696. doi: 10.1038/sj.leu.2403879.

Nangalia, J., C. E. Massie, E. J. Baxter, F. L. Nice, G. Gundem, D. C. Wedge, E. Avezov, J. Li, K. Kollmann, D. G. Kent, A. Aziz, A. L. Godfrey, J. Hinton, I. Martincorena, P. Van Loo, A. V. Jones, P. Guglielmelli, P. Tarpey, H. P. Harding, J. D. Fitzpatrick, C. T. Goudie, C. A. Ortmann, S. J. Loughran, K. Raine, D. R. Jones, A. P. Butler, J. W. Teague, S. O'Meara, S. McLaren, M. Bianchi, Y. Silber, D. Dimitropoulou, D. Bloxham, L. Mudie, M. Maddison, B. Robinson, C. Keohane, C. Maclean, K. Hill, K. Orchard, S. Tauro, M. Q. Du, M. Greaves, D. Bowen, B. J. P. Huntly, C. N. Harrison, N. C. P. Cross, D. Ron, A. M. Vannucchi, E. Papaemmanuil, P. J. Campbell, and A. R. Green. 2013. "Somatic CALR mutations in myeloproliferative neoplasms with nonmutated JAK2." *N Engl J Med* 369 (25):2391–2405. doi: 10.1056/NEJMoa1312542.

Navarro, A., T. Diaz, A. Martinez, A. Gaya, A. Pons, B. Gel, C. Codony, G. Ferrer, C. Martinez, E. Montserrat, and M. Monzo. 2009. "Regulation of JAK2 by miR-135a: prognostic impact in classic Hodgkin lymphoma." *Blood* 114 (14):2945–2951. doi: 10.1182/blood-2009-02-204842.

Nebral, K., D. Denk, A. Attarbaschi, M. Konig, G. Mann, O. A. Haas, and S. Strehl. 2009. "Incidence and diversity of PAX5 fusion genes in childhood acute lymphoblastic leukemia." *Leukemia* 23 (1):134–143. doi: 10.1038/leu.2008.306.

Nelson, E. A., S. R. Walker, A. Kepich, L. B. Gashin, T. Hideshima, H. Ikeda, D. Chauhan, K. C. Anderson, and D. A. Frank. 2008. "Nifuroxazide inhibits survival of multiple myeloma cells by directly inhibiting STAT3." *Blood* 112 (13):5095–5102. doi: 10.1182/blood-2007-12-129718.

Nelson, E. A., S. R. Walker, E. Weisberg, M. Bar-Natan, R. Barrett, L. B. Gashin, S. Terrell, J. L. Klitgaard, L. Santo, M. R. Addorio, B. L. Ebert, J. D. Griffin, and D. A. Frank. 2011. "The STAT5 inhibitor pimozide decreases survival of chronic myelogenous leukemia cells resistant to kinase inhibitors." *Blood* 117 (12):3421–3429. doi: 10.1182/blood-2009-11-255232.

Nelson, E. A., S. R. Walker, M. Xiang, E. Weisberg, M. Bar-Natan, R. Barrett, S. Liu, S. Kharbanda, A. L. Christie, M. Nicolais, J. D. Griffin, R. M. Stone, A. L. Kung, and D. A. Frank. 2012. "The STAT5 inhibitor pimozide displays efficacy in models of acute myelogenous leukemia driven by FLT3 mutations." *Genes Cancer* 3 (7–8):503–511. doi: 10.1177/1947601912466555.

Nielsen, C., S. E. Bojesen, B. G. Nordestgaard, K. F. Kofoed, and H. S. Birgens. 2014. "JAK2V617F somatic mutation in the general population: myeloproliferative neoplasm development and progression rate." *Haematologica* 99 (9):1448–1455. doi: 10.3324/haematol.2014.107631.

Ok, C. Y., J. Chen, Z. Y. Xu-Monette, A. Tzankov, G. C. Manyam, L. Li, C. Visco, S. Montes-Moreno, K. Dybkaer, A. Chiu, A. Orazi, Y. Zu, G. Bhagat, K. L. Richards, E. D. Hsi, W. W. Choi, J. H. van Krieken, J. Huh, X. Zhao, M. Ponzoni, A. J. Ferreri, F. Bertoni, J. P. Farnen, M. B. Moller, M. A. Piris, J. N. Winter, L. J. Medeiros, and K. H. Young. 2014. "Clinical implications of phosphorylated STAT3 expression in De Novo diffuse large B-cell lymphoma." *Clin Cancer Res* 20 (19):5113–5123. doi: 10.1158/1078-0432.CCR-14-0683.

O'Shea, J. J., D. M. Schwartz, A. V. Villarino, M. Gadina, I. B. McInnes, and A. Laurence. 2015. "The JAK-STAT pathway: impact on human disease and therapeutic intervention." *Annu Rev Med* 66:311–328. doi: 10.1146/annurev-med-051113-024537.

Padron, E., A. Dezern, M. Andrade-Campos, K. Vaddi, P. Scherle, Q. Zhang, Y. Ma, M. E. Balasis, S. Tinsley, H. Ramadan, C. Zimmerman, D. P. Steensma, G. J. Roboz, J. E. Lancet, A. F. List, M. A. Sekeres, R. S. Komrokji, and Consortium Myelodysplastic Syndrome Clinical Research. 2016. "A multi-institution phase i trial of ruxolitinib in patients with chronic myelomonocytic Leukemia (CMML)." *Clin Cancer Res* 22 (15):3746–3754. doi: 10.1158/1078-0432.CCR-15-2781.

Padron, E., J. S. Painter, S. Kunigal, A. W. Mailloux, K. McGraw, J. M. McDaniel, E. Kim, C. Bebbington, M. Baer, G. Yarranton, J. Lancet, R. S. Komrokji, O. Abdel-Wahab, A. F. List, and P. K. Epling-Burnette. 2013. "GM-CSF-dependent pSTAT5 sensitivity is a feature with therapeutic potential in chronic myelomonocytic leukemia." *Blood* 121 (25):5068–5077. doi: 10.1182/blood-2012-10-460170.

Pan, Y. R., C. C. Chen, Y. T. Chan, H. J. Wang, F. T. Chien, Y. L. Chen, J. L. Liu, and M. H. Yang. 2018. "STAT3-coordinated migration facilitates the dissemination of diffuse large B-cell lymphomas." *Nat Commun* 9 (1):3696. doi: 10.1038/s41467-018-06134-z.

Pardanani, A., R. R. Laborde, T. L. Lasho, C. Finke, K. Begna, A. Al-Kali, W. J. Hogan, M. R. Litzow, A. Leontovich, M. Kowalski, and A. Tefferi. 2013. "Safety and efficacy of CYT387, a JAK1 and JAK2 inhibitor, in myelofibrosis." *Leukemia* 27 (6):1322–1327. doi: 10.1038/leu.2013.71.

Pardanani, A., T. Lasho, G. Smith, C. J. Burns, E. Fantino, and A. Tefferi. 2009. "CYT387, a selective JAK1/JAK2 inhibitor: in vitro assessment of kinase selectivity and preclinical studies using cell lines and primary cells from polycythemia vera patients." *Leukemia* 23 (8):1441–1445. doi: 10.1038/leu.2009.50.

Pardanani, A. D., R. L. Levine, T. Lasho, Y. Pikman, R. A. Mesa, M. Wadleigh, D. P. Steensma, M. A. Elliott, A. P. Wolanskyj, W. J. Hogan, R. F. McClure, M. R. Litzow, D. G. Gilliland, and A. Tefferi. 2006. "MPL515 mutations in myeloproliferative and other myeloid disorders: a study of 1182 patients." *Blood* 108 (10):3472–3476. doi: 10.1182/blood-2006-04-018879.

Passamonti, F., M. Griesshammer, F. Palandri, M. Egyed, G. Benevolo, T. Devos, J. Callum, A. M. Vannucchi, S. Sivgin, C. Bensasson, M. Khan, N. Mounedji, and G. Saydam. 2017. "Ruxolitinib for the treatment of inadequately controlled polycythaemia vera without splenomegaly (RESPONSE-2): a randomised, open-label, phase 3b study." *Lancet Oncol* 18 (1):88–99. doi: 10.1016/S1470-2045(16)30558-7.

Patnaik, M. M., R. A. Knudson, N. Gangat, C. A. Hanson, A. Pardanani, A. Tefferi, and R. P. Ketterling. 2010. "Chromosome 9p24 abnormalities: prevalence, description of novel JAK2 translocations, JAK2V617F mutation analysis and clinicopathologic correlates." *Eur J Haematol* 84 (6):518–524. doi: 10.1111/j.1600-0609.2010.01428.x.

Pianko, M. J., Y. Liu, S. Bagchi, and A. M. Lesokhin. 2017. "Immune checkpoint blockade for hematologic malignancies: a review." *Stem Cell Investig* 4:32. doi: 10.21037/sci.2017.03.04.

Podar, K., D. Chauhan, and K. C. Anderson. 2009. "Bone marrow microenvironment and the identification of new targets for myeloma therapy." *Leukemia* 23 (1):10–24. doi: 10.1038/leu.2008.259.

Poitras, J. L., P. Dal Cin, J. C. Aster, D. J. Deangelo, and C. C. Morton. 2008. "Novel SSBP2-JAK2 fusion gene resulting from a t(5;9)(q14.1;p24.1) in pre-B acute lymphocytic leukemia." *Genes Chromosomes Cancer* 47 (10):884–889. doi: 10.1002/gcc.20585.

Poulsen, A., A. William, S. Blanchard, A. Lee, H. Nagaraj, H. Wang, E. Teo, E. Tan, K. C. Goh, and B. Dymock. 2012. "Structure-based design of oxygen-linked macrocyclic kinase inhibitors: discovery of SB1518 and SB1578, potent inhibitors of Janus kinase 2 (JAK2) and Fms-like tyrosine kinase-3 (FLT3)." *J Comput Aided Mol Des* 26 (4):437–450. doi: 10.1007/s10822-012-9572-z.

Rampal, R., F. Al-Shahrour, O. Abdel-Wahab, J. P. Patel, J. P. Brunel, C. H. Mermel, A. J. Bass, J. Pretz, J. Ahn, T. Hricik, O. Kilpivaara, M. Wadleigh, L. Busque, D. G. Gilliland, T. R. Golub, B. L. Ebert, and R. L. Levine. 2014. "Integrated genomic analysis illustrates the central role of JAK-STAT pathway activation in myeloproliferative neoplasm pathogenesis." *Blood* 123 (22):e123–e133. doi: 10.1182/blood-2014-02-554634.

Reiter, A., C. Walz, A. Watmore, C. Schoch, I. Blau, B. Schlegelberger, U. Berger, N. Telford, S. Aruliah, J. A. Yin, D. Vanstraelen, H. F. Barker, P. C. Taylor, A. O'Driscoll, F. Benedetti, C. Rudolph,

H. J. Kolb, A. Hochhaus, R. Hehlmann, A. Chase, and N. C. Cross. 2005. "The t(8;9)(p22;p24) is a recurrent abnormality in chronic and acute leukemia that fuses PCM1 to JAK2." *Cancer Res* 65 (7):2662–2667. doi: 10.1158/0008-5472.CAN-04-4263.

Rosenwald, A., G. Wright, K. Leroy, X. Yu, P. Gaulard, R. D. Gascoyne, W. C. Chan, T. Zhao, C. Haioun, T. C. Greiner, D. D. Weisenburger, J. C. Lynch, J. Vose, J. O. Armitage, E. B. Smeland, S. Kvaloy, H. Holte, J. Delabie, E. Campo, E. Montserrat, A. Lopez-Guillermo, G. Ott, H. K. Muller-Hermelink, J. M. Connors, R. Braziel, T. M. Grogan, R. I. Fisher, T. P. Miller, M. LeBlanc, M. Chiorazzi, H. Zhao, L. Yang, J. Powell, W. H. Wilson, E. S. Jaffe, R. Simon, R. D. Klausner, and L. M. Staudt. 2003. "Molecular diagnosis of primary mediastinal B cell lymphoma identifies a clinically favorable subgroup of diffuse large B cell lymphoma related to Hodgkin lymphoma." *J Exp Med* 198 (6):851–862. doi: 10.1084/jem.20031074.

Saharinen, P., and O. Silvennoinen. 2002. "The pseudokinase domain is required for suppression of basal activity of Jak2 and Jak3 tyrosine kinases and for cytokine-inducible activation of signal transduction." *J Biol Chem* 277 (49):47954–47963. doi: 10.1074/jbc.M205156200.

Saharinen, P., M. Vihinen, and O. Silvennoinen. 2003. "Autoinhibition of Jak2 tyrosine kinase is dependent on specific regions in its pseudokinase domain." *Mol Biol Cell* 14 (4):1448–1459. doi: 10.1091/mbc.e02-06-0342.

Sansone, P., and J. Bromberg. 2012. "Targeting the interleukin-6/Jak/stat pathway in human malignancies." *J Clin Oncol* 30 (9):1005–1014. doi: 10.1200/JCO.2010.31.8907.

Scott, L. M., P. J. Campbell, E. J. Baxter, T. Todd, P. Stephens, S. Edkins, R. Wooster, M. R. Stratton, P. A. Futreal, and A. R. Green. 2005. "The V617F JAK2 mutation is uncommon in cancers and in myeloid malignancies other than the classic myeloproliferative disorders." *Blood* 106 (8):2920–2921. doi: 10.1182/blood-2005-05-2087.

Seif, F., M. Khoshmirsafa, H. Aazami, M. Mohsenzadegan, G. Sedighi, and M. Bahar. 2017. "The role of JAK-STAT signaling pathway and its regulators in the fate of T helper cells." *Cell Commun Signal* 15 (1):23. doi: 10.1186/s12964-017-0177-y.

Seth, P. P., G. Vasquez, C. A. Allerson, A. Berdeja, H. Gaus, G. A. Kinberger, T. P. Prakash, M. T. Migawa, B. Bhat, and E. E. Swayze. 2010. "Synthesis and biophysical evaluation of 2′,4′-constrained 2′O-methoxyethyl and 2′,4′-constrained 2′O-ethyl nucleic acid analogues." *J Org Chem* 75 (5):1569–1581. doi: 10.1021/jo902560f.

Silvennoinen, O., and S. R. Hubbard. 2015. "Targeting the inactive conformation of JAK2 in hematological malignancies." *Cancer Cell* 28 (1):1–2. doi: 10.1016/j.ccell.2015.06.010.

Skoda, R. C., A. Duek, and J. Grisouard. 2015. "Pathogenesis of myeloproliferative neoplasms." *Exp Hematol* 43 (8):599–608. doi: 10.1016/j.exphem.2015.06.007.

Solary, E. 2016. "Unplugging JAK/STAT in chronic myelomonocytic Leukemia." *Clin Cancer Res* 22 (15):3707–3709. doi: 10.1158/1078-0432.CCR-16-0372.

Steensma, D. P., R. F. McClure, J. E. Karp, A. Tefferi, T. L. Lasho, H. L. Powell, G. W. DeWald, and S. H. Kaufmann. 2006. "JAK2 V617F is a rare finding in de novo acute myeloid leukemia, but STAT3 activation is common and remains unexplained." *Leukemia* 20 (6):971–978. doi: 10.1038/sj.leu.2404206.

Syed, Y. Y. 2017. "Durvalumab: first global approval." *Drugs* 77 (12):1369–1376. doi: 10.1007/s40265-017-0782-5.

Tyner, J. W., T. G. Bumm, J. Deininger, L. Wood, K. J. Aichberger, M. M. Loriaux, B. J. Druker, C. J. Burns, E. Fantino, and M. W. Deininger. 2010. "CYT387, a novel JAK2 inhibitor, induces hematologic responses and normalizes inflammatory cytokines in murine myeloproliferative neoplasms." *Blood* 115 (25):5232–5240. doi: 10.1182/blood-2009-05-223727.

Vainchenker, W., and S. N. Constantinescu. 2013. "JAK/STAT signaling in hematological malignancies." *Oncogene* 32 (21):2601–2613. doi: 10.1038/onc.2012.347.

Van Den Neste, E., M. Andre, T. Gastinne, A. Stamatoullas, C. Haioun, A. Belhabri, O. Reman, O. Casasnovas, H. Ghesquieres, G. Verhoef, M. J. Claessen, H. A. Poirel, M. C. Copin, R. Dubois, P. Vandenberghe, I. A. Stoian, A. S. Cottereau, S. Bailly, L. Knoops, and F. Morschhauser. 2018. "A phase II study of the oral JAK1/JAK2 inhibitor ruxolitinib in advanced relapsed/refractory Hodgkin lymphoma." *Haematologica* 103 (5):840–848. doi: 10.3324/haematol.2017.180554.

Vannucchi, A. M. 2015. "Ruxolitinib versus standard therapy for the treatment of polycythemia vera." *N Engl J Med* 372 (17):1670–1671. doi: 10.1056/NEJMc1502524.

Vannucchi, A. M., S. Verstovsek, P. Guglielmelli, M. Griesshammer, T. C. Burn, A. Naim, D. Paranagama, M. Marker, B. Gadbaw, and J. J. Kiladjian. 2017. "Ruxolitinib reduces JAK2 p. V617F allele burden in patients with polycythemia vera enrolled in the RESPONSE study." *Ann Hematol* 96 (7):1113–1120. doi: 10.1007/s00277-017-2994-x.

Van Roosbroeck, K., L. Cox, T. Tousseyn, I. Lahortiga, O. Gielen, B. Cauwelier, P. De Paepe, G. Verhoef, P. Marynen, P. Vandenberghe, C. De Wolf-Peeters, J. Cools, and I. Wlodarska. 2011. "JAK2 rearrangements, including the novel SEC31A-JAK2 fusion, are recurrent in classical Hodgkin lymphoma." *Blood* 117 (15):4056–4064. doi: 10.1182/blood-2010-06-291310.

Velazquez, L., M. Fellous, G. R. Stark, and S. Pellegrini. 1992. "A protein tyrosine kinase in the interferon alpha/beta signaling pathway." *Cell* 70 (2):313–322. doi: 10.1016/0092-8674(92)90105-l.

Verstovsek, S., J. Gotlib, R. A. Mesa, A. M. Vannucchi, J. J. Kiladjian, F. Cervantes, C. N. Harrison, R. Paquette, W. Sun, A. Naim, P. Langmuir, T. Dong, P. Gopalakrishna, and V. Gupta. 2017. "Long-term survival in patients treated with ruxolitinib for myelofibrosis: COMFORT-I and -II pooled analyses." *J Hematol Oncol* 10 (1):156. doi: 10.1186/s13045-017-0527-7.

Verstovsek, S., R. A. Mesa, J. Gotlib, R. S. Levy, V. Gupta, J. F. DiPersio, J. V. Catalano, M. Deininger, C. Miller, R. T. Silver, M. Talpaz, E. F. Winton, J. H. Harvey, Jr., M. O. Arcasoy, E. Hexner, R. M. Lyons, R. Paquette, A. Raza, K. Vaddi, S. Erickson-Viitanen, I. L. Koumenis, W. Sun, V. Sandor, and H. M. Kantarjian. 2012. "A double-blind, placebo-controlled trial of ruxolitinib for myelofibrosis." *N Engl J Med* 366 (9):799–807. doi: 10.1056/NEJMoa1110557.

Verstovsek, S., A. M. Vannucchi, M. Griesshammer, T. Masszi, S. Durrant, F. Passamonti, C. N. Harrison, F. Pane, P. Zachee, K. Kirito, C. Besses, M. Hino, B. Moiraghi, C. B. Miller, M. Cazzola, V. Rosti, I. Blau, R. Mesa, M. M. Jones, H. Zhen, J. Li, N. Francillard, D. Habr, and J. J. Kiladjian. 2016. "Ruxolitinib versus best available therapy in patients with polycythemia vera: 80-week follow-up from the RESPONSE trial." *Haematologica* 101 (7):821–829. doi: 10.3324/haematol.2016.143644.

Wardrop, D., and D. P. Steensma. 2009. "Is refractory anaemia with ring sideroblasts and thrombocytosis (RARS-T) a necessary or useful diagnostic category?." *Br J Haematol* 144 (6):809–817. doi: 10.1111/j.1365-2141.2008.07526.x.

Weisberg, E., Q. Liu, E. Nelson, A. L. Kung, A. L. Christie, R. Bronson, M. Sattler, T. Sanda, Z. Zhao, W. Hur, C. Mitsiades, R. Smith, J. F. Daley, R. Stone, I. Galinsky, J. D. Griffin, and N. Gray. 2012. "Using combination therapy to override stromal-mediated chemoresistance in mutant FLT3-positive AML: synergism between FLT3 inhibitors, dasatinib/multi-targeted inhibitors and JAK inhibitors." *Leukemia* 26 (10):2233–2244. doi: 10.1038/leu.2012.96.

Xu, X., Q. Zhang, J. Luo, S. Xing, Q. Li, S. B. Krantz, X. Fu, and Z. J. Zhao. 2007. "JAK2(V617F): prevalence in a large Chinese hospital population." *Blood* 109 (1):339–342. doi: 10.1182/blood-2006-03-009472.

Younes, A., J. Romaguera, M. Fanale, P. McLaughlin, F. Hagemeister, A. Copeland, S. Neelapu, L. Kwak, J. Shah, S. de Castro Faria, S. Hart, J. Wood, R. Jayaraman, K. Ethirajulu, and J. Zhu. 2012. "Phase I study of a novel oral Janus kinase 2 inhibitor, SB1518, in patients with relapsed lymphoma: evidence of clinical and biologic activity in multiple lymphoma subtypes." *J Clin Oncol* 30 (33):4161–4167. doi: 10.1200/JCO.2012.42.5223.

Yu, H., D. Pardoll, and R. Jove. 2009. "STATs in cancer inflammation and immunity: a leading role for STAT3." *Nat Rev Cancer* 9 (11):798–809. doi: 10.1038/nrc2734.

Zhang, Q., H. Bai, X. H. Yu, B. Wu, Y. Z. Pan, C. B. Wang, L. P. Zhao, W. B. Li, F. Xu, and J. Zhang. 2018. "[Changes of IL-21 and its mediated JAK/STAT signaling pathway in patients with immune thrombocytopenia]." *Zhongguo Shi Yan Xue Ye Xue Za Zhi* 26 (3):859–865. doi: 10.7534/j.issn.1009-2137.2018.03.038.

Zhang, Y., and J. M. Kolesar. 2011. "Eltrombopag: an oral thrombopoietin receptor agonist for the treatment of idiopathic thrombocytopenic purpura." *Clin Ther* 33 (11):1560–1576. doi: 10.1016/j.clinthera.2011.10.004.

13

Immunodeficiency: Consequences of Mutations in JAK-STAT Signaling

Daniel Silberger and Duy Pham
Department of Pathology
University of Alabama at Birmingham
Birmingham, Alabama

13.1 Overview JAK-STAT Signaling

The JAK-STAT is a classical pathway that plays an essential role in immunity, cell division, cell death, and tumor formation (Stark and Darnell 2012). JAK-STAT signaling starts with the interaction between extracellular stimuli such as cytokines or growth factors with their corresponding transmembrane receptors (Murray 2007). Cytokine-receptor interaction induces receptor dimerization, which promotes transactivation of receptor-associated JAKs by bringing it into close proximity and by prompting conformational changes that separate their kinase domains and inhibitory pseudo-kinase domains (Leonard 2002; Rawlings 2004; Schindler 1999). Activated JAKs then phosphorylate the receptor's tyrosine residues to create docking sites for SH2 domain-associated STAT proteins (Meyer et al. 2004). Subsequently, JAKs phosphorylate receptor-bound STATs, resulting in STAT dissociation from the receptor. Activated STATs can form homo- or heterodimers with other STATs, translocate to the cell nucleus, bind DNA, and promote transcription of target genes (**Figure 13.1**).

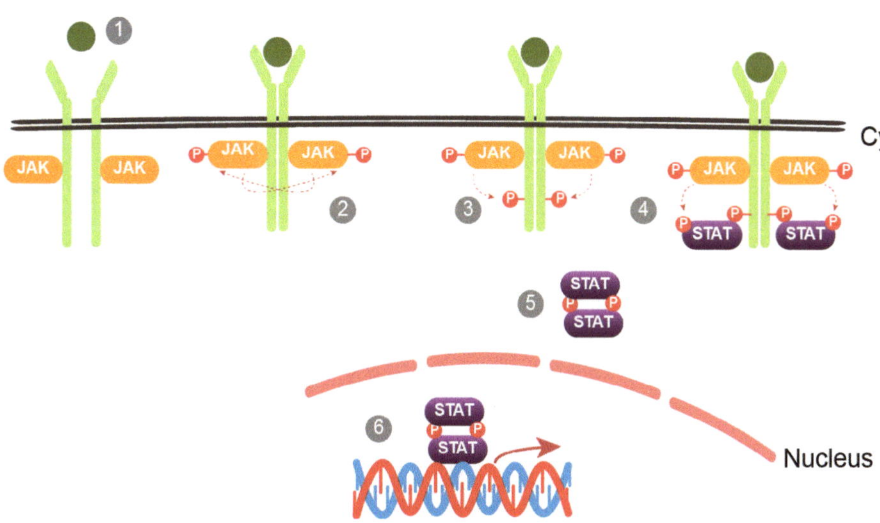

FIGURE 13.1 JAK-STAT signaling cascade. (**1**) Cytokine-receptor interaction and receptor dimerization (**2**) transactivation of receptor-associated JAKs (**3**) JAK phosphorylation of receptor (**4**) STAT binds phosphorylated receptor and is phosphorylated by JAK (**5**) STAT dimerization, and (**6**) STAT dimers translocate to nucleus, bind DNA, and modulate gene expression

In mammals, there are four JAKs (JAK1, JAK2, JAK3, and TYK2) and seven STATs (STAT1, STAT2, STAT3, STAT4, STAT5a, STAT5b, and STAT6) that are associated with the signaling of more than 50 cytokines and growth factors (Bousoik and Aliabadi 2018; Darnell 1997). Importantly, the specificity of JAK-STAT signaling greatly depends on cell lineages, qualitative differences in the duration and or intensity, heterogeneous signaling involving in multiple STATs, and hierarchical order of STATs in response to stimuli (O'Shea and Murray 2008; Seidel et al. 1995; Villarino et al. 2003).

As mentioned, STATs not only form homodimers and heterodimers in a stimuli-dependent manner, but also form higher order tetramers that are poorly understood (Lin et al. 2017, 2012). In contrast to dimerization that occurs in the cytoplasm, tetramerization assembles in the nucleus upon binding of STAT dimers and DNA (John et al. 1999). For example, while Type I IFN promotes STAT1/STAT2 heterodimers, Type II IFN elicits STAT1 homodimers (Au-Yeung et al. 2014). IL-10 also induces heterodimers between STAT1 and STAT3 at the sis oncogene (SIE) sequence of the c-fos promoter (Wehinger et al. 1999). Another non-canonical model of JAK-STAT signaling is JAK-independent activation of STAT (Nan, Wu, and Zhang 2018; Singh 2011). It has been shown that certain receptor tyrosine kinases can bypass JAK activity and mediate the phosphorylation of STAT on its own (Zhang et al. 2000). Flt-3 receptor and STING have been demonstrated to activate STAT5 and STAT6, respectively, in a JAK-independent manner (Zhang et al. 2000). In addition to tyrosine phosphorylation, other post-translational modifications regulate STAT function, including serine phosphorylation, which influences the DNA-binding affinity of STATs and the interaction of STATs with other factors (Lim and Cao 2006). STAT function can also be altered by other modifications such as acetylation, methylation, and sumoylation, that can either promote or limit STAT function (Mowen et al. 2001; Ungureanu et al. 2003; Yuan 2005). It has been shown that both lysine acetylation and arginine methylation of STAT proteins can enhance its function (Mowen et al. 2001; Zhuang 2013). In contrast, sumoylation has been strongly associated with repressing the activity of STAT, including STAT1 and STAT5 (Begitt et al. 2011; Van Nguyen et al. 2012). Another non-canonical characteristic of JAK-STAT is the location of its function. Although STATs have been shown to localize to the nucleus and regulate gene expression, recent studies have demonstrated its function elsewhere (Meier and Larner 2014). For example, all other STATs with the exception of STAT4 have been reported to translocate to the mitochondrion, but their function within this compartment remains poorly understood (Meier and Larner 2014).

STAT proteins act as classical transcription factors (TFs) with the ability to recognize and bind to consensus DNA sequence known as GAS elements (Villarino et al. 2015). In T helper cells, STAT binding has been mapped to various regions of the genome including proximal to transcription start sites, enhancers, and other cis regulatory elements (Vahedi et al. 2012). However, each STAT has some selectivity in the core sequence that results in variable DNA-binding affinity and overlapping binding sites between individual STATs (O'Shea et al. 2011). This partially explains for common genes that are regulated by multiple STAT family members (Villarino et al. 2015). In addition, STAT proteins are capable of recruiting co-activators and other TFs to promote epigenetic changes and enhance transcription of target genes (Shuai 2000; Vahedi et al. 2012). STATs have been shown to associate with CBP, p300, and histone acetyltransferases (HATs) that promote active histone marks thus enhancing gene expression (Shuai 2000). In addition, STATs integrate with other TFs to create regulatory networks that have been dissected in detail in T cell biology (O'Shea et al. 2011; Vahedi et al. 2012). For example, STAT3 involves in a transcriptional network including Irf4, Batf, and Rorγt that plays a crucial role in T cell biology (Ciofani et al. 2012). Other groups show the cooperation of STATs with lineage-specific TFs including T-bet, Gata-3, and Foxp3 (Maier et al. 2012; Passerini et al. 2008; Thieu et al. 2008). Whether STATs have a central role in the network or are just components remain to be explored.

13.2 Regulation of JAK-STAT Signaling

Given the importance of the JAK-STAT pathway, every step of the signaling is strictly monitored. One key component of the JAK-STAT signaling pathway is the phosphorylation of tyrosine residues of JAKs, STATs, and upstream receptors. Protein tyrosine phosphatases (PTPs), such as SHP1, SHP2, and CD45, can hydrolyze phosphorylated tyrosine residues thus inhibiting JAK-STAT signaling (Böhmer and Friedrich 2014; Shuai and Liu 2003). There are other PTPs including PTPRD, PTPRT, TCPTP, PTP1B, and DUSP2 (Lu et al. 2015). Another key regulatory module is protein inhibitors of activated STATs (PIAS) including four-member protein family: PIAS1, PIAS3, PIASx, and PIASγ (Niu et al. 2018). These PIAS exert a variety of mechanisms to suppress STAT activation, such as adding small ubiquitin-like modifier (SUMO) onto JAK and STAT to modify their function, directly interacting with STATs to block their DNA-binding activity, and recruiting histone deacetylases (HDACs) that modulate gene expression (Niu et al. 2018). Another essential regulatory component comprises proteins of the suppressor of cytokine signaling (SOCS) family including cytokine-inducible SH2 domain-containing protein (CISH), SOCS1, SOCS2, SOCS3, SOCS4, SOCS5, SOCS6, and SOCS7, but only the first four members are linked with the JAK-STAT pathway in mammals (Croker et al. 2008; Palmer and Restifo 2009). SOCS proteins have a SH2 domain that allows receptor binding and directly suppresses JAK activation (Croker et al. 2008). Since STATs can induce the expression of SOCS, they act as negative feedback loops to modulate both quantitative and qualitative aspects of JAK-STAT signaling. Last, but not least, STATs are maintained at a limiting resource via self-regulated gene expression, post-transcriptional regulation, and post-translational regulation, such that minor changes can have significant effects in its function (Qi and Yang 2014).

13.3 JAK-STAT in Immunology

13.3.1 T Helper Cell Development

T helper cells secrete cytokines that are important in mediating immune responses, and JAK-STAT signaling is an essential component for the development of T cell lineages (**Figure 13.2**).

Th1 cells require T-bet, IFN-γ-STAT1, and IL-12-STAT4 for its development, which produce IFN-γ and mediate immune response against intracellular pathogens (Athie-Morales et al. 2004; G Ler et al. 1996; Ma et al. 2010). Th2 cells produce IL-4, IL-5, and IL-13 that are important in mounting immune responses against parasites and contributing to allergic reactions (Walker and McKenzie 2017). IL-4/STAT6 is required for the differentiation of Th2 cells and inhibition of Th1 cell differentiation (Kaplan et al. 1996; Wurtz 2004). Th17 cells produce IL-17A, IL-17F, IL-21, and IL-22 that are critical in immunity against extracellular bacteria and antifungal responses (Becher and Segal 2011). T follicular helper cells (Tfh) produce IL-4 and IL-21 and are important in the maintenance of germinal center responses (Crotty 2011). Both Th17 and Tfh cells require IL-6-STAT3 signaling for their development (Foley 2007; H. Wu et al. 2016). Regulatory T cells (Treg) are essential to suppress immune response to maintain homeostasis and self-tolerance (Bluestone and Tang 2005). IL-2-STAT5 is required for the generation of Treg cells, by promoting the expression of Foxp3, a Treg-defining factor (Mahmud et al. 2013). Th9 cells secrete IL-9 and are essential in protecting against helminth infections, promoting allergic responses, autoimmunity, and tumor suppression (Kaplan et al. 2015). Several STAT proteins are required to modulate Th9 cell differentiation including STAT5 and STAT6 that act as positive regulators, while STAT3 inhibits Th9 differentiation (Goswami et al. 2012; Olson et al. 2016; Ulrich et al. 2017). STATs play a complicated role in the regulation of T helper cell differentiation. In fact, multiple STAT signals are integrated in order to regulate the development of T cells (O'Shea et al. 2011). IL-2-STAT5 is also required for Th1 and Th2 cell differentiation by promoting T-bet expression and IL4Rα-induced GATA3 expression, respectively (Liao et al. 2008). In addition, STAT3 has been shown to be required for Th2 cell differentiation (Stritesky et al. 2011). Type I IFN activates STAT1 and STAT2 that can suppress Treg cell development (Srivastava et al. 2014). In addition, Type I IFN can induce STAT4 activation with lesser potency compared to IL-12 and might regulate Th1 cell differentiation (Ramos et al. 2007). Taken together, JAK-STAT signaling plays a key role in T helper cell differentiation but the interplay between STAT proteins is still poorly understood.

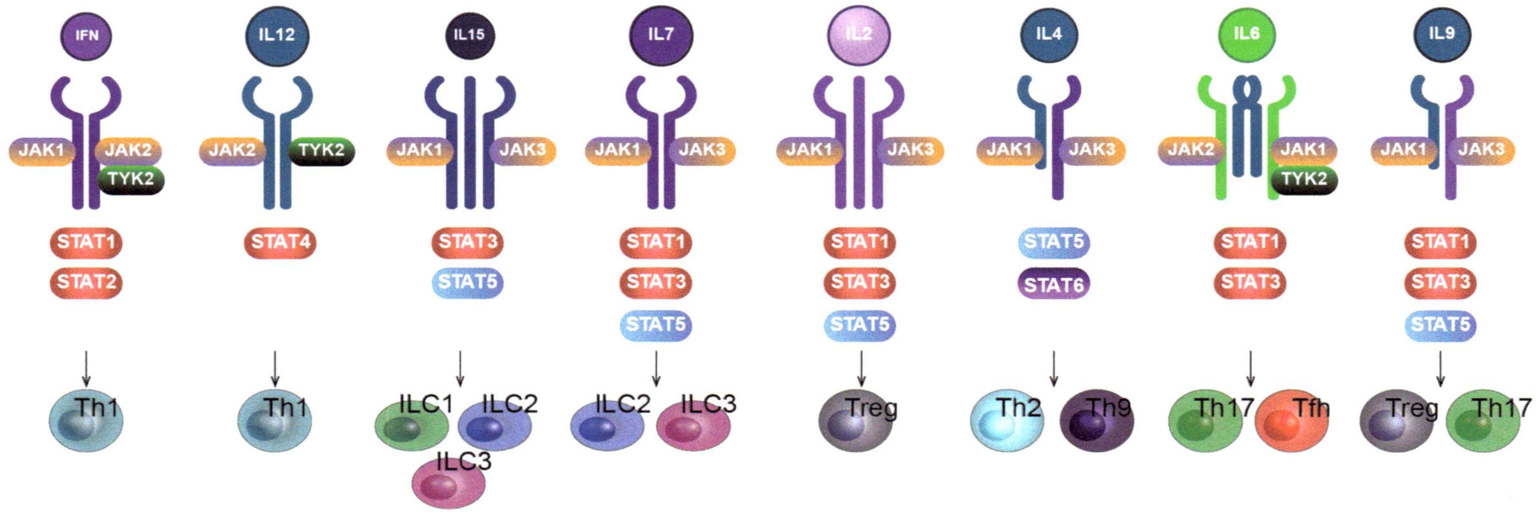

FIGURE 13.2 The contribution of JAK-STAT signaling in the development of T helper cell and ILC subsets. Cytokines activate JAK-STAT pathways that are essential in the differentiation of immune cells

13.3.2 Innate Lymphoid Cell Development

JAK-STAT signaling also plays an essential role in the development of innate lymphoid cells (ILC) (**Figure 13.2**) (Stabile et al. 2018). ILC1 cells are required for protective immunity but are also involved in the homeostatic regulation of immunity. IL-7 and IL-15 provide critical signals that are necessary for the development of ILC1 cells (Villarino et al. 2017b). This is further demonstrated by mice with deficiencies in IL-15 or IL-2Rγ, which have reduced ILC1 subsets (Lodolce et al. 1998; Ranson et al. 2003). In addition, IL-7 signaling is pivotal for the development of Type 2 (ILC2) and type 3 (ILC3) ILCs cells (Vonarbourg and Diefenbach 2012). Mice with deficiency in IL-7 signaling, including *Il7ra*$^{-/-}$ or *JAK3*$^{-/-}$ mice, have defect in lymphoid organ development (Cao et al. 1995; Park et al. 1995). Additionally, both IL-7 an IL-15 utilize the common y chain receptor for their signaling and patients with severe combined immunodeficiency (SCID), who carry mutations of *IL2RG* or *JAK3*, are deficient in ILC subsets; further demonstrating the importance of JAK-STAT pathway in ILC development and homeostasis (Vély et al. 2016). IL-7 and IL-15 signal through JAK3–STAT5 via two STAT5 isoforms including STAT5A and STAT5B. STAT5B has an important role in regulating development and activation of ILCs (Imada et al. 1998; Villarino et al. 2017b).

Both IL-12-STAT4 and Type I IFN-STAT1 signaling are important for NK cell development, though the role of the latter remains controversial (Nguyen et al. 2000; Thierfelder et al. 1996). In particular, STAT4 is indispensable for the differentiation and proliferation of specific subsets of NK cells (Madera et al. 2018). In addition, ILC1 cells require IL-12-STAT4 activity to promote their pro-inflammatory functions against pathogens, especially tissue-resident ILC (Weizman et al. 2017).

ILC2 cells require transcription factors RORα and GATA3 for their development, and secrete IL-5, IL-9, and IL-13 that are indispensable in immune response to parasitic worm infections and allergic inflammation (Walker and McKenzie 2013). Although IL-4/STAT6 signaling does not play an essential role for the differentiation of ILC2, it is still important for ILC2 function in specific inflammatory conditions. For example, reduced IL-13-producing ILC2 were observed in STAT6-deficient mice during *Nippostrongylus brasiliensis* infection and in a mouse model of lung inflammation induced by the fungal allergen Alternaria (Doherty et al. 2012; Liang et al. 2011).

ILC3 cells require Rorγt and Ahr for their development and production of IL-17A, IL-17F, and IL-22 cytokines that regulate intestinal homeostasis, infection, and inflammation (Melo-Gonzalez and Hepworth 2017). IL-23/STAT3 signaling is required to promote ILC3 functions (Geremia et al. 2011; Takatori et al. 2009). Similar to T helper cell development, a combination of multiple STATs are involved in the development of ILC subsets (Stabile et al. 2018). While STAT3 is important for ILC3 functions, it is also essential for NK cell effector function (Zhu et al. 2014). In addition, IL-23 activates STAT4 and is required for the development of the IFN-γ-producing NCR+ILC3 cell population (Mikami et al. 2018).

13.4 Mutations of JAKs and STATs Associated with Primary Immunodeficiency

Given the important role of JAK-STAT, defects or mutation of JAK-STAT have been linked to several disorders of primary immunodeficiency. JAK1 mutations are found in 20% of down syndrome-associated B-ALL and 9% of patients with hepatitis B-associated hepatocellular carcinoma (Hammarén et al. 2018). Two homozygous missense germline mutations, including P733L and P832S, of JAK1 have been identified (Eletto et al. 2016). Patients with the mutations have diminished expression of IFN-target genes due to reduced JAK1–STAT signaling, which results in enhanced susceptibility to atypical mycobacterial infection and metastatic bladder carcinoma (Eletto et al. 2016). In addition, frameshift mutations in JAK1 have been identified that cause increased mutation burden and microsatellite instability (Albacker et al. 2017). Gene expression analysis from biopsy samples of patients with JAK1 frameshift mutations reveals defects in multiple interferon and anti-tumor immune response signatures (Stelloo et al. 2016).

Altogether, JAK1 mutations lead to the loss of multiple IFN-mediated immune responses against tumor and bacterial infections.

JAK2 has a high mutation frequency of 38.7% (Villarino et al. 2017a). Erythropoietin and thrombopoietin-mediated JAK2 signaling pathway is important for the expansion of erythrocytes and megakaryocytes (Ihle and Gilliland 2007). The most common JAK2 mutation is V617F of pseudokinase domain that is responsible for its inhibitory function (Nielsen et al. 2011).

JAK3 is commonly expressed in T cells, NK cells, and intestinal epithelial cells (Kawamura et al. 1994; Musso et al. 1995). JAK3 signaling involves receptors containing the common gamma chain (γc) of type I cytokine receptor family including IL-2R, IL-4R, IL-7R, IL-9R, IL-15R, and IL-21R (Babon et al. 2014). Because JAK3 signaling is associated with γc, its mutation has been found to be similar to X-link severe combined immunodeficiency (SCID) underlying by γc mutation (Pesu et al. 2005). The diagnosis of SCID is generally based on recurrent severe infection, diarrhea, and atopic dermatitis (Pesu et al. 2005). Patients with JAK3 or γc mutation are unable to mount a proper immune responses due to nonfunctional T cells, defective B cell activation, and impaired immunoglobulin production (Pesu et al. 2005).

Tyrosine kinase 2 (TYK2) signals through the receptors of type I IFN, IL-6, IL-10, IL-12, and IL-23 (Strobl 2011). A frameshift at codon 90 disrupts the translation of TYK2 protein and results in mutation (Nemoto et al. 2018). TYK2 mutation was discovered based on a patient with symptoms of hyper-IgE syndrome (HIES), that developed infection with bacteria including Staphylococcus and mycobacteria and/or viruses (Minegishi et al. 2006). In a mild case, a recessive partial TYK2 deficiency with heterozygous mutations (c.209_212delGCTT/c.691C > T, p.Cys70Serfs*21/p.Arg231Trp) was found in patients with T-cell lymphopenia who developed Epstein–Barr virus (EBV)-associated B-cell lymphoma (Nemoto et al. 2018). Unlike the case of complete TYK2 mutation, patients with partial TYK2 deficiency have impaired T cell responses to IL-23 (Nemoto et al. 2018). Genome-wide association study (GWAS) studies show an association between polymorphisms of TYK2 with various autoimmune diseases including inflammatory bowel disease, systemic lupus erythematosus, multiple sclerosis, psoriasis, and rheumatoid arthritis (Dendrou et al. 2016).

STAT1 mutations include both gain-of-function (GOF) and loss-of-function (LOF) mutations (Villarino et al. 2017a). GOF mutation in STAT1 is associated with chronic mucocutaneous candidiasis (CMC) characterized by persistent *Candida albicans* infections of skin, nails, and mucosa (Carey et al. 2018). Moreover, STAT1 GOF mutations have been linked with autoimmunity and IPEX-like syndrome (Uzel et al. 2013). The mechanisms underlying these abnormalities are due to enhanced IFN-γ signaling, thus inhibiting IL-17 production and increasing susceptibility to fungal infections (Uzel et al. 2013). In contrast, patients with LOF mutation in STAT1, have increased susceptibility to mycobacterial infections due to a reduction in type II IFN signaling (Boisson-Dupuis et al. 2012). Another variant of STAT1 LOF have impaired type I and type II IFN signaling, thus causing reduced anti-viral response (O'Shea et al. 2013).

Type I IFN signaling through STAT2 is important for immunity against viral infection (Morrison and García-Sastre 2014). Patients with STAT2 mutations have been identified including STAT2 c.381+5 G>C and c.1836 C>A (Hambleton et al. 2013; Shahni et al. 2015). STAT2 c.381+5 G>C results in a splice mutation, and patients with this mutation are susceptible to severe viral infection (Hambleton et al. 2013). In the latter case, STAT2 mutation was identified using whole exome sequencing from patient with severe neurological deterioration following viral infection (Shahni et al. 2015). Patients with STAT2 c.1836 C>A have elongated mitochondria due to inactivation of fission protein DRP1 (encoded by DNM1L) (Shahni et al. 2015). Thus, STAT2 mutation can lead to defects in mitochondrial dynamics and could be the cause of neurological disease (Hambleton et al. 2013).

IL-6 and other gp130 cytokines signal through STAT3 to induce gene expression and are critical in various cellular processes including cell growth and apoptosis (Hillmer et al. 2016). Dominant negative mutation of STAT3 is associated with hyperimmunoglobulin E syndrome (HIES, or Job's syndrome) that is characterized by recurring and severe cutaneous and sinopulmonary bacterial infections, increased serum immunoglobulin E (IgE), and connective tissue abnormalities (Holland et al. 2009). Because IL-6-STAT3 signaling plays a key role in the development of Th17 cells that secrete IL-17 and IL-22 cytokines to mediate protective immune responses against bacterial and fungal infections at barrier sites, mutations in STAT3 signaling are almost always Th17/IL-17-related defects (Foley 2007; Weaver et al. 2013; Zindl et al. 2013). Patients with STAT3 mutation often experience atopic dermatitis, staphylococcal skin abscesses, and mucocutaneous candidiasis typical of the disease (Mogensen 2013). STAT3 is also required for CD8 memory T cells; thus, impaired STAT3 function results in fragile immune response against viral infections (Siegel et al. 2011). In contrast to STAT3 LOF that causes immunodeficiency, GOF mutations in STAT3 result in defective STAT1 and STAT5 phosphorylation and regulatory T cell development, and are associated with autoimmune diseases (Milner et al. 2015; Nabhani et al. 2017). Patients with STAT3 GOF were diagnosed with lymphadenopathy, autoimmune cytopenias, multi-organ autoimmunity, and infections (Milner et al. 2015).

IL-12 activation of STAT4 is important for the development of Th1 and Th17 cell differentiation, monocyte activation, and IFN-γ production (Athie-Morales et al. 2004). Although germline GOF and LOF in STAT4 have not been reported, a number of GWAS studies have identified an association of *STAT4* variants with autoimmune diseases including inflammatory bowel disease, systemic lupus erythematosus, rheumatoid arthritis, primary biliary cirrhosis, Sjögren's syndrome, and systemic sclerosis (Jiang et al. 2013; Korman et al. 2008; Remmers et al. 2007).

STAT5 exists in two highly identical variants, STAT5A and STAT5B, at the amino acid level (Heltemes-Harris et al. 2011; Rani and Murphy 2016). STAT5A and STAT5B expression has been detected in mammary tissue and muscle/liver, respectively (Mui et al. 1995; Rani and Murphy 2016). Both STAT5A and STAT5B are activated by various cytokines and hormones such as growth hormone, erythropoietin, prolactin, and interleukins, and are essential for the development and homeostasis of lymphoid cells (Rani and Murphy 2016). Because of the importance of STAT5 in growth hormone signaling, defect in STAT5 function especially autosomal recessive STAT5B mutations, can result in dwarfism, immunodeficiency, and autoimmunity (Hwa et al. 2011; Klammt et al. 2018; Kofoed et al. 2003). In addition, STAT5B mutations have been linked with increased susceptibility to bacterial and viral infections, severe eczema, immunodeficiency, pulmonary inflammation, and immunodeficiency (Leonard and O'Shea 1998).

STAT6 is activated in response to the cytokine IL-4, and is important in Th2 cell development and anti-apoptotic activity (Kaplan et al. 1996). Although GOF and LOF in STAT6 have not been reported, GWAS studies have discovered *STAT6* variants that are associated with atopy, asthma, and eosinophilic esophagitis (Sleiman et al. 2014). Additionally, a novel STAT6 mutation hotspot (p.419D/G, p.419D/A, and p.419D/H) is responsible for 11% of follicular lymphoma (FL), a common non-Hodgkin lymphoma (Yildiz et al. 2015).

13.5 Therapeutic Targeting of JAK-STAT Signaling

Because of the importance of JAK-STAT signaling in autoimmune disease and malignancy, therapeutic approaches targeting the pathway have been intensively developed (**Figure 13.3**).

One of the approaches is to target the upstream activators including recombinant cytokines, engineered cytokines, agonist cytokine–antibody complexes, and agonist cytokine–antibody fusions that have had some success in animal models and clinical trials (O'Shea et al. 2015). For example, Type I IFN has been used to treat hepatitis C and multiple sclerosis (Reder and Feng 2014; Rong and Perelson 2010). Low dose of IL-2 has been used to target certain autoimmune diseases (Collison 2018). Moreover, FDA-approved cytokine- and receptor-blocking antibodies, such as anti-IL2Rα, anti-IL-6, anti-IL-6R, anti-IL-12, and anti-IL-23, have shown promising results in immunotherapies (Niederreiter et al. 2013; Rider et al. 2016). Given the vital role of the kinase domain for JAK activity, it has become an attractive target for therapeutic approaches (O'Shea et al. 2015). The first FDA-approved JAK inhibitor (JAKinibs) is Ruxolitinib that targets both JAK1 and JAK2 (Mesa 2010). It has been used for treatment of primary myelofibrosis, and has some encouraging success in psoriasis and in rheumatoid arthritis (RA) (Mesa 2010). Tofacitinib was developed to block JAK3, that is used for treatment of rheumatoid arthritis,

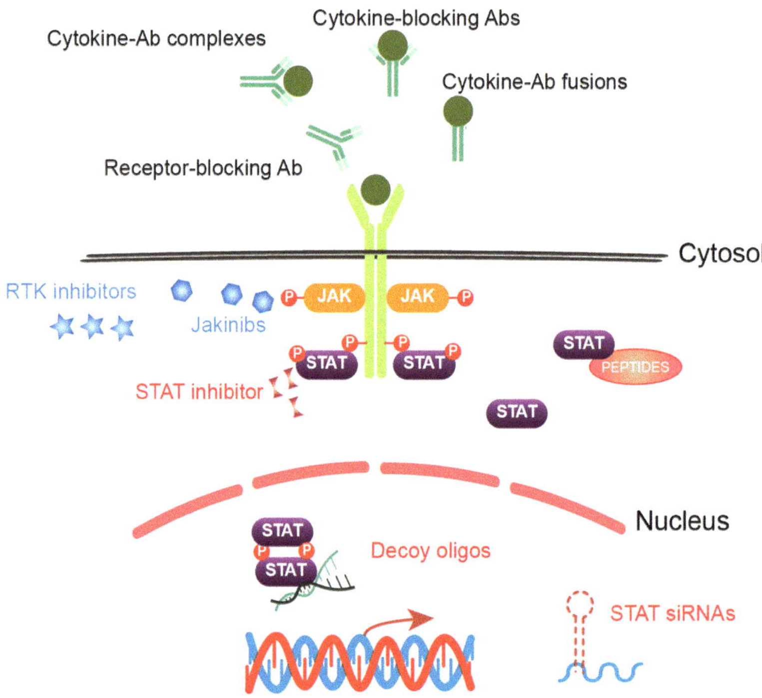

FIGURE 13.3 Current therapeutic approaches targeting JAK-STAT. (**1**, black) Inhibitors target upstream activators including recombinant cytokines, engineered cytokines, agonist cytokine–antibody complexes, and agonist cytokine–antibody fusions (**2**, blue) JAK inhibitors (JAKinibs) and receptor tyrosine kinase (RTK) inhibitors (**3**, red) STAT inhibitors

psoriasis, IBD, transplant rejection, juvenile arthritis, and many others (Ellis et al. 2015; Furumoto and Gadina 2013). JAK3 was thought to be an attractive therapeutic target due to its essential role in lymphocytes, and that mutations in JAK3 does not affect non-immunologic organs or tissues (Babon et al. 2014). However, tofacitinib also targets, to a lesser extent, JAK1 and JAK2, thus causing adverse effect on patients. In fact, mild anemia with neutropenia and increased malignancy risk were observed in patients treated with tofacitinib (Clark et al. 2014; Riese et al. 2010). Thus, further studies will be needed to clarify the risk of tofacitinib during the course of treatment. Another JAK inhibitor, oclacitinib, does not target a specific JAK, and this has raised the question of the drug's effectiveness and side effects (Gonzales et al. 2014). However, it has shown promising results in treating allergic skin disease and could be a potential therapeutic approach to treat immune-mediated dermatologic conditions in patients (Damsky and King 2017). Other JAK inhibitors have been developed for various diseases such as asthma, malignancies, and autoimmune conditions but are currently at different stages of clinical trials (O'Shea et al. 2015). Baricitinib and momelotinib inhibit both JAK1 and JAK2, and have shown encouraging results in the treatment of autoinflammatory diseases and myelofibrosis, respectively (Fleischmann et al. 2017; Tefferi et al. 2018). Although the use of JAK inhibitors has yielded some successes, there are remaining issues that need to be addressed and rigorously interrogated, including the development of more selective JAK inhibitors and the optimal dosing during treatment.

In contrast to JAK inhibitors that specifically target kinase domains, STAT proteins do not have catalytic activity and pose more of a challenge for therapeutic targets. STAT inhibitors have been developed by blocking phosphorylation and thus preventing SH2-mediated binding to receptors and dimerization or inhibiting STAT DNA-binding activity. Several strategies have been developed to target STATs including oligonucleotide-based STAT inhibitors, small molecule inhibitors, and STAT-binding intrabodies (Howell et al. 2018; Koo et al. 2014; Sen and Grandis 2014). These STAT inhibitors are still at various stages of the development (O'Shea et al. 2015). For example, oligonucleotide-based STAT inhibitors are being examined in several malignancies (Fagard et al. 2014). OPB-51,062 and OPB-31,121.6 have been developed to target STAT3 and have shown promising results in animal models of autoimmune disease (O'Shea et al. 2015).

13.6 Concluding Remarks

The JAK-STAT represents an ancient signaling pathway that regulates cellular communication and gene expression. Mutations in the JAK-STAT pathway are frequently associated with primary immunodeficiencies. Because JAK-STAT plays an important role in the development and homeostasis of T and ILC cells, defects in any of its signaling components results in severe immune dysregulation; therefore, therapeutic targeting of JAK-STAT is challenging due to the requirement of high specificity. Recent studies show the involvement of JAK-STAT in cell biology expanding beyond immunology and cancer biology and emerging to the field of stem cell biology and metabolomics. Thus, JAK-STAT signaling pathway acts as a central hub for the regulation of molecular and cellular functions. As a result, it is required to interrogate in depth the regulation of JAK-STAT pathway and its role in regulating chromatin biology and cellular differentiation in a cell-type specific manner. In addition, the precise mechanisms of how JAK-STAT inhibitors work in different clinical settings need to be investigated in detail. Thus, a better understanding of the molecular function of JAK-STAT using advance technology for genetic manipulation, such as CRISPR-Cas9, would provide tremendous mechanistic insights for improved diagnosis and therapeutic options for JAK-STAT-related diseases.

REFERENCES

Albacker, Lee A, Jeremy Wu, Peter Smith, Markus Warmuth, Philip J Stephens, Ping Zhu, Lihua Yu, and Juliann Chmielecki. 2017. Loss of Function JAK1 Mutations Occur at High Frequency in Cancers with Microsatellite Instability and are Suggestive of Immune Evasion. *PLoS One* 12 (11):e0176181.

Athie-Morales, Veronica, Hermelijn H Smits, Doreen A Cantrell, and Catharien M U Hilkens. 2004. Sustained IL-12 Signaling Is Required for Th1 Development. *J Immunol* 172 (1):61–9.

Au-Yeung, Nancy, Roli Mandhana, and Curt M Horvath. 2014. Transcriptional Regulation by STAT1 and STAT2 in the Interferon JAK–STAT Pathway. *JAKStat* 2 (3):e23931.

Babon, Jeffrey J, Isabelle S Lucet, James M Murphy, Nicos A Nicola, and Leila N Varghese. 2014. "the Molecular Regulation of Janus Kinase (JAK) Activation. *Biochemical J* 462 (1):1–13.

Becher, Burkhard, and Benjamin M Segal. 2011. TH17 Cytokines in Autoimmune Neuro-Inflammation. *Curr Opinion Immunol* 23 (6):707–12.

Begitt, Andreas, Mathias Droescher, Klaus-Peter Knobeloch, and Uwe Vinkemeier. 2011. "SUMO Conjugation of STAT1 Protects Cells from Hyperresponsiveness to IFNγ. *Blood* 118 (4):1002–7.

Bluestone, Jeffrey A, and Qizhi Tang. 2005. How Do CD4+CD25+ Regulatory T Cells Control Autoimmunity? *Curr Opinion in Immunol* 17 (6):638–42.

Böhmer, Frank-D, and Karlheinz Friedrich. 2014. Protein Tyrosine Phosphatases as Wardens of STAT Signaling. *JAKStat* 3 (1):e28087.

Boisson-Dupuis, Stephanie, Xiao-Fei Kong, Satoshi Okada, Sophie Cypowyj, Anne Puel, Laurent Abel, and Jean-Laurent Casanova. 2012. Inborn Errors of Human STAT1: Allelic Heterogeneity Governs the Diversity of Immunological and Infectious Phenotypes. *Curr Opinion in Immunol* 24 (4):364–78.

Bousoik, Emira, and Hamidreza Montazeri Aliabadi. 2018. Do We Know Jack' about JAK? a Closer Look at JAK/STAT Signaling Pathway. *Front in Oncol* 8:338.

Cao, X, E W Shores, J Hu-Li, M R Anver, B L Kelsall, S M Russell, J Drago, M Noguchi, A Grinberg, and E T Bloom. 1995. Defective Lymphoid Development in Mice Lacking Expression of the Common Cytokine Receptor Gamma Chain. *Immunity* 2 (3):223–38.

Carey, Barbara, Jonathan Lambourne, Stephen Porter, and Tim Hodgson. 2018. Chronic Mucocutaneous Candidiasis Due to Gain-of-Function Mutation in STAT1. *Oral Diseases* 17 (4). (10.1111):399.

Ciofani, Maria, Aviv Madar, Carolina Galan, MacLean Sellars, Kieran Mace, Florencia Pauli, Ashish Agarwal et al. 2012. A Validated Regulatory Network for Th17 Cell Specification. *Cell* 151 (2):289–303.

Clark, James D, Mark E Flanagan, and Jean-Baptiste Telliez. 2014. Discovery and Development of Janus Kinase (JAK) Inhibitors for Inflammatory Diseases. *J Med Chem* 57 (12):5023–38.

Collison, Joanna. 2018. Low-Dose IL-2 Therapy for Autoimmune Diseases. *Nat Rev Rheumatol* 15 (1):2.

Croker, Ben A, Hiu Kiu, and Sandra E Nicholson. 2008. SOCS Regulation of the JAK/STAT Signalling Pathway. *Semin Cell Dev Biol* 19 (4):414–22.

Crotty, Shane. 2011. Follicular Helper CD4 T Cells (T FH). *Annu Rev Immunol* 29 (1):621–63.

Damsky, William, and Brett A King. 2017. JAK Inhibitors in Dermatology: The Promise of a New Drug Class. *J Am Acad Dermatol* 76 (4):736–44.

Darnell, J E, Jr. 1997. STATs and Gene Regulation. *Science* 277 (5332):1630–35.

Dendrou, Calliope A, Adrian Cortes, Lydia Shipman, Hayley G Evans, Kathrine E Attfield, Luke Jostins, Thomas Barber et al. 2016. Resolving TYK2 Locus Genotype-to-Phenotype Differences in Autoimmunity. *Sci Transl Med* 8 (363):363ra149.

Doherty, Taylor A, Naseem Khorram, Jinny E Chang, Hee-Kyoo Kim, Peter Rosenthal, Michael Croft, and David H Broide. 2012. STAT6 Regulates Natural Helper Cell Proliferation during Lung Inflammation Initiated by Alternaria. *Am J Physiol Lung Cell Mol Physiol* 303 (7):L577–88.

Eletto, Davide, Siobhan O Burns, Ivan Angulo, Vincent Plagnol, Kimberly C Gilmour, Frances Henriquez, James Curtis et al. 2016. Biallelic JAK1 Mutations in Immunodeficient Patient with Mycobacterial Infection. *Nat Commun* 7 (1):13992.

Ellis, Martin H, Noa Lavi, Elena Mishchenko, Najib Dally, David Lavie, Anna Courevitch, Odit Gutwein et al. 2015. Ruxolitinib Treatment for Myelofibrosis: Efficacy and Tolerability in Routine Practice. *Leukemia Research* 39 (11):1154–58.

Fagard, Remi, Valeri Metelev, Inès Souissi, and Fanny Baran-Marszak. 2014. STAT3 Inhibitors for Cancer Therapy. *JAKStat* 2 (1):e22882.

Fleischmann, Roy, Jahangir Alam, Vipin Arora, John Bradley, Douglas E Schlichting, David Muram, and Josef S Smolen. 2017. Safety and Efficacy of Baricitinib in Elderly Patients with Rheumatoid Arthritis. *RMD Open* 3 (2):e000546.

Foley, J F. 2007. STAT3 Regulates the Generation of Th17 Cells. *Science's STKE* 2007 (380):tw113.

Furumoto, Yasuko, and Massimo Gadina. 2013. The Arrival of JAK Inhibitors: Advancing the Treatment of Immune and Hematologic Disorders. *BioDrugs* 27 (5):431–8.

Geremia, Alessandra, Carolina V Arancibia-Cárcamo, Myles P P Fleming, Nigel Rust, Baljit Singh, Neil J Mortensen, Simon P L Travis, and Fiona Powrie. 2011. IL-23–Responsive Innate Lymphoid Cells Are Increased in Inflammatory Bowel Disease. *J Exp Med* 208 (6):1127–33.

G Ler, M L, J D Gorham, C S Hsieh, A J Mackey, R G Steen, W F Dietrich, and K M Murphy. 1996. Genetic Susceptibility to Leishmania: IL-12 Responsiveness in TH1 Cell Development. *Science* 271 (5251):984–7.

Gonzales, A J, J W Bowman, G J Fici, M Zhang, D W Mann, and M Mitton-Fry. 2014. Oclacitinib (APOQUEL ®) Is a Novel Janus Kinase Inhibitor with Activity against Cytokines Involved in Allergy. *J Vet Pharmacol Ther* 37 (4):317–24.

Goswami, Ritobrata, Rukhsana Jabeen, Ryoji Yagi, Duy Pham, Jinfang Zhu, Shreevrat Goenka, and Mark H Kaplan. 2012. "STAT6-dependent Regulation of Th9 Development. *J Immunol* 188 (3):968–75.

Hambleton, Sophie, Stephen Goodbourn, Dan F Young, Paul Dickinson, Siti M B Mohamad, Manoj Valappil, Naomi McGovern et al. 2013. STAT2 Deficiency and Susceptibility to Viral Illness in Humans. *Proc Natl Acad Sci U S A* 110 (8):3053–8.

Hammarén, Henrik M, Anniina T Virtanen, Juuli Raivola, and Olli Silvennoinen. 2018. The Regulation of JAKs in Cytokine Signaling and Its Breakdown in Disease. *Cytokine* 118:48–63 .

Heltemes-Harris, Lynn M, Mark J L Willette, Kieng B Vang, and Michael A Farrar. 2011. The Role of STAT5 in the Development, Function, and Transformation of B and T Lymphocytes. *Ann N Y Acad Sci* 1217 (1):18.

Hillmer, Emily J, Huiyuan Zhang, Haiyan S Li, and Stephanie S Watowich. 2016. STAT3 Signaling in Immunity. *Cytokine Growth Factor Rev* 31:1–15.

Holland, Steven M, Frank R DeLeo, Houda Z Elloumi, Amy P Hsu, Gulbu Uzel, Nina Brodsky, Alexandra F Freeman et al. 2009. STAT3 Mutations in the Hyper-IgE Syndrome. *Dx.Doi.org* 357 (16):1608–19.

Howell, Michael D, Carolyn Fitzsimons, and Paul A Smith. 2018. JAK/STAT Inhibitors and Other Small Molecule Cytokine Antagonists for the Treatment of Allergic Disease. *Ann Allergy Asthma Immunol* 120 (4):367–75.

Hwa, Vivian, Kari Nadeau, Jan M Wit, and Ron G Rosenfeld. 2011. STAT5b Deficiency: Lessons From STAT5b Gene Mutations. *Best Pract Res Clin Endocrinol Metab* 25 (1):61–75.

Ihle, James N, and D Gary Gilliland. 2007. JAK2: Normal Function and Role in Hematopoietic Disorders. *Curr Opin Genet Dev* 17 (1):8–14.

Imada, Kazunori, Eda T Bloom, Hiroshi Nakajima, Judith A Horvath-Arcidiacono, Garry B Udy, Helen W Davey, and Warren J Leonard. 1998. Stat5b Is Essential for Natural Killer Cell–Mediated Proliferation and Cytolytic Activity. *J Exp Med* 188 (11):2067–74.

Jiang, De-Ke, Jielin Sun, Guangwen Cao, Yao Liu, Dongxin Lin, Yu-Zhen Gao, Wei-Hua Ren et al. 2013. Genetic Variants in STAT4 and HLA-DQ Genes Confer Risk of Hepatitis B Virus-Related Hepatocellular Carcinoma. *Nat Genet* 45 (1):72–5.

John, S, U Vinkemeier, E Soldaini, J E Darnell, and W J Leonard. 1999. The Significance of Tetramerization in Promoter Recruitment by Stat5. *Mol Cell Biol* 19 (3):1910–18.

Kaplan, Mark H, Matthew M Hufford, and Matthew R Olson. 2015. The Development and in Vivo Function of T Helper 9 Cells. *Nat Rev Immunol* 15 (5):295–307.

Kaplan, Mark H, Ulrike Schindler, Stephen T Smiley, and Michael J Grusby. 1996. Stat6 Is Required for Mediating Responses to IL-4 and for the Development of Th2 Cells. *Immunity* 4 (3):313–19.

Kawamura, M, D W McVicar, J A Johnston, T B Blake, Y Q Chen, B K Lal, A R Lloyd, D J Kelvin, J E Staples, and J R Ortaldo. 1994. Molecular Cloning of L-JAK, a Janus Family Protein-Tyrosine Kinase Expressed in Natural Killer Cells and Activated Leukocytes. *Proc Natl Acad Sci U S A* 91 (14):6374–8.

Klammt, Jürgen, David Neumann, Evelien F Gevers, Shayne F Andrew, I David Schwartz, Denise Rockstroh, Roberto Colombo et al. 2018. Dominant-Negative STAT5B Mutations Cause Growth Hormone Insensitivity with Short Stature and Mild Immune Dysregulation. *Nat Commun* 9 (1):472.

Kofoed, Eric M, Vivian Hwa, Brian Little, Katie A Woods, Caroline K Buckway, Junko Tsubaki, Katherine L Pratt et al. 2003. Growth Hormone Insensitivity Associated with a STAT5bMutation. *New Engl J Med* 349 (12):1139–47.

Koo, Mi Young, Jiyoung Park, Jung Mi Lim, Sei Yoon Joo, Seung-Pil Shin, Hyun Bo Shim, Junho Chung, Dongmin Kang, Hyun Ae Woo, and Sue Goo Rhee. 2014. Selective Inhibition of the Function of Tyrosine-Phosphorylated STAT3 with a Phosphorylation Site-Specific Intrabody.". *Proc Natl Acad Sci U S A* 111 (17):6269–74.

Korman, B D, M I Alba, J M Le, I Alevizos, J A Smith, N P Nikolov, D L Kastner, E F Remmers, and G G Illei. 2008. Variant Form of STAT4 Is Associated with Primary Sjögren's Syndrome. *Genes Immun* 9 (3):267–70.

Leonard, Warren J. 2002. The JAK–STAT Pathway. *In Hormone Signaling* 17:103–120.

Leonard, Warren J, and John J O'Shea. 1998. JAKS and STATS: Biological Implications. *Annu Rev Immunol* 16 (1):293–322.

Liang, Hong-Erh, R Lee Reinhardt, Jennifer K Bando, Brandon M Sullivan, I-Cheng Ho, and Richard M Locksley. 2011. Divergent Expression Patterns of IL-4 and IL-13 Define Unique Functions in Allergic Immunity. *Nat Immunol* 13 (1):58–66.

Liao, Wei, Dustin E Schones, Jangsuk Oh, Yongzhi Cui, Kairong Cui, Tae-Young Roh, Keji Zhao, and Warren J Leonard. 2008. Priming for T Helper Type 2 Differentiation by Interleukin 2-Mediated Induction of Interleukin 4 Receptor Alpha-Chain Expression. *Nat Immunol* 9 (11):1288–96. doi: 10.1038/ni.1656.

Lim, Cheh Peng, and Xinmin Cao. 2006. Structure, Function, and Regulation of STAT Proteins. *Mol BioSyst* 2 (11):536–50.

Lin, Jian-Xin, Ning Du, Peng Li, Majid Kazemian, Tesfay Gebregiorgis, Rosanne Spolski, and Warren J Leonard. 2017. Critical Functions for STAT5 Tetramers in the Maturation and Survival of Natural Killer Cells. *Nat Commun* 8 (1):293.

Lin, Jian-Xin, Peng Li, Delong Liu, Hyun Tak Jin, Jianping He, Mohammed Ata Ur Rasheed, Yrina Rochman et al. 2012. Critical Role of STAT5 Transcription Factor Tetramerization for Cytokine Responses and Normal Immune Function. *Immunity* 36 (4):586–99.

Lodolce, James P, David L Boone, Sophia Chai, Rachel E Swain, Themistocles Dassopoulos, Shanthi Trettin, and Averil Ma. 1998. IL-15 Receptor Maintains Lymphoid Homeostasis by Supporting Lymphocyte Homing and Proliferation. *Immunity* 9 (5):669–76.

Lu, Dan, Liang Liu, Xin Ji, Yanan Gao, Xi Chen, Yu Liu, Yang Liu et al. 2015. The Phosphatase DUSP2 Controls the Activity of the Transcription Activator STAT3 and Regulates TH17 Differentiation. *Nat Immunol* 16 (12):1263–73.

Ma, Da, Hua Huang, and Zan Huang. 2010. STAT1 Signaling Is Required for Optimal Th1 Cell Differentiation in Mice. *Chin Sci Bull* 55 (11):1032–40. doi: 10.1007/s11434-010-0030-9.

Madera, Sharline, Clair D Geary, Colleen M Lau, Olga Pikovskaya, Steven L Reiner, and Joseph C Sun. 2018. Cutting Edge: Divergent Requirement of T-Box Transcription Factors in Effector and Memory NK Cells. *J Immunol* 200 (6):1977–81.

Mahmud, Shawn A, Luke S Manlove, and Michael A Farrar. 2013. Interleukin-2 and STAT5 in Regulatory T Cell Development and Function. *JAKStat* 2 (1):e23154.

Maier, Elisabeth, Albert Duschl, and Jutta Horejs-Hoeck. 2012. STAT6-Dependent and -Independent Mechanisms in Th2 Polarization. *Eur J Immunol* 42 (11):2827–33.

Meier, Jeremy A, and Andrew C Larner. 2014. Toward a New STATe: The Role of STATs in Mitochondrial Function. *Semin Immunol* 26 (1):20–8.

Melo-Gonzalez, Felipe, and Matthew R Hepworth. 2017. Functional and Phenotypic Heterogeneity of Group 3 Innate Lymphoid Cells. *Immunology* 150 (3):265–75.

Mesa, Ruben A. 2010. Ruxolitinib, a Selective JAK1 and JAK2 Inhibitor for the Treatment of Myeloproliferative Neoplasms and Psoriasis. *IDrugs* 13 (6):394–403.

Meyer, Thomas, Lisa Hendry, Andreas Begitt, Susan John, and Uwe Vinkemeier. 2004. A Single Residue Modulates Tyrosine Dephosphorylation, Oligomerization, and Nuclear Accumulation of Stat Transcription Factors. *J Biol Chem* 279 (18):18998–19007.

Mikami, Yohei, Gianluca Scarno, Beatrice Zitti, Han-Yu Shih, Yuka Kanno, Angela Santoni, John J O'Shea, and Giuseppe Sciumè. 2018. NCR+ ILC3 Maintain Larger STAT4 Reservoir via T-BET to Regulate Type 1 Features upon IL-23 Stimulation in Mice. *Eur J Immunol* 48 (7):1174–80.

Milner, Joshua D, Tiphanie P Vogel, Lisa Forbes, Chi A Ma, Asbjørg Stray-Pedersen, Julie E Niemela, Jonathan J Lyons et al. 2015. Early-Onset Lymphoproliferation and Autoimmunity Caused by Germline STAT3 Gain-of-Function Mutations. *Blood* 125 (4):591–9.

Minegishi, Yoshiyuki, Masako Saito, Tomohiro Morio, Ken Watanabe, Kazunaga Agematsu, Shigeru Tsuchiya, Hidetoshi Takada et al. 2006. Human Tyrosine Kinase 2 Deficiency Reveals Its Requisite Roles in Multiple Cytokine Signals Involved in Innate and Acquired Immunity. *Immunity* 25 (5):745–55.

Mogensen, Trine H. 2013. STAT3 and the Hyper-IgE Syndrome: Clinical Presentation, Genetic Origin, Pathogenesis, Novel Findings and Remaining Uncertainties. *JAKStat* 2 (2):e23435.

Morrison, Juliet, and Adolfo García-Sastre. 2014. STAT2 Signaling and Dengue Virus Infection. *JAKStat* 3 (1):e27715.

Mowen, Kerri A, Jie Tang, Wei Zhu, Brandon T Schurter, Ke Shuai, Harvey R Herschman, and Michael David. 2001. Arginine Methylation of STAT1 Modulates IFNα/B-Induced Transcription. *Cell* 104 (5):731–41.

Mui, A L, H Wakao, A M O'Farrell, N Harada, and A Miyajima. 1995. Interleukin-3, Granulocyte-Macrophage Colony Stimulating Factor and Interleukin-5 Transduce Signals through Two STAT5 Homologs. *Embo J* 14 (6):1166–75.

Murray, P J. 2007. The JAK–STAT Signaling Pathway: Input and Output Integration. *J Immunol* 178 (5):2623–29.

Musso, T, J A Johnston, D Linnekin, L Varesio, T K Rowe, J J O'Shea, and D W McVicar. 1995. Regulation of JAK3 Expression in Human Monocytes: Phosphorylation in Response to Interleukins 2, 4, and 7. *J Exp Med* 181 (4):1425–31.

Nabhani, Schafiq, Cyrill Schipp, Hagit Miskin, Carina Levin, Sergey Postovsky, Tal Dujovny, Ariel Koren et al. 2017. STAT3 Gain-of-Function Mutations Associated with Autoimmune Lymphoproliferative

Syndrome like Disease Deregulate Lymphocyte Apoptosis and Can Be Targeted by BH3 Mimetic Compounds. *Clin Immunol* 181:32–42.

Nan, Yuchen, Chunyan Wu, and Yan-Jin Zhang. 2018. Interferon Independent Non-Canonical STAT Activation and Virus Induced Inflammation. *Viruses* 10 (4):196.

Nemoto, Michiko, Hiroyoshi Hattori, Naoko Maeda, Nobuhiro Akita, Hideki Muramatsu, Suzuko Moritani, Tomonori Kawasaki et al. 2018. Compound Heterozygous TYK2 Mutations Underlie Primary Immunodeficiency with T-Cell Lymphopenia. *Sci Rep* 8 (1):515.

Nguyen, Khuong B, Leslie P Cousens, Lesley A Doughty, Gary C Pien, Joan E Durbin, and Christine A Biron. 2000. Interferon A/B-Mediated Inhibition and Promotion of Interferon Γ: STAT1 Resolves a Paradox. *Nat Immunol* 1 (1):70–6.

Niederreiter, Lukas, Timon Adolph, and Arthur Kaser. 2013. Anti-IL-12/23 in Crohn'S Disease: Bench and Bedside. *Curr Drug Targets* 14 (12):1379–84.

Nielsen, Camilla, Henrik S Birgens, Børge G Nordestgaard, Lasse Kjaer, and Stig E Bojesen. 2011. The JAK2 V617F Somatic Mutation, Mortality and Cancer Risk in the General Population. *Haematologica* 96 (3). Haematologica:450–3.

Niu, Guo-Juan, Ji-Dong Xu, Wen-Jie Yuan, Jie-Jie Sun, Ming-Chong Yang, Zhong-Hua He, Xiao-Fan Zhao, and Jin-Xing Wang. 2018. Protein Inhibitor of Activated STAT (PIAS) Negatively Regulates the JAK/STAT Pathway by Inhibiting STAT Phosphorylation and Translocation. *Front Immunol* 9:2392.

O'Shea, John J, Steven M Holland, and Louis M Staudt. 2013. JAKs and STATs in Immunity, Immunodeficiency, and Cancer. *N Engl J Med* 368 (2):161–70.

O'Shea, John J, Riitta Lahesmaa, Golnaz Vahedi, Arian Laurence, and Yuka Kanno. 2011. Genomic Views of STAT Function in CD4+ T Helper Cell Differentiation. *Nat Rev Immunol* 11 (4):239–50.

O'Shea, John J, and Peter J Murray. 2008. Cytokine Signaling Modules in Inflammatory Responses. *Immunity* 28 (4):477–87.

O'Shea, John J, Daniella M Schwartz, Alejandro V Villarino, Massimo Gadina, Iain B McInnes, and Arian Laurence. 2015. The JAK–STAT Pathway: Impact on Human Disease and Therapeutic Intervention. *Annu Rev Med* 66 (1):311–28.

Olson, Matthew R, Felipe Fortino Verdan, Matthew M Hufford, Alexander L Dent, and Mark H Kaplan. 2016. STAT3 Impairs STAT5 Activation in the Development of IL-9-Secreting T Cells. *J Immunol* 196 (8):3297–304.

Palmer, Douglas C, and Nicholas P Restifo. 2009. Suppressors of Cytokine Signaling (SOCS) in T Cell Differentiation, Maturation, and Function. *Trend Immunol* 30 (12):592–602.

Park, S Y, K Saijo, T Takahashi, M Osawa, H Arase, N Hirayama, K Miyake, H Nakauchi, T Shirasawa, and T Saito. 1995. Developmental Defects of Lymphoid Cells in JAK3 Kinase-Deficient Mice. *Immunity* 3 (6):771–82.

Passerini, Laura, Sarah E Allan, Manuela Battaglia, Sara Di Nunzio, Alicia N Alstad, Megan K Levings, Maria G Roncarolo, and Rosa Bacchetta. 2008. STAT5-Signaling Cytokines Regulate the Expression of FOXP3 in CD4+CD25+ Regulatory T Cells and CD4+CD25- Effector T Cells. *Int Immunol* 20 (3):421–31.

Pesu, Marko, Fabio Candotti, Matthew Husa, Sigrun R Hofmann, Luigi D Notarangelo, and John J O'Shea. 2005. JAK3, Severe Combined Immunodeficiency, and a New Class of Immunosuppressive Drugs. *Immunol Rev* 203 (1):127–42.

Qi, Qian-Rong, and Zeng-Ming Yang. 2014. Regulation and Function of Signal Transducer and Activator of Transcription 3. *World J Biol Chem* 5 (2):231–9.

Ramos, Hilario J, Ann M Davis, Thaddeus C George, and J David Farrar. 2007. IFN-Alpha Is Not Sufficient to Drive Th1 Development Due to Lack of Stable T-Bet Expression. *J Immunol* 179 (6):3792–803.

Rani, Aradhana, and John J Murphy. 2016. STAT5 in Cancer and Immunity. *J Interferon Cytokine Res* 36 (4):226–37.

Ranson, Thomas, Christian A J Vosshenrich, Erwan Corcuff, Odile Richard, Werner Müller, and James P Di Santo. 2003. IL-15 Is an Essential Mediator of Peripheral NK-Cell Homeostasis. *Blood* 101 (12):4887–93.

Rawlings, J S. 2004. The JAK/STAT Signaling Pathway. *J Cell Sci* 117 (8):281–3.

Reder, Anthony T, and Xuan Feng. 2014. How Type I Interferons Work in Multiple Sclerosis and Other Diseases: Some Unexpected Mechanisms. *J Interferon Cytokine Res* 34 (8):589–99.

Remmers, Elaine F, Robert M Plenge, Annette T Lee, Robert R Graham, Geoffrey Hom, Timothy W Behrens, Paul I W de Bakker et al. 2007. STAT4 and the Risk of Rheumatoid Arthritis and Systemic Lupus Erythematosus. *N Engl J Med* 357 (10):977–86.

Rider, Peleg, Yaron Carmi, and Idan Cohen. 2016. Biologics for Targeting Inflammatory Cytokines, Clinical Uses, and Limitations. *Int J Cell Biol* 2016 (7):1–11.

Riese, Richard J, Sriram Krishnaswami, and Joel Kremer. 2010. Inhibition of JAK Kinases in Patients with Rheumatoid Arthritis: Scientific Rationale and Clinical Outcomes. *Best Pract Res Clin Rheumatol* 24 (4):513–26.

Rong, Libin, and Alan S Perelson. 2010. Treatment of Hepatitis C Virus Infection with Interferon and Small Molecule Direct Antivirals: Viral Kinetics and Modeling. *Crit Rev Immunol* 30 (2):131–48.

Schindler, Christian. 1999. Cytokines and JAK–STAT Signaling. *Exp Cell Res* 253 (1):7–14.

Seidel, H M, L H Milocco, P Lamb, J E Darnell, R B Stein, and J Rosen. 1995. Spacing of Palindromic Half Sites as a Determinant of Selective STAT (Signal Transducers and Activators of Transcription) DNA Binding and Transcriptional Activity. *Proc Natl Acad Sci U S A* 92 (7):3041–5.

Sen, Malabika, and Jennifer R Grandis. 2014. Nucleic Acid-Based Approaches to STAT Inhibition. *JAKStat* 1 (4):285–91.

Shahni, Rojeen, Catherine M Cale, Glenn Anderson, Laura D Osellame, Sophie Hambleton, Thomas S Jacques, Yehani Wedatilake et al. 2015. Signal Transducer and Activator of Transcription 2 Deficiency Is a Novel Disorder of Mitochondrial Fission. *Brain* 138:2834–46.

Shuai, K. 2000. Modulation of STAT Signaling by STAT-Interacting Proteins. *Oncogene* 19 (21):2638–44.

Shuai, Ke, and Bin Liu. 2003. Regulation of JAK–STAT Signalling in the Immune System. *Nat Rev Immunol* 3 (11):900–11.

Siegel, Andrea M, Jennifer Heimall, Alexandra F Freeman, Amy P Hsu, Erica Brittain, Jason M Brenchley, Daniel C Douek et al. 2011. A Critical Role for STAT3 Transcription Factor Signaling in the Development and Maintenance of Human T Cell Memory. *Immunity* 35 (5):806–18.

Singh, Rekha. 2011. JAK2-Independent Activation of Stat3 by Intracellular Angiotensin II in Human Mesangial Cells. *J Signal Transduct* 2011 (6):257862–10.

Sleiman, Patrick M A, Mei-Lun Wang, Antonella Cianferoni, Seema Aceves, Nirmala Gonsalves, Kari Nadeau, Albert J Bredenoord, Glenn T Furuta, Jonathan M Spergel, and Hakon Hakonarson. 2014. GWAS Identifies Four Novel Eosinophilic Esophagitis Loci. *Nat Commun* 5 (1):5593.

Srivastava, Shivani, Meghan A Koch, Marion Pepper, and Daniel J Campbell. 2014. Type I Interferons Directly Inhibit Regulatory T Cells to Allow Optimal Antiviral T Cell Responses during Acute LCMV Infection. *J Exp Med* 211 (5):961–74.

Stabile, Helena, Gianluca Scarno, Cinzia Fionda, Angela Gismondi, Angela Santoni, Massimo Gadina, and Giuseppe Sciumè. 2018. JAK/STAT Signaling in Regulation of Innate Lymphoid Cells: The Gods before the Guardians. *Immunol Rev* 286 (1):148–59.

Stark, George R, and James E Darnell Jr. 2012. The JAK–STAT Pathway at Twenty. *Immunity* 36 (4):503–14.

Stelloo, Ellen, Marco A Versluis, Hans W Nijman, Marco de Bruyn, Annechien Plat, Elisabeth M Osse, Reinhardt H van Dijk et al. 2016. Microsatellite Instability Derived JAK1 Frameshift Mutations are Associated with Tumor Immune Evasion in Endometrioid Endometrial Cancer. *Oncotarget* 7 (26):39885–93.

Stritesky, Gretta L, Rajarajeswari Muthukrishnan, Sarita Sehra, Ritobrata Goswami, Duy Pham, Jared Travers, Evelyn T Nguyen, David E Levy, and Mark H Kaplan. 2011. The Transcription Factor STAT3 Is Required for T Helper 2 Cell Development. *Immunity* 34 (1):39–49.

Strobl, Birgit. 2011. Tyrosine Kinase 2 (TYK2) in Cytokine Signalling and Host Immunity. *Front Biosci (Landmark Ed)* 16 (1):3214–32.

Takatori, Hiroaki, Yuka Kanno, Wendy T Watford, Cristina M Tato, Greta Weiss, Ivaylo I Ivanov, Dan R Littman, and John J O'Shea. 2009. Lymphoid Tissue Inducer-Like Cells are an Innate Source of IL-17 and IL-22. *J Exp Med* 206 (1):35–41.

Tefferi, Ayalew, Daniela Barraco, Terra L Lasho, Sahrish Shah, Kebede H Begna, Aref Al-Kali, William J Hogan et al. 2018. Momelotinib Therapy for Myelofibrosis: A 7-Year Follow-Up. *Blood Cancer J* 8 (3):1322.

Thierfelder, William E, Jan M van Deursen, Koh Yamamoto, Ralph A Tripp, Sally R Sarawar, Richard T Carson, Mark Y Sangster et al. 1996. Requirement for Stat4 in Interleukin-12-Mediated Responses of Natural Killer and T Cells. *Nature* 382 (6587):171–4.

Thieu, Vivian T, Qing Yu, Hua-Chen Chang, Norman Yeh, Evelyn T Nguyen, Sarita Sehra, and Mark H Kaplan. 2008. Signal Transducer and Activator of Transcription 4 Is Required for the Transcription Factor T-Bet to Promote T Helper 1 Cell-Fate Determination. *Immunity* 29 (5):679–90.

Ulrich, Benjamin J, Felipe Fortino Verdan, Andrew N J McKenzie, Mark H Kaplan, and Matthew R Olson. 2017. STAT3 Activation Impairs the Stability of Th9 Cells. *J Immunol* 198 (6):2302–9.

Ungureanu, Daniela, Sari Vanhatupa, Noora Kotaja, Jie Yang, Saara Aittomaki, Olli A Jänne, Jorma J Palvimo, and Olli Silvennoinen. 2003. PIAS Proteins Promote SUMO-1 Conjugation to STAT1. *Blood* 102 (9):3311–3.

Uzel, Gulbu, Elizabeth P Sampaio, Monica G Lawrence, Amy P Hsu, Mary Hackett, Morna J Dorsey, Richard J Noel et al. 2013. Dominant Gain-of-Function STAT1 Mutations in FOXP3 Wild-Type Immune Dysregulation-Polyendocrinopathy-Enteropathy-X-Linked-Like Syndrome. *J Allergy Clin Immunol* 131 (6):1611–23.

Vahedi, Golnaz, Hayato Takahashi, Shingo Nakayamada, Hong-wei Sun, Vittorio Sartorelli, Yuka Kanno, and John J O'Shea. 2012. STATs Shape the Active Enhancer Landscape of T Cell Populations. *Cell* 151 (5):981–93.

Van Nguyen, Thang, Pornpimon Angkasekwinai, Hong Dou, Feng-Ming Lin, Long-Sheng Lu, Jinke Cheng, Y Eugene Chin, Chen Dong, and Edward T H Yeh. 2012. SUMO-Specific Protease 1 Is Critical for Early Lymphoid Development through Regulation of STAT5 Activation. *Mol Cell* 45 (2):210–21.

Vély, Frédéric, Vincent Barlogis, Blandine Vallentin, Bénédicte Neven, Christelle Piperoglou, Mikael Ebbo, Thibaut Perchet et al. 2016. Evidence of Innate Lymphoid Cell Redundancy in Humans. *Nat Immunol* 17 (11):1291–9.

Villarino, Alejandro, Linda Hibbert, Linda Lieberman, Emma Wilson, Tak Mak, Hiroki Yoshida, Robert A Kastelein, Christiaan Saris, and Christopher A Hunter. 2003. The IL-27R (WSX-1) Is Required to Suppress T Cell Hyperactivity during Infection. *Immunity* 19 (5):645–55.

Villarino, Alejandro V, Yuka Kanno, John R Ferdinand, and John J O'Shea. 2015. Mechanisms of JAK/STAT Signaling in Immunity and Disease. *J Immunol* 194 (1):21–7.

Villarino, Alejandro V, Yuka Kanno, and John J O'Shea. 2017a. Mechanisms and Consequences of JAK–STAT Signaling in the Immune System. *Nat Immunol* 18 (4):374–84.

Villarino, Alejandro V, Giuseppe Sciumè, Fred P Davis, Shigeru Iwata, Beatrice Zitti, Gertraud W Robinson, Lothar Hennighausen, Yuka Kanno, and John J O'Shea. 2017b. Subset- and Tissue-Defined STAT5 Thresholds Control Homeostasis and Function of Innate Lymphoid Cells. *Journal Exp Med* 214 (10):2999–3014.

Vonarbourg, Cedric, and Andreas Diefenbach. 2012. Multifaceted Roles of Interleukin-7 Signaling for the Development and Function of Innate Lymphoid Cells. *Sem Immunol* 24 (3):165–74.

Walker, Jennifer A, and Andrew N J McKenzie. 2013. Development and Function of Group 2 Innate Lymphoid Cells. *Curr Opin Immunol* 25 (2):148–55.

Walker, Jennifer A, and Andrew N J McKenzie. 2017. TH2 Cell Development and Function. *Nat Rev Immunol* 18 (2):121–33.

Weaver, Casey T, Charles O Elson, Lynette A Fouser, and Jay K Kolls. 2013. The Th17 Pathway and Inflammatory Diseases of the Intestines, Lungs, and Skin. *Annu Rev Pathol* 8 (1):477–512.

Wehinger, Jens, Fabrice Gouilleux, Bernd Groner, Juergen Finke, Roland Mertelsmann, and Renate Maria Weber-Nordt. 1999. IL-10 Induces DNA Binding Activity of Three STAT Proteins (stat1, Stat3, and Stat5) and Their Distinct Combinatorial Assembly in the Promoters of Selected Genes. *FEBS Lett* 394 (3):365–70.

Weizman, Orr-El, Nicholas M Adams, Iona S Schuster, Chirag Krishna, Yuri Pritykin, Colleen Lau, Mariapia A Degli-Esposti, Christina S Leslie, Joseph C Sun, and Timothy E O'Sullivan. 2017. ILC1 Confer Early Host Protection at Initial Sites of Viral Infection. *Cell* 171 (4):795–808.e12.

Wu, Hao, Markus M Xie, Hong Liu, and Alexander L Dent. 2016. Stat3 Is Important for Follicular Regulatory T Cell Differentiation. Edited by Derya Unutmaz. *PLoS One* 11 (5):e0155040.

Wurtz, O. 2004. IL-4-Mediated Inhibition of IFN-Gamma Production by CD4+ T Cells Proceeds by Several Developmentally Regulated Mechanisms. *Int Immunol* 16 (3):501–8.

Yildiz, M, H Li, D Bernard, N A Amin, P Ouillette, S Jones, K Saiya-Cork et al. 2015. Activating STAT6 Mutations in Follicular Lymphoma. *Blood* 125 (4):668–79.
Yuan, Z L. 2005. Stat3 Dimerization Regulated by Reversible Acetylation of a Single Lysine Residue. *Science* 307 (5707):269–73.
Zhang, S, S Fukuda, Y Lee, G Hangoc, S Cooper, R Spolski, W J Leonard, and H E Broxmeyer. 2000. Essential Role of Signal Transducer and Activator of Transcription (stat)5a but Not Stat5b for Flt3-Dependent Signaling. *J Exp Med* 192 (5):719–28.
Zhu, Shiguo, Prasad V Phatarpekar, Cecele J Denman, Vladimir V Senyukov, Srinivas S Somanchi, Hoainam T Nguyen-Jackson, Emily M Mace et al. 2014. Transcription of the Activating Receptor NKG2D in Natural Killer Cells Is Regulated by STAT3 Tyrosine Phosphorylation. *Blood* 124 (3):403–11.
Zhuang, Shougang. 2013. Regulation of STAT Signaling by Acetylation. *Cell Signal* 25 (9):1924–31.
Zindl, Carlene L, Jen-Feng Lai, Yun Kyung Lee, Craig L Maynard, Stacey N Harbour, Wenjun Ouyang, David D Chaplin, and Casey T Weaver. 2013. IL-22-Producing Neutrophils Contribute to Antimicrobial Defense and Restitution of Colonic Epithelial Integrity during Colitis. *Proc Natl Acad Sci U S A* 110 (31):12768–73.

14

Targeting JAK-STAT Pathway for Various Inflammatory Diseases and Viral Infections

Christina Gavegnano and Raymond F. Schinazi
Center for AIDS Research, Laboratory of Biochemical Pharmacology
Department of Pediatrics, Emory University
Atlanta, GA

14.1 Introduction

The (Janus activating kinase) JAK–signal transducer and activator of transcription (STAT) pathway has been implicated as a key driver in many diseases, including myelofibrosis and polycythemia vera (overexpression of JAKs), rheumatoid and psoriatic arthritis, inflammatory bowel disease, ulcerative colitis, uveitis, systemic lupus erythematosus (SLE), and psoriasis. Previous treatment for these pathologies centered around symptom relief, rather than selective targeting of the pathway that is responsible for the dysregulated inflammatory response driving the disease. With the onset of small molecule inhibitors that can selectively and specifically target key JAKs involved in controlling downstream inflammation and immunoregulatory cytokines, exploration for their utility across a broad range of diseases has become a rapidly expanding field.

JAK inhibitors present with unique advantages over non-steroidal anti-inflammatories (NSAIDs) such as aspirin, or steroidal anti-inflammatories such as prednisone and cortisone based agents. For example, NSAIDs, including aspirin, target the cyclooxygenase enzyme pathway, rendering it inactive upon binding; the end result is multifactorial and weakly selective to COX-1 *versus* COX-2 (1). Block of COX signaling primarily confers its anti-inflammatory effects by non-specific events, including block of prostaglandins, which are paracrine hormones that confer a wide array of chemical functions including temperature regulation, homeostasis, and inflammation. Further, block of COX-2 has been associated with severe adverse effects, including cardiac events leading to death; it is thought that excessive block of COX-2 within the endothelial cells that comprise the microvasculature within the heart can result in increased thrombosis, leading to heart attack and death (2,3). Additionally, COX blockade can also result in decreased clotting, which, in turn, can lead to detrimental long-term side effects systemically (2).

Steroidal anti-inflammatories also present with lack of specificity, and are often associated with immunosuppression, which can lead to increased opportunistic infections; broad blunting of T cell, B cell, and antigen-presenting cell responses to pathogens; and increased malignancies due to blunted ability to sense eliminate potential malignant cells. Together, the onset of selective, specific inhibitors of the JAK-STAT pathway presents with a unique opportunity to treat many diseases, or prevent disease progression. Specific targeting significantly reduces off-target effects, while simultaneously blocking the pathway that produces the inflammatory dysregulation, as opposed to a generic block of non-specific events, some of which are anti-inflammatory. Together, the discovery of JAK inhibitors to block production of excessive inflammation at the root provides an attractive and still-blossoming field. The long-term ramifications as they relate to approval for disease indications, the impact of JAK inhibitors long-term within distinct populations, and the ability of these agents to control chronic inflammation is still being explored. Despite the relatively new discovery of JAK inhibitors, and the ongoing research to better understand how these agents can mitigate disease, multiple approvals of JAK inhibitors have transpired, and several novel JAK inhibitors are in late-phase clinical investigation. This review focuses on the details of each agent, the phase of investigation, the target indication, and the mechanistic block of each JAK as it relates to the downstream disease and inflammatory factors targeted. The dynamic interaction between these events is discussed, with a goal of broadening the overall knowledge of JAK inhibitors toward treatment, prevention, or eradication of diseases driven by inflammatory dysregulation.

14.1.1 JAK-STAT Pathway and Downstream Cytokine Production

A wide variety of ligands bind to type 1/2 cytokine receptors, including growth factors/colony-stimulating factors, interferons, and HIV-1 envelope gp120 (4–10) (Figure 14.1). The resulting activation of these receptors by bound ligand results in activation of JAKs, comprised of JAK1, JAK2, JAK3, and Tyk2 (Figure 14.2). JAKs then engage in cross-phosphorylation (one JAK phosphorylates the other JAK), along with phosphorylation of the intracellular domain of the

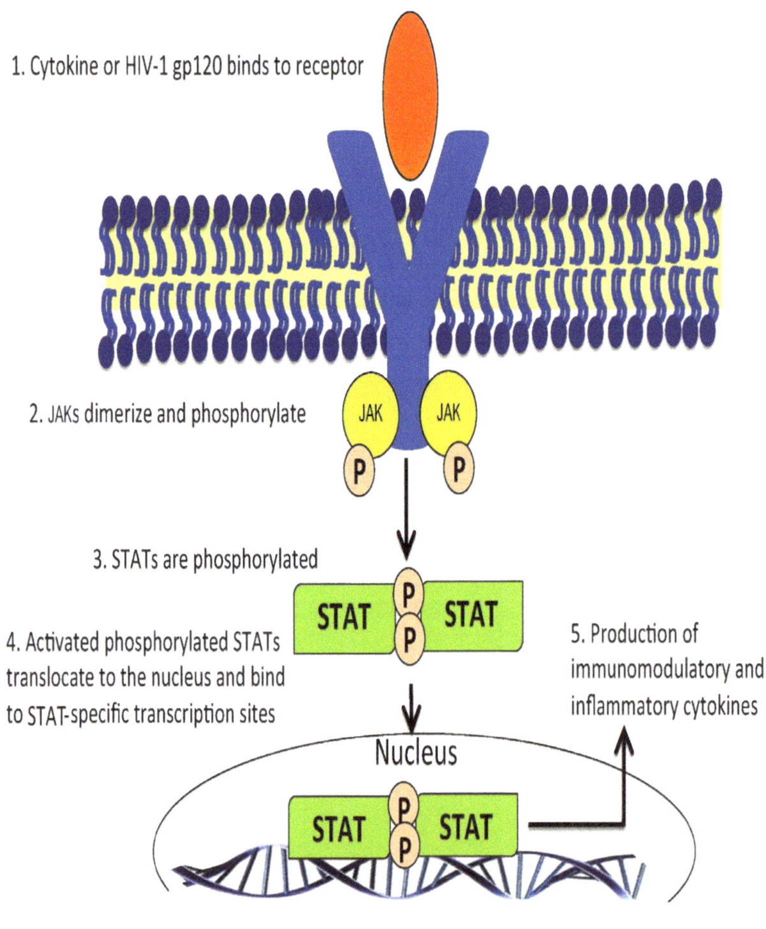

FIGURE 14.1 JAK-STAT pathway signaling pathway and downstream activation of proinflammatory cytokines.

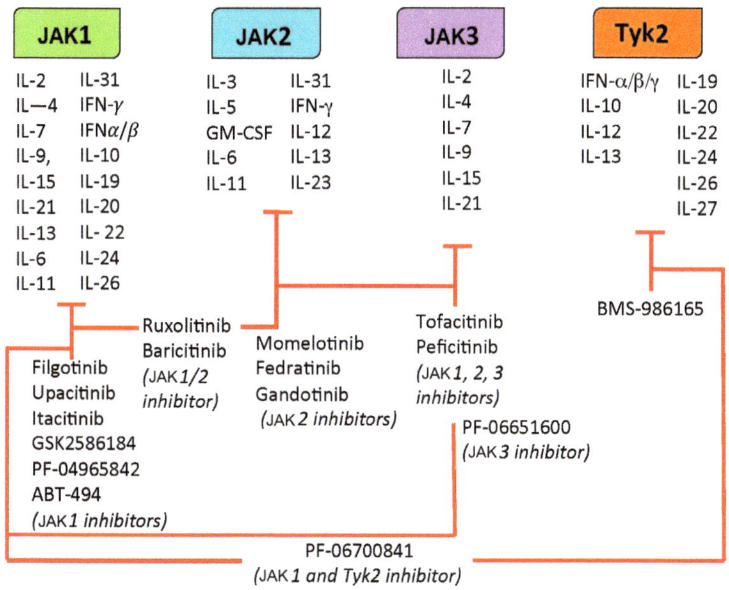

FIGURE 14.2 Summary of JAK inhibitors, specificity for JAKs, and corresponding block on proinflammatory cytokines.

receptor, the latter of which results in docking sites for various recruited signaling molecules (5–10). Activated phosphorylated JAKs then phosphorylate STATs (Figure 14.2). The STAT family is comprised of STAT1, STAT2, STAT3, STAT4, STAT5a, STAT5b, and STAT6 (5–10). Activated, dimerized phosphorylated STATs then translocate to the nucleus, and bind to specific transcription sites within the DNA that confer transcription of gene products for a wide array pro-inflammatory and immunomodulatory cytokines. Different receptors can trigger with specific JAKs, thereby agonist engagement at the receptor level can lead to different downstream cascades of eventual cytokine production. Therefore, based on the desired blockade of cytokines, selective inhibition of a specific JAK can be sought. Nonetheless, there is some overlap in cytokines that are eventually produced from phosphorylation of JAKs. For example, JAK1 and JAK2 phosphorylation can result in production of IL-6, and IL-7 is produced after JAK1 or JAK3 phosphorylation (Figure 14.2). A summary of JAK phosphorylation and downstream production of cytokines appears in Figure 14.2. Notably, autocrine and paracrine activation by cytokines produced from JAK-STAT triggering can further activate JAK-STAT signaling, recruitment of bystander cells, and promote a systemic or tissue-localized inflammatory response (5–10), underscoring the importance of this pathway in inflammatory disease pathologies.

14.1.2 Diseases Targeted with JAKinibs

Many diseases are hallmarked by chronic inflammation, driven by a dysregulated autoimmune phenotype, including rheumatoid arthritis, psoriasis, and systemic lupus Erythmeus, Castleman's disease (pediatrics), and Chron's diseases (7–18). Additionally, dysregulated JAK-STAT signaling, governed by overexpressing JAKs, have been reported in myeloproliferative malignancies including myelofibrosis and polycythemia vera (9,16–18).

Until recently, dysregulated autoimmune diseases were treated with non-specific steroid-based regimens or immunosuppressants intended to blunt the inflammatory response. These agents are blunt tools that systemically reduce the ability of the adaptive and innate immunity to mount a response to pathogens, and are also present with off-target toxicities and malignancies (8,10,12,19,20). A major unmet clinical need is specific blockade of the pro-inflammatory cytokines that are dysregulated and overexpressed in these conditions. Further, prior to the use of JAK inhibitors, JAK-overexpressing cancers, such as myelofibrosis and polycythemia vera, were considered incurable cancers with little more than palliative care as therapeutic interventions. The discovery of JAK inhibitors to treat inflammation and also block JAK overexpression that drives cancer has provided the field with a revolutionary modality to tackle previously incurable diseases with significantly reduced toxicity and an increased quality of life. These discoveries have lead to exploratory consideration for JAK inhibitors for more than 47 autoimmune and inflammatory-based disorders (clinicaltrials.gov), including chronic inflammation in HIV-infected individuals (clinicaltrials.gov/ruxolitinib.a5336; actg.gov/a5336). A closer look at the role of JAK inhibitors in treating or reversing specific disease pathologies are discussed below.

14.1.3 Arthritis

JAK-STAT dysregulation transpires in arthritis, including rheumatoid arthritis (RA) and psoriatic arthritis (PsA). PsA is hallmarked by musculoskeletal inflammation, resulting in chronic inflammation of joints and the systemic musculoskeletal system (10,21–24). Pathogenesis is associated with multiple factors, including mutations to the human leukocyte antigens (HLA-Cw0602 and HLA-AB*27 alleles); these mutations are associated with increased dysregulation of inflammatory cytokines, including tumor necrosis factor alpha (TNF-α, IL-18, IL-17, IL-α/β, and IL-6). TNF-α, in particular, confers autocrine and paracrine signaling across myeloid and lymphoid cells, promoting production of chemotactic factors that induce localization and recruitment of a Th1 cytokine response and induction of antigen-presenting cell (APC) inflammatory responses (10,21–24). Production of these cytokines is triggered by activation of the JAK-STAT pathway; therefore, blocking its activation with JAK inhibitors represents an attractive and specific modality to prevent clinical manifestation of symptoms in individuals with PsA. RA is an autoimmune arthritis found in 1% of the total population, hallmarked by systemic joint inflammation conferred by JAK-STAT activation and production of IL-6, TNF-α, IL-α/β, D-dimer, C reactive protein (CRP), and other pro-inflammatory cytokines (21,23,24). JAK-STAT signaling is a primary modulator of production of these cytokines (Figure 14.2). Prior to the discovery of JAK inhibitors, treatment for RA centered around blunt steroidal anti-inflammatory regimens, without blocking the root cause of the signaling dysregulation. More recently, biologics targeting circulating inflammatory cytokines, such as anti-TNF-α monoclonal antibody (mAb) or anti-IL-1 mAb, remained a more attractive option versus a systemic and non-specific blunting of systemic immune function. Biologics present with limitations, however, including lack of oral bioavailability and eventual tolerance from the immune system, were compounded by lack of multi-cytokine targeting. These limitations led to pursuit of orally available, targeted inhibitors of multiple pro-inflammatory cytokines by blocking JAK phosphorylation. Tofacitinib was the first JAK inhibitor approved for this indication, and is a JAK3 inhibitor that also demonstrates block of other JAKs with lower specificity (summarized in Figure 14.2). JAK3 blockade has been associated with reduced neutrophil and natural killer cell counts; therefore, selective JAK1 or 2 blockade without significant JAK3 block has begun to be considered. To date, baricitinib, a JAK1/2 inhibitor, is also FDA approved for treatment of RA; baricitinib is a qd-dosed orally bioavailable agent that is also approved for use in children (summarized in Figure 14.2) (8,10). Multiple investigational JAK inhibitors are also under investigation for rheumatoid arthritis, including peficitinib (phase 3; pan JAK inhibitor), filgotinib (phase 3; JAK1 inhibitor), and upadcitinib (phase 3; JAK1 inhibitor); summarized in Figure 14.2 (8,10).

14.1.4 Inflammatory Bowel Disease

Inflammatory bowel disease (IBD) is delineated into two different disease pathologies: Crohn's disease (CD) and ulcerative colitis (UC) (19). Both are comprised of localized inflammation within the intestine that results in clinical manifestation of symptoms including diarrhea, abdominal pain, and fistulas/abscesses (12,19). Polymorphisms in JAK2, Tyk2, and STAT3 genes have been associated with increased risk of developing IBD, and these JAKs and STATs are crucial parts of downstream signaling that produces inflammatory cytokines that result in IBD. Further, the inflammatory responses are conferred in T cells and T cell subsets in response to JAK-STAT-mediated inflammatory induction compounds', Th1, Th2, and Th17, dysregulation and inflammation within the intestine of IBD patients (12,19). Therefore, safe, specific blockade of JAK-STAT signaling in the context of IBD represents an attractive modality to alleviate symptoms of this disease. To date, the JAK3 inhibitor, tofacitinib, is approved for this indication, and multiple JAK inhibitors are under investigation, including filgotinib (phase 3; JAK1 inhibitor for Crohn's disease and ulcerative colitis), upadcitinib (phase 2; JAK1 inhibitor for Crohn's disease and ulcerative colitis), PF-06651600 (phase 2; JAK3 inhibitor for Crohn's disease and ulcerative colitis), and PF-06700841 (JAK1 and Tyk2 inhibitor; phase 2 for Crohn's disease and ulcerative colitis) (8,10).

14.1.5 Myeloproliferative Disorders

Polycythemia vera (PV) and myelofibrosis (MF) are two distinct malignancies that are characterized by BCR-ABL1-negative myeloproliferations associated with hematopoietic mutations and hyperactivation of the JAK-STAT pathway (14,16). Dysregulated JAK-STAT signaling disrupts hematopoietic growth factors, including proliferation of stem cells and production of erythropoietin/thrombopoietin (14,16); this is hallmarked in both MF and PV. Further, a gain of function mutation in the JAK2 gene (JAK2V617F) promotes constitutive activation of JAK2, resulting in elevated levels of inflammatory cytokines, including IL-6, TNF-α, IL-1, CRP, and D-dimer (14,16). Presence of abnormally high levels of these cytokines is thought to further promote autocrine and paracrine stimulation to produce these cytokines in non-JAK2 mutated cells, further compounding the hyperstimulatory profile of MF and PV. This chronic triggering of pro-inflammatory cytokines is thought to drive the clinical manifestation of symptoms in patients with MF and PV, further underscoring necessity of JAK-STAT blockade in treatment of myeloproliferative disorders. To date, ruxolitinib, a JAK1/2 inhibitor, is the only FDA-approved JAK inhibitor for treatment of PV and MF (16,18,25), but multiple other JAK inhibitors are under various phases of clinical investigation for myeloproliferative disorders, including momelotinib (phase 3; JAK2 inhibitor), fedratinib (phase 2; JAK2 inhibitor), and gandotinib (phase 2; JAK2 inhibitor), summarized in Figure 14.2 (8,10).

14.1.6 Psoriasis and Autoimmune Skin Diseases

Inflammatory skin diseases are hallmarked by cutaneous pro-inflammatory milieu in localized APCs (dendritic cells, macrophages, etc.), and cells of the adaptive immune response (Th cells and B cells) (20). Localization of these cells to the skin provides a robust innate and adaptive immune response, but can become dysregulated in chronic inflammatory skin diseases, often by autoimmune-based mechanisms. This results in elevated pro-inflammatory cytokines, localized within the dermis, including elevated levels of IL-17, and IL-22 (psoriasis) or IL-4 (atopic dermatitis) (20). Most inflammatory skin diseases, independent of origin, were treated up until recent with topical steroids or oral systemic immunosuppressive agents designed to blunt the chronic inflammation. These treatments do not treat the root signaling cascades that result in production of inflammatory cytokines, but instead only treat the phenotype of inflamed skin. Recent advances have defined the critical role of the JAK-STAT pathway in (1) producing the inflammatory milieu found in inflammatory skin disorders (8,20), and (2) dysregulation and genetic alterations found in autoimmune skin disorders (8,10). Together, these findings have resulted in exploration of JAK inhibitors for treatment of inflammatory skin disorders, including psoriasis. To date, tofacitinib (JAK3 inhibitor with some specificity for other JAKs) is FDA approved for psoriatic arthritis (21), and multiple other JAK inhibitors are under various phases of clinical investigation for inflammatory skin disorders, including ruxolitinib (phase 2; JAK1/2 inhibitor for atopic dermatitis), baricitinib (JAK1/2 inhibitor for atopic dermatitis), PF-06263276 (pan JAK inhibitor; phase 1 for psoriasis), upadcitinib (JAK1 inhibitor; phase 2 atopic dermatitis), and PF-04965842 (JAK1 inhibitor; phase 2 for psoriasis, phase 3 for atopic dermatitis), and BMS- 986165 (Tyk2 inhibitor; phase 2 for psoriasis), summarized in Figure 14.2 (8,10).

14.1.7 Systemic Lupus Erythematosus (SLE)

SLE is a systemic autoimmune disorder affecting ~5 million people worldwide, which is characterized by high levels of circulating interferons that produce systemic inflammatory immune dysregulation. Together, this can result in massive autoimmune inflammatory responses against major tissues and organs, including skin and vital organs; left unchecked during flare-ups, this disease can lead to death (8,10,20). Treatments for this condition previously included systemic immunosuppressive agents, including cyclosporins and methotrexate, each of which is present with significant toxicities and an overall immunosuppressive phenotype due to lack of specificity for the disease pathology. A major unmet clinical need for SLE is safe, specific potent block of pro-inflammatory cytokine production, making JAK inhibitors, in particular JAK1/2 inhibitors that block the IFN response, very attractive. To that end, in December 2018, it was announced that baricitinib, an orally bioavailable qd-dosed JAK1/2 inhibitor (currently FDA-approved for rheumatoid arthritis), received fast-track status from the FDA for treatment of SLE (fda.gov; elililly.com). Data from this trial will inform the field about the impact of JAK1/2 inhibition in downregulating the chronic inflammation hallmarking SLE, and provide a foundation for further studies toward careful evaluation of JAK inhibitors for this indication.

14.1.8 Alopecia and Hair Loss

Alopecia areata (AA) is an autoimmune disease with no currently FDA-approved therapies; the condition is characterized by non-scarring hair loss (26). The full mechanism(s) responsible for this phenotype is/are not well understood; however, recent discoveries elucidated over active JAK-STAT signaling in patients with AA (26). Recently, a multi-center study (ClinicalTrials.gov NCT02197455 and NCT02312882) evaluated the ability of tofacitinib, a pan JAK inhibitor with JAK3 specificity, to reverse AA (27,28). Data demonstrated that tofacitinib was well-tolerated, safe, and effective for patients with severe AA; however, the treatment length did not demonstrate a durable and sustained reversal of AA (27,28). Nonetheless, this important study demonstrated encouraging results, and provided a foundation for future studies with other JAK inhibitors, including JAK1/2 inhibitors, to evaluate their ability to reverse AA. In particular, a recent report was described for a single patient entering a trial for baricitinib for treatment of CANDLE (chronic atypical neutrophilic dermatitis with lipodystrophy and elevated temperature). The patient who entered the study also had AA; the report describes a reversal of AA in this individual (29). Although this report is a sample size of one, it underscores the potential for JAK inhibitors toward reversal of AA or hair loss, especially in light of the key role of JAK-STAT signaling in conferring the disease pathology in vivo.

14.1.9 Closer Look at JAKinibs Toward HIV-1 Eradication

Despite major advances in treatment of HIV infection, including sustained viral suppression with highly active antiretroviral therapy (HAART), purge of the HIV reservoir remains challenging. A major barrier to eradication of HIV is the maintenance of the HIV reservoir (30–33). Recent reports demonstrate that elevated levels of inflammatory and immunomodulatory cytokines that trigger through the JAK-STAT pathway, including IL-7 and IL-15 promote proliferation of the HIV reservoir (homeostatic proliferation) (31,34). Further, other inflammatory cytokines that are produced from the JAK-STAT pathway remain elevated in HIV-infected individuals, even with well-controlled viremia, and are associated with non-AIDS-related morbidity/mortality, HIV persistence, progression to AIDS, and death (13,35–40). Recent reports demonstrate that JAK inhibitors, tofacitinib (JAK3, some specificity for JAK1 and 2) and ruxolitinib (JAK1/2), block HIV reservoir establishment, maintenance, and expansion in vitro and ex vivo (41). Further, these inhibitors downregulate key markers of HIV persistence ex vivo, underscoring the potential of JAK inhibitors as an add-on therapy to block key events that prevent HIV eradication in vivo, which could result in an eventual purge of the HIV reservoir in vivo (Figure 14.3A) (41). The impact of ruxolitinib on inflammation associated with HIV infection, which plays an important role in seeding viral reservoirs, viral persistence, and ongoing low-level replication in sanctuary sites is currently being evaluated in an NIH-ACTG sponsored 21-site Phase 2a study in humans (A5336; https://actgnetwork.org/newsletter/april2016/A5336); data from this study will be reported at the CROI meeting in 2019. Briefly, the population studied are "otherwise healthy" HIV-infected individuals, with CD4 T cell counts >350 mm^3, no underlying disease or co-infections, and no elevated inflammatory markers at onset of study. This population was selected because of the "proof-of-concept" nature of A5336; the goal of the study was to evaluate safety of an immunomodulator-based agent in the context of HIV infection, and further to quantify any potential decreases in inflammatory or HIV-persistence markers within the population. It is well established that chronic inflammation is associated with end-organ disease and mortality for people living with HIV (PLWH); and ruxolitinib demonstrates reduction in biomarkers of systemic inflammation in HIV-uninfected individuals and HIV reservoir and persistence markers ex vivo. The goal of this trial was to determine the safety and efficacy of ruxolitinib in treated HIV disease. In a highly selected cohort of HIV-positive adults on suppressive ART, ruxolitinib was safe and well tolerated, and demonstrated a trend in reduction of IL-6 levels within the short duration of treatment (5-week intervention). These data were coupled with a statistically significant

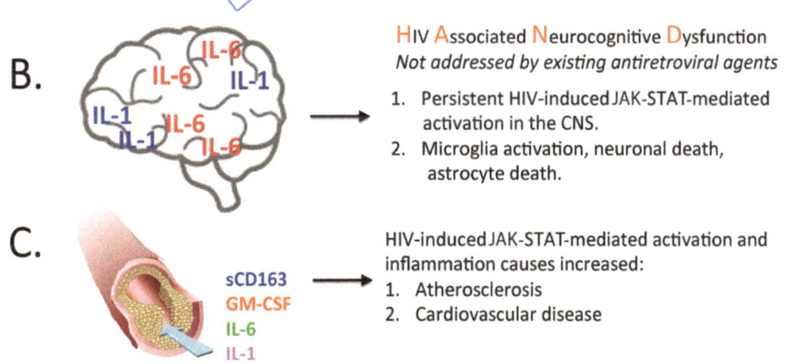

FIGURE 14.3 Role of the JAK-STAT pathway in HIV-1 persistence, HIV-associated neurocognitive dysfunction, and HIV-associated cardiovascular disease. Elevated pSTATS correlates with increased HIV reservoirs, HIV persistence markers, and progression to AIDS in HIVinfected individuals (A). Increased JAK-STAT activation is associated with HIV-associated neurocognitive dysfunction (HAND), which is hallmarked by CNS inflammation that promotes HAND (B). Increased inflammation in myeloid cells in HIV-infected individuals results in increased atherosclerosis; key markers including sCD163 are associated with increased risk of CVD in HIV-infected individuals (C).

decrease in sCD14 levels, and an increase in circulating T cells through mechanisms undefined. This proof-of-concept trial provides a rationale for future studies of JAK inhibitors in PLWH, who have residual inflammation or immune dysfunction despite long-term suppressive ART, including in individuals with HIV-associated neurocognitive dysfunction, and HIV-associated cardiovascular abnormalities and atherosclerosis.

Dysregulation of the JAK-STAT pathway has also been associated with HAND (6,15,38,42–45), and ruxolitinib demonstrated amelioration of HIV-induced inflammation in a murine model (46). Further, recent reports demonstrate that baricitinib crosses the blood–brain barrier and reverses behavioral abnormalities associated with HAND in a murine model, and reverses key pathological markers of HAND (47) (Figure 14.3B). Given these findings, JAK inhibitors, especially baricitinib, which demonstrates a safe, qd dosing in humans with FDA approval in children, could be used with HAART to reverse HAND and key events driving HIV persistence; further studies in humans are warranted for this indication.

Increased inflammation that persists in HIV-infected individuals on otherwise effective therapy have a higher-than-expected risk of a number of "non-AIDS" conditions, including cardiovascular disease (6,35,36,38–40,42,48–50). Chronic inflammation predicts and presumably causes this effect, and elevated activation of the JAK-STAT pathway, which drives a pro-inflammatory milieu, has been reported during HIV infection and is associated with HIV persistence and cardiovascular disease in HIV-infected individuals (summarized in Figure 14.3) (6,35,36,38–40,42,48–50). Mitigating increased risk of CVD in HIV-infected individuals, which data suggests is driven by increased, persistent elevated levels of inflammation, is an unmet clinical need. Safe, specific potent inhibition of these events with a well-tolerated, safe add-on therapy could reduce the risk of CVD in HIV-infected individuals. Given the body of data regarding the potent, specific block that JAK inhibitors confer on inflammatory cytokines, and their anti-HIV properties, it is possible that use of JAK inhibitors to reduce CVD in HIV-infected individuals could represent an attractive modality for a currently unaddressed clinical indication.

14.1.10 Metabolism and Future Indications

Metabolism of JAK inhibitors relative to their clearance and potential drug–drug interactions is an important factor for consideration of these agents when co-administered with other agents. Agents that are orally administered have two major routes of metabolism; renal or hepatic (51,52). Hepatic clearance often presents with drug–drug interactions, due to potential competition for key metabolizing enzymes such as CYP3A4 and CYP2B6; occupation of the CYP-binding sites for metabolism by a JAK inhibitor can prevent another drug from binding, and therefore prevent its metabolism (51,52). The end result of such an interaction is a potentially toxic concentration of an agent, when co-administered in the presence of a JAK inhibitor. Ruxolitinib is a hepatic cleared JAK inhibitor, for example (18,53,54); therefore, careful consideration for co-administration of agents that also require hepatic metabolism and clearance by CYP3A4 and CYP2B6 must be carefully monitored, or avoided altogether. Second generation JAK inhibitors, such as baricitinib (Olumiant, Eli Lilly), are renally cleared, which avoid the potential drug–drug interaction at the level of CYP metabolism (53,54), therefore significantly limiting the list of agents that may have drug–drug interactions with baricitinib *versus* ruxolitinib (or another hepatically cleared JAK inhibitor). Further, recent reports have begun to evaluate the pharmacokinetics of JAK inhibitors across different compartments, including the CNS (46,47), providing the beginning of exploration of use of these agents for tissue or compartment-specific pathologies, including neuroinflammatory diseases.

An additional indication that is not currently being explored clinically is atherosclerosis, which is driven by a pro-inflammatory milieu within cardiac tissue, including sCD163, sCD14, and localized inflammation from resident myeloid cells within the heart (39). Decreasing persistent tissue-specific inflammation for this indication could provide an add-on therapy to reverse or prevent cardiac events that lead to morbidity and mortality in the aging population. Atherosclerosis is increased in the aging HIV-positive population, and key markers of soluble inflammation have been correlated with increased morbidity and mortality from non-AIDS-related events including heart attacks (39), further highlighting the potential for this indication across the ~39 million HIV-infected individuals worldwide.

14.1.11 Summary

The discovery of safe, JAK-specific inhibitors that are orally bioavailable and well-tolerated has provided a new chapter toward harnessing of the JAK-STAT pathway in many diseases. Myeloproliferative disorders previously had no FDA-approved therapies prior to JAK inhibitors, resulting in rapid progression of disease and death. Other conditions including many autoimmune disorders hallmarked by chronic uncontrolled inflammation were treated with blunt tools, such as systemic immunosuppressants, which resulted in significant toxicities. The onset of JAK inhibitors to treat inflammatory conditions marks the beginning of a burgeoning new field. Exploration of the effect of safe, specific inhibition of JAKs toward immunomodulation of a wide array of inflammatory-driven diseases, including viral infections, will continue to be elucidated. New information gained from clinical trials will guide eventual additional FDA approvals of novel, well-tolerated, and effective JAK inhibitors.

ACKNOWLEDGMENTS

This work was funded in part by NIH grant 1RO1-MH-116695 and Emory's CFAR NIH grant P30-AI-050409.

REFERENCES

1. Lei J, Zhou Y, Xie D, Zhang Y. 2015. Mechanistic insights into a classic wonder drug–aspirin. *J Am Chem Soc* **137**:70–73.
2. Ornelas A, Zacharias-Millward N, Menter DG, Davis JS, Lichtenberger L, Hawke D, Hawk E, Vilar E, Bhattacharya P, Millward S. 2017. Beyond COX-1: the effects of aspirin on platelet biology and potential mechanisms of chemoprevention. *Cancer Metastasis Rev* **36**:289–303.
3. Shim YK, Kim N. 2016. [Nonsteroidal anti-inflammatory drug and aspirin-induced peptic ulcer disease]. *Korean J Gastroenterol* **67**:300–312.
4. Kohler JJ, Tuttle DL, Coberley CR, Sleasman JW, Goodenow MM. 2003. Human immunodeficiency virus type 1 (HIV-1) induces activation of multiple STATs in CD4+ cells of lymphocyte or monocyte/macrophage lineages. *J Leukoc Biol* **73**:407–416.
5. Benczik M, Gaffen SL. 2004. The interleukin (IL)-2 family cytokines: survival and proliferation signaling pathways in T lymphocytes. *Immunol Invest* **33**:109–142.
6. Chaudhuri A, Yang B, Gendelman HE, Persidsky Y, Kanmogne GD. 2008. STAT1 signaling modulates HIV-1-induced inflammatory responses and leukocyte transmigration across the blood-brain barrier. *Blood* **111**:2062–2072.
7. Darnell JE, Jr., Kerr IM, Stark GR. 1994. JAKJAK–STAT pathways and transcriptional activation in response to IFNs and other extracellular signaling proteins. *Science* **264**:1415–1421.
8. Gadina M, Johnson C, Schwartz D, Bonelli M, Hasni S, Kanno Y, Changelian P, Laurence A, O'Shea JJ. 2018. Translational and clinical advances in JAK-STAT biology: the present and future of jakinibs. *J Leukoc Biol* **104**:499–514.
9. McLornan DP, Khan AA, Harrison CN. 2015. Immunological consequences of JAK inhibition: friend or foe? *Curr Hematol Malig Rep* **10**:370–379.
10. O'Shea JJ, Schwartz DM, Villarino AV, Gadina M, McInnes IB, Laurence A. 2015. The JAK-STAT pathway: impact on human disease and therapeutic intervention. *Annu Rev Med* **66**:311–328.
11. Bovolenta C, Camorali L, Lorini AL, Ghezzi S, Vicenzi E, Lazzarin A, Poli G. 1999. Constitutive activation of STATs upon in vivo human immunodeficiency virus infection. *Blood* **94**:4202–4209.
12. Coskun M, Salem M, Pedersen J, Nielsen OH. 2013. Involvement of JAK/STAT signaling in the pathogenesis of inflammatory bowel disease. *Pharmacol Res* **76**:1–8.

13. Gavegnano C, Detorio M, Montero C, Bosque A, Planelles V, Schinazi RF. 2014. Ruxolitinib and tofacitinib are potent and selective inhibitors of HIV-1 replication and virus reactivation in vitro. *Antimicrob Agents Chemother* **58**:1977–1986.
14. Griesshammer M, Sadjadian P. 2017. The BCR-ABL1-negative myeloproliferative neoplasms: a review of JAK inhibitors in the therapeutic armamentarium. *Expert Opin Pharmacother* **18**:1929–1938.
15. Nabavi SM, Ahmed T, Nawaz M, Devi KP, Balan DJ, Pittala V, Arguelles-Castilla S, Testai L, Khan H, Sureda A, de Oliveira MR, Vacca RA, Xu S, Yousefi B, Curti V, Daglia M, Sobarzo-Sanchez E, Filosa R, Nabavi SF, Majidinia M, Dehpour AR, Shirooie S. 2018. Targeting STATs in neuroinflammation: the road less traveled! *Pharmacol Res* **141**:73–84.
16. Bryan JC, Verstovsek S. 2016. Overcoming treatment challenges in myelofibrosis and polycythemia vera: the role of ruxolitinib. *Cancer Chemother Pharmacol* **77**:1125–1142.
17. Massa M, Rosti V, Campanelli R, Fois G, Barosi G. 2014. Rapid and long-lasting decrease of T-regulatory cells in patients with myelofibrosis treated with ruxolitinib. *Leukemia* **28**:449–451.
18. Verstovsek S, Mesa RA, Gotlib J, Levy RS, Gupta V, DiPersio JF, Catalano JV, Deininger M, Miller C, Silver RT, Talpaz M, Winton EF, Harvey JH, Jr., Arcasoy MO, Hexner E, Lyons RM, Paquette R, Raza A, Vaddi K, Erickson-Viitanen S, Koumenis IL, Sun W, Sandor V, Kantarjian HM. 2015. A double-blind, placebo-controlled trial of ruxolitinib for myelofibrosis. *N Engl J Med* **366**:799–807.
19. Chan HC, Ng SC. 2017. Emerging biologics in inflammatory bowel disease. *J Gastroenterol* **52**:141–150.
20. Shreberk-Hassidim R, Ramot Y, Zlotogorski A. 2017. Janus kinase inhibitors in dermatology: a systematic review. *J Am Acad Dermatol* **76**:745–753e719.
21. Costa L, Del Puente A, Peluso R, Tasso M, Caso P, Chimenti MS, Sabbatino V, Girolimetto N, Benigno C, Bertolini N, Del Puente A, Perricone R, Scarpa R, Caso F. 2017. Small molecule therapy for managing moderate to severe psoriatic arthritis. *Expert Opin Pharmacother* **18**:1557–1567.
22. Song GG, Bae SC, Lee YH. 2014. Efficacy and safety of tofacitinib for active rheumatoid arthritis with an inadequate response to methotrexate or disease-modifying antirheumatic drugs: a meta-analysis of randomized controlled trials. *Korean J Intern Med* **29**:656–663.
23. Tanaka Y. 2016. [Tofacitinib for the treatment of rheumatoid arthritis]. *Nihon Rinsho* **74**:974–980.
24. Winthrop KL. 2017. The emerging safety profile of JAK inhibitors in rheumatic disease. *Nat Rev Rheumatol* **13**:234–243.
25. Chen X, Williams WV, Sandor V, Yeleswaram S. 2013. Population pharmacokinetic analysis of orally-administered ruxolitinib (INCB018424 Phosphate) in patients with primary myelofibrosis (PMF), post-polycythemia vera myelofibrosis (PPV-MF) or post-essential thrombocythemia myelofibrosis (PET MF). *J Clin Pharmacol* **53**:721–730.
26. Xing L, Dai Z, Jabbari A, Cerise JE, Higgins CA, Gong W, de Jong A, Harel S, DeStefano GM, Rothman L, Singh P, Petukhova L, Mackay-Wiggan J, Christiano AM, Clynes R. 2014. Alopecia areata is driven by cytotoxic T lymphocytes and is reversed by JAK inhibition. *Nat Med* **20**:1043–1049.
27. Kennedy Crispin M, Ko JM, Craiglow BG, Li S, Shankar G, Urban JR, Chen JC, Cerise JE, Jabbari A, Winge MC, Marinkovich MP, Christiano AM, Oro AE, King BA. 2016. Safety and efficacy of the JAK inhibitor tofacitinib citrate in patients with alopecia areata. *JCI Insight* **1**:e89776.
28. Samadi A, Ahmad Nasrollahi S, Hashemi A, Nassiri Kashani M, Firooz A. 2017. Janus kinase (JAK) inhibitors for the treatment of skin and hair disorders: a review of literature. *J Dermatolog Treat* **28**:476–483.
29. Jabbari A, Dai Z, Xing L, Cerise JE, Ramot Y, Berkun Y, Sanchez GA, Goldbach-Mansky R, Christiano AM, Clynes R, Zlotogorski A. 2015. Reversal of alopecia areata following treatment with the JAK1/2 inhibitor baricitinib. *EBioMedicine* **2**:351–355.
30. Barouch DH, Deeks SG. 2014. Immunologic strategies for HIV-1 remission and eradication. *Science* **345**:169–174.
31. Chomont N, El-Far M, Ancuta P, Trautmann L, Procopio FA, Yassine-Diab B, Boucher G, Boulassel MR, Ghattas G, Brenchley JM, Schacker TW, Hill BJ, Douek DC, Routy JP, Haddad EK, Sekaly RP. 2009. HIV reservoir size and persistence are driven by T cell survival and homeostatic proliferation. *Nat Med* **15**:893–900.
32. Gavegnano C, Schinazi RF. 2009. Antiretroviral therapy in macrophages: implication for HIV eradication. *Antivir Chem Chemother* **20**:63–78.

33. Honeycutt JB, Thayer WO, Baker CE, Ribeiro RM, Lada SM, Cao Y, Cleary RA, Hudgens MG, Richman DD, Garcia JV. 2017. HIV persistence in tissue macrophages of humanized myeloid-only mice during antiretroviral therapy. *Nat Med* **23**:638–643.
34. Vandergeeten C, Fromentin R, DaFonseca S, Lawani MB, Sereti I, Lederman MM, Ramgopal M, Routy JP, Sekaly RP, Chomont N. 2013. Interleukin-7 promotes HIV persistence during antiretroviral therapy. *Blood* **121**:4321–4329.
35. Bastard JP, Soulie C, Fellahi S, Haim-Boukobza S, Simon A, Katlama C, Calvez V, Marcelin AG, Capeau J. 2016. Circulating interleukin-6 levels correlate with residual HIV viraemia and markers of immune dysfunction in treatment-controlled HIV-infected patients. *Antivir Ther* **17**:915–919.
36. Borges AH, O'Connor JL, Phillips AN, Baker JV, Vjecha MJ, Losso MH, Klinker H, Lopardo G, Williams I, Lundgren JD, Group ISS, Group ES, Committee SS. 2014. Factors associated with D-dimer levels in HIV-infected individuals. *PLoS One* **9**:e90978.
37. Chakrabarti LA, Boucherie C, Bugault F, Cumont MC, Roussillon C, Breton G, Patey O, Chene G, Richert L, Lortholary O, Anrs 129 Bkvir-Cytok Study Group. 2014. Biomarkers of CD4+ T-cell activation as risk factors for tuberculosis-associated immune reconstitution inflammatory syndrome. *AIDS* **28**:1593–1602.
38. Chaudhuri A, Duan F, Morsey B, Persidsky Y, Kanmogne GD. 2008. HIV-1 activates proinflammatory and interferon-inducible genes in human brain microvascular endothelial cells: putative mechanisms of blood-brain barrier dysfunction. *J Cereb Blood Flow Metab* **28**:697–711.
39. Frustaci A, Petrosillo N, Vizza D, Francone M, Badagliacca R, Verardo R, Fedele F, Ippolito G, Chimenti C. 2014. Myocardial and microvascular inflammation/infection in patients with HIV-associated pulmonary artery hypertension. *AIDS* **28**:2541–2549.
40. Tenorio AR, Zheng Y, Bosch RJ, Krishnan S, Rodriguez B, Hunt PW, Plants J, Seth A, Wilson CC, Deeks SG, Lederman MM, Landay AL. 2014. Soluble markers of inflammation and coagulation but not T-cell activation predict non-AIDS-defining morbid events during suppressive antiretroviral treatment. *J Infect Dis* **210**:1248–1259.
41. Gavegnano C, Brehm JH, Dupuy FP, Talla A, Ribeiro SP, Kulpa DA, Cameron C, Santos S, Hurwitz SJ, Marconi VC, Routy JP, Sabbagh L, Schinazi RF, Sekaly RP. 2017. Novel mechanisms to inhibit HIV reservoir seeding using JAK inhibitors. *PLoS Pathog* **13**:e1006740.
42. Clayton KL, Garcia JV, Clements JE, Walker BD. 2017. HIV infection of macrophages: implications for pathogenesis and cure. *Pathog Immun* **2**:179–192.
43. Dunfee R, Thomas ER, Gorry PR, Wang J, Ancuta P, Gabuzda D. 2006. Mechanisms of HIV-1 neurotropism. *Curr HIV Res* **4**:267–278.
44. Sippy BD, Hofman FM, Wallach D, Hinton DR. 1995. Increased expression of tumor necrosis factor-alpha receptors in the brains of patients with AIDS. *J Acquir Immune Defic Syndr Hum Retrovirol* **10**:511–521.
45. Tyor WR, Glass JD, Griffin JW, Becker PS, McArthur JC, Bezman L, Griffin DE. 1992. Cytokine expression in the brain during the acquired immunodeficiency syndrome. *Ann Neurol* **31**:349–360.
46. Haile WB, Gavegnano C, Tao S, Jiang Y, Schinazi RF, Tyor WR. 2016. The Janus kinase inhibitor ruxolitinib reduces HIV replication in human macrophages and ameliorates HIV encephalitis in a murine model. *Neurobiol Dis* doi:10.1016/j.nbd.2016.02.007.
47. Gavegnano C, Haile W, Koneru R, Hurwitz S, Tao S, Tyor WR, Schinazi RF. 2018. Baricitinib reverses HAND behavioral phenotype in vivo and reservoir seeding in vitro. Keystone Symposium, Whistler, BC, April 2018.
48. Buzon MJ, Massanella M, Llibre JM, Esteve A, Dahl V, Puertas MC, Gatell JM, Domingo P, Paredes R, Sharkey M, Palmer S, Stevenson M, Clotet B, Blanco J, Martinez-Picado J. 2010. HIV-1 replication and immune dynamics are affected by raltegravir intensification of HAART-suppressed subjects. *Nat Med* **16**:460–465.
49. Kamat A, Misra V, Cassol E, Ancuta P, Yan Z, Li C, Morgello S, Gabuzda D. 2015. A plasma biomarker signature of immune activation in HIV patients on antiretroviral therapy. *PLoS One* **7**: e30881.
50. Katlama C, Deeks SG, Autran B, Martinez-Picado J, van Lunzen J, Rouzioux C, Miller M, Vella S, Schmitz JE, Ahlers J, Richman DD, Sekaly RP. 2013. Barriers to a cure for HIV: new ways to target and eradicate HIV-1 reservoirs. *Lancet* **381**:2109–2117.

51. Knights KM, Rowland A, Miners JO. 2013. Renal drug metabolism in humans: the potential for drug-endobiotic interactions involving cytochrome P450 (CYP) and UDP-glucuronosyltransferase (UGT). *Br J Clin Pharmacol* **76**:587–602.
52. Konig J, Muller F, Fromm MF. 2013. Transporters and drug-drug interactions: important determinants of drug disposition and effects. *Pharmacol Rev* **65**:944–966.
53. Shi JG, Chen X, Emm T, Scherle PA, McGee RF, Lo Y, Landman RR, McKeever EG, Jr., Punwani NG, Williams WV, Yeleswaram S. 2014. The effect of CYP3A4 inhibition or induction on the pharmacokinetics and pharmacodynamics of orally administered ruxolitinib (INCB018424 phosphate) in healthy volunteers. *J Clin Pharmacol* **52**:809–818.
54. Shi JG, Chen X, McGee RF, Landman RR, Emm T, Lo Y, Scherle PA, Punwani NG, Williams WV, Yeleswaram S. 2014. The pharmacokinetics, pharmacodynamics, and safety of orally dosed INCB018424 phosphate in healthy volunteers. *J Clin Pharmacol* **51**:1644–1654.

Index

J

K

L

T